2010

POWER ELECTRONICS AND AC DRIVES

B. K. Bose

General Electric Company
Corporate Research and Development
Schenectady, New York

Prentice-Hall, Englewood Cliffs, New Jersey 07632

Library of Congress Cataloging-in-Publication Data

BOSE, BIMAL K.
  Power electronics and ac drives.

  Bibliography: p.
  Includes index.
  1. Power electronics.   2. Electric machinery—
Alternating current.   I. Title.
TK7881.15.B67   1987   621.31′7   85-28184
ISBN 0-13-686882-7

o 81892

Editorial/production supervision and
    interior design: Denise Gannon
Cover design: 20/20 Services, Inc.
Manufacturing buyer: Gordon Osbourne

*The book is dedicated to those people whose contributions
have made power electronics and ac drives
technology so rich today*

Printed in the United States of America

10 9 8 7 6 5 4 3 2 1

ISBN 0-13-686882-7 025

PRENTICE-HALL INTERNATIONAL (UK) LIMITED, *London*
PRENTICE-HALL OF AUSTRALIA PTY. LIMITED, *Sydney*
PRENTICE-HALL CANADA INC., *Toronto*
PRENTICE-HALL HISPANOAMERICANA, S.A., *Mexico*
PRENTICE-HALL OF INDIA PRIVATE LIMITED, *New Delhi*
PRENTICE-HALL OF JAPAN, INC., *Tokyo*
PRENTICE-HALL OF SOUTHEAST ASIA PTE. LTD., *Singapore*
EDITORA PRENTICE-HALL DO BRASIL, LTDA., *Rio de Janeiro*
WHITEHALL BOOKS LIMITED, *Wellington, New Zealand*

CONTENTS

2   AC MACHINES   28

3   PHASE-CONTROLLED CONVERTERS AND CYCLOCONVERTERS   68

4 *VOLTAGE-FED INVERTERS*   *121*

7   CONTROL OF INDUCTION MACHINES   232

PREFACE

Power electronics and ac drives are an interdisciplinary technology and embrace many areas, including power semiconductor devices, ac machines, converter circuits, control theory, and signal electronics. More recently, the advent of microcomputers and VLSI circuits has advanced the frontier of this technology. After nearly two decades of intense technological evolution, power electronics and ac drives are finding widespread application in industry. This momentum will continue to grow in the future with reductions in cost and improvements in performance.

Unfortunately, because of the interdisciplinary character of the subject, power electronics and ac drives as a whole are complex and have been difficult for most of the readers to comprehend. The literature in this area has grown immensely in recent years, but a motivated reader who is trying to understand the subject systematically gathers only frustration. Although a number of books are available which deal with specific areas of power electronics, such as rectifiers, inverters, and cycloconverters, few books focus on all aspects of the subject.

This book, which gives an integrated treatment of the entire subject, is intended to fill this need. The book begins with an introduction to power semiconductor devices. The salient features of diodes, thyristors, triacs, asymmetrical thyristors, GTOs, power transistors, power MOSFETs, and hybrid devices are reviewed. Then the performance of induction and synchronous machines is reviewed in Chapter 2. This includes a discussion of d-q theory, which is useful to explain the field-oriented or vector control principles discussed later. The understanding of machines is so important in adjustable speed ac drives that its discussion is distributed into several subsequent chapters. Phase-controlled converters and cycloconverters, both widely discussed topics in power electronics, are reviewed in Chapter 3 with an emphasis on the circuits and waveforms. Then voltage-fed and current-fed inverters, the principal power circuits that interface the machines, are discussed in Chapters 4 and 5, respectively. Adjustable speed drives based on slip power control are discussed in Chapter 6. In Chapters 7 and 8, the control of induction and synchronous machines is treated in detail. Both scalar control and vector control are reviewed, and vector control particularly is discussed in depth. Chapter 7 also includes a general review of classical control, state variable control, and adaptive control principles that can be applied to ac machines. A special feature of this book is that it contains two chapters on microcomputer control. Chapter 9 covers microcomputers fundamentals and discusses some state-of-the-art microcomputers and peripheral chips. Chapter 10 describes the basic control functions of microcomputers and includes general discussions on systems, hardware, software and diagnostic design methodologies.

Throughout this book, physical description is emphasized in explaining a technical topic and mathematics are used to supplement the understanding. Control circuits, modeling, simulation, converter-machine interface and practical performance aspects are discussed in appropriate places. Numerical examples are also included. Each chapter ends with a list of selected references.

The book will be useful to college and university students as well as engineers working in industry. Selected portions of the book can be considered for a one-semester senior/graduate course. Students are required to have a basic knowledge of power devices, circuits, machines, control theory, and electronics.

The author expresses gratitude to his numerous colleagues at the General Electric Research and Development Center whose contributions and helpful discussions have made this book possible. I am particularly thankful to F. G. Turnbull, T. M. Jahns, K. D. T. Ngo, J. N. Park, and P. M. Szczesny for reviewing all the chapters of the manuscript. Special gratitude goes to Professor S. K. Ghandhi, former chairman of Electro-Physics Division of Rensselaer Polytechnic Institute, Troy, New York, who invited me to organize RPI's power electronics program and encouraged me to write a book in this area. I am also grateful to our secretary, Sue Folus, for typing the entire manuscript. Finally, I must acknowledge my family's cooperation throughout the ordeal of three long years.

B. K. BOSE

LIST OF PRINCIPAL SYMBOLS

Symbols are generally defined where they are used. Lowercase letters normally refer to instantaneous quantities and uppercase letters refer to constants, or average, rms, or peak values. Subscripts and superscripts relate to particular circuits or systems concerned. Sometimes, one symbol represents more than one quantity.

f Frequency (Hz)
I_d Dc current
I_f Machine field current (dc)
I_L Rms load current
I_m Rms magnetizing current
I_P Rms active current
I_Q Rms reactive current
I_r Rms rotor current (referred to stator)
I_s Rms stator current
i_D Instantaneous diode current
i_Q Instantaneous thyristor current
i_{dr} Instantaneous d-axis rotor current
i_{ds} Instantaneous d-axis stator current
i_{qr} Instantaneous q-axis rotor current
i_{qs} Instantaneous q-axis stator current
T Moment of inertia
$V_m,$

V_g	Rms air gap voltage
V_R	Rectifier dc voltage (also thyristor reverse voltage)
v	Instantaneous supply voltage
v_d	Instantaneous dc voltage
v_D	Instantaneous diode voltage
v_f	Instantaneous field voltage
v_Q	Instantaneous thyristor voltage
v_{dr}	Instantaneous d-axis rotor voltage
v_{ds}	Instantaneous d-axis stator voltage
v_{qr}	Instantaneous q-axis rotor voltage
v_{qs}	Instantaneous q-axis stator voltage
W	Digital word
X_c	Commutating reactance
X_r	Rotor reactance
X_s	Synchronous reactance
X_{ds}	d-Axis synchronous reactance
X_{lr}	Rotor leakage reactance
X_{ls}	Stator leakage reactance
X_{qs}	q-Axis synchronous reactance
α	Firing angle
β	Advance angle
γ	Turn-off angle
δ	Power angle of synchronous machine (also duty cycle ratio)
Θ	Thermal impedance (also angle or torque angle)
θ_e	Angle of synchronously rotating frame ($\omega_e t$)
θ_r	Rotor angle
θ_{sl}	Slip angle
μ	Overlap angle
τ	Time constant
L_c	Commutating inductance
L_d	Dc link filter inductance
L_m	Magnetizing inductance
L_r	Rotor inductance
L_s	Stator inductance
L_{lr}	Rotor leakage inductance
L_{ls}	Stator leakage inductance
L_{dm}	d-Axis magnetizing inductance
L_{qm}	q-Axis magnetizing inductance
n	Turns ratio
Ne	Stator sychronous speed (rpm)
Nr	Rotor electrical speed (rpm)
P	Number of poles (also active power)
P_g	Air gap power
P_m	Mechanical output power

P_{sl} Slip power

Q Reactive power

R_r Rotor resistance

R_s Stator resistance

S Slip per unit (also the Laplace operator)

T Time period (also temperature)

T_e Developed torque

T_L Load torque

T_s Sampling time

t_{off} Turn-off time, impressed ($t_{off} > t_q$)

t_q Device turn-off time

V Rms supply voltage

V_d Dc voltage

V_f Induced emf (rms)

ϕ Displacement factor angle

ψ_a Armature reaction flux linkage

ψ_f Field flux linkage

ψ_m Air gap flux linkage

ψ_s Stator flux linkage

ψ_{dr} d-Axis rotor flux linkage

ψ_{ds} d-Axis stator flux linkage

ψ_{qr} q-Axis rotor flux linkage

ψ_{qs} q-Axis stator flux linkage

ω_e Stator frequency (rad/s)

ω_m Rotor mechanical speed (rad/s)

ω_r Rotor electrical speed [$(P/2)\,\omega_m$] (rad/s)

ω_{sl} Slip frequency (rad/s)

$\hat{\ }$ Peak value of a phasor

1

POWER SEMICONDUCTOR DEVICES

1.0 INTRODUCTION

The electronic power conversion circuits that convert and control electrical power use power semiconductor devices. The devices operate in the switching mode, which causes the losses to be reduced and therefore the conversion efficiency to be improved. However, the disadvantages of switching-mode operation are the generation of harmonics and the fact that the converter system model tends to be more complex.

Power semiconductor devices can generally be classified as follows:

- Diode rectifier
- Thyristor
- Power transistor
- Power MOSFET

This chapter reviews the characteristics of the devices listed above. It should be mentioned here that the cost and reliability of power conversion circuits are profoundly affected by the power semiconductor devices and therefore, a thorough knowledge of the devices' influence on the factors noted is mandatory.

The power semiconductors of today have grown by a dynamic evolutionary process in the last two and a half decades. Silicon is the principal material used and will possibly remain so in the foreseeable future. The advances in processing technology and new packaging concepts are helping power circuit integration and integration of the power circuit with the control, thereby minimizing the size and cost.

1.1 DIODE RECTIFIER

Power diodes provide uncontrolled rectification of ac-to-dc power, and generally, feedback and free-wheeling functions in inverters. The power diodes have the P-I-N structure and are fabricated by a diffusion process. Line-frequency rectifiers have slow-recovery characteristics and are available in high voltage and current ratings, such as 5000 V and 2000 A. The application of these devices includes electric traction, battery charging, electroplating, electromechanical processing, power supplies, welding, and uninterruptible power supply (UPS) systems. Fast-recovery rectifiers are used in high-frequency rectification and switching applications where the recovery current and recovery time are critically important in circuit applications. In this class of diodes, the carrier lifetime, and consequently the recovery time, are controlled by gold or platinum diffusion, which causes increase of conduction voltage drop. The fast-recovery diodes are available in intermediate voltage and current ranges and are used in commutation circuits, feedback and free-wheeling functions in inverters, switching-mode power supplies, induction heating, and so on. Schottky rectifiers are formed by metal semiconductor junctions and are therefore defined as majority-carrier devices. These have the special attributes of low conduction voltage and minimal switching times, but have the limitations of high reverse leakage current and low blocking voltage capability. The application of these devices includes high-frequency instrumentation and switching power supplies.

The electrical and thermal characteristics of power diodes are somewhat analogous to those of thyristors, and these are reviewed in the next section.

1.2 THYRISTOR

The thyristor, also known as a silicon-controlled rectifier (SCR), is the main workhorse for bulk power conversion and control in industry. A thyristor has a P-N-P-N structure with the three terminals shown in Fig. 1.1 and is fabricated by diffusion process. If forward voltage is applied to the device, the middle junction becomes reverse biased and the depletion region extends mainly into the N_1 region. The N_1 layer is lightly doped and has a large width to withstand the high voltage. For reverse voltage on the device, the end junctions become reverse biased, and the N_1 layer again absorbs most of the voltage.

Two-Transistor Theory

The principle of thyristor operation is normally explained by the classical two-transistor analogy shown in Fig. 1.1. If an oblique section is taken on the device, a thyristor can be represented by P-N-P and N-P-N transistors in regenerative

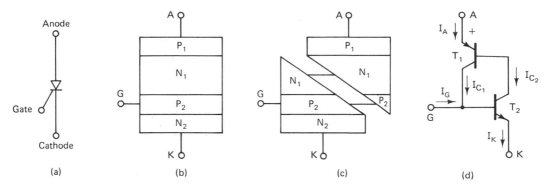

Figure 1.1 Two-transistor analogy of thyristor.

feedback. If gate current I_G is injected to the base of T_2 transistor, its collector current I_{C_2}, which constitutes the base current of transistor T_1, amplifies the collector current I_{C_1}, and therefore reinforces the original gate current I_G. Eventually, T_1 and T_2 go into complete saturation and all the junctions become forward biased. In the off condition the following equations can be written:

$$I_{C_1} = \alpha_1 I_A + I_{CBO_1} \tag{1.1}$$

$$I_{C_2} = \alpha_2 I_K + I_{CBO_2} \tag{1.2}$$

$$I_K = I_A + I_G \tag{1.3}$$

where α_1 and α_2 are the common-base current gain and I_{CBO_1} and I_{CBO_2} are the common-base leakage current of transistors T_1 and T_2, respectively. Combining equations (1.1) to (1.3) yields

$$I_A = \frac{\alpha_2 I_G + I_{CBO_1} + I_{CBO_2}}{1 - (\alpha_1 + \alpha_2)} \tag{1.4}$$

A silicon transistor has the common property that α is very low at low emitter current, and as the emitter current builds up, α rises rapidly. In the normally off condition, $I_G = 0$ and $\alpha_1 + \alpha_2$ is very low, and therefore the leakage current will be somewhat higher than the sum of the individual leakage currents. If by some mechanism the emitter current of component transistors can be increased so that $\alpha_1 + \alpha_2$ approaches 1, then I_A approaches infinity and the device triggers into saturation. Physically, the external load will limit the anode current as the device goes into conduction. There are several mechanisms by which a thyristor can be triggered into conduction, and these can be explained as follows.

Gate current.   If gate current is injected, the emitter currents of the component transistors are increased by normal transistor action and the device goes into saturation. The gate current does not have any control once the device goes into conduction (except for the gate-turn-off thyristor, discussed later).

Voltage effect.   If the forward anode blocking voltage is gradually increased to a high value, the minority-carrier leakage current at the middle junction, which is also the collector junction current of the component transistors, increases due to an avalanche effect. The amplification of the leakage current due to regenerative action eventually causes switching action.

dV/dt effect. If the anode voltage rises at a certain rate, the depletion layer capacitance C at the middle junction will create displacement current $i = C\,dV/dt$ which will induce emitter current of the component transistors and will eventually cause switching action.

Temperature effect. At a higher junction temperature, the leakage current of the component transistors increases and eventually causes switching action.

Light firing. Direct irradiation of light on silicon creates electron–hole pairs, which under the influence of an electric field, produce current that triggers the thyristor.

Although in actual practice, a number of the foregoing effects in combination influence switching on of the device, gate firing is the most usual method for triggering a thyristor. Light-fired thyristors have recently been developed for high-voltage direct-current (HVDC) applications. In this application, a number of devices are connected in series–parallel combination, and light firing provides the advantage of electrical isolation between the control and power circuits.

Volt–Ampere Characteristics

The volt–ampere characteristics of a thyristor are shown in Fig. 1.2. With $I_G = 0$, if forward voltage is applied across the device, there will be a leakage current, as explained earlier. If the voltage exceeds a critical limit, the device switches into conduction. With increasing magnitude of gate current, the forward breakover voltage is reduced, and eventually at I_{G3} the device behaves as a diode with the entire forward blocking voltage removed. During conduction, if the gate current is zero and the anode current falls below a critical limit called the holding current, the device reverts to the forward blocking state. With reverse voltage, the end P-N junctions of the device become reverse biased and the V–I curve becomes essentially similar to that of a diode rectifier.

Switching Characteristics

A thyristor, like other power semiconductor switches, is a delicate device and its static and dynamic characteristics are to be taken into consideration for economical and reliable design of converter equipment. The thyristors can be classified as being of converter or phase-control type and of fast-switching or inverter type. It is in the latter category that the switching characteristics require far more critical consideration. Figure 1.3 shows the switching characteristics of a thyristor. Initially, when forward voltage is applied across the device, the off-state or static dv/dt has to be limited so that it does not switch on spuriously. In the forward

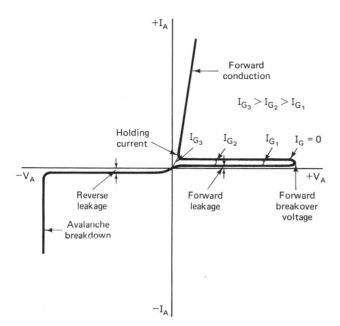

Figure 1.2 Volt–ampere characteristics of thyristor.

blocking condition, if gate current is applied, there is a finite delay time t_d before the anode current starts building up. The delay time is normally a fraction of a microsecond and is determined by the carrier transition delay before thyristor action starts.   Then the device turns on and the current rises to the full value I_T with a slope di/dt in the rise time t_r.   The physical mechanism of turn-on is quite complex but can be explained briefly as follows.   The initial turn-on occurs in the gate–cathode periphery and then the turn-on area spreads across the whole cross section with a finite velocity.   If I_A builds up at a faster rate than the spreading velocity, the current density increases, which might eventually cause localized hot-spot burn-out of the device.   During conduction, the middle N_1, and P_2 regions remain heavily saturated with minority carriers and the gate loses its control, as mentioned earlier.   To recover the forward voltage blocking capability, these minority carriers are to be cleaned out so that the middle junction can hold the reverse voltage.   In the conducting condition, a thyristor can be turned off or commutated by tem-porarily applying reverse voltage from the external circuit.   Methods of thyristor commutation are discussed in Chapter 4.   When reverse voltage is applied, the forward current first goes to zero and then the current builds up in the reverse direction with a commutation di/dt slope determined by the circuit leakage in-ductance.   The reverse recovery current flows due to sweeping out of the minority carriers across the end junctions, which remain forward biased during the initial turn-off process.   At maximum recovery current I_{RM}, the junctions begin to block, causing decay of recovery current.   The fast decay of recovery current causes a voltage overshoot V_{RRM} across the device due to a leakage inductance effect.   At zero current, the middle junction continues to remain forward biased and the minority-carrier electrons and holes in the vicinity should be allowed to disappear by a recombination process.   Eventually the middle junction regains its voltage

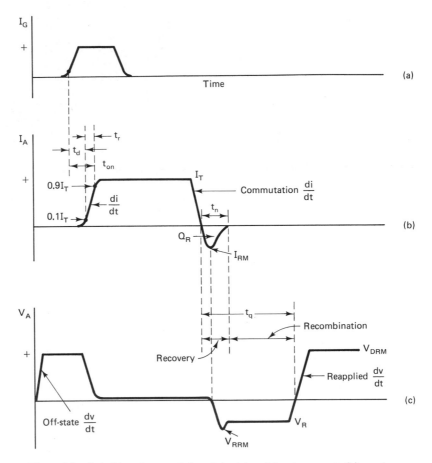

Figure 1.3 Switching characteristics of thyristor: (a) gate current; (b) anode current; (c) anode voltage.

blocking capability and forward voltage can be successfully reapplied. The reapplied dv/dt has to be limited so that its effect, in combination with the stored minority-carrier leakage current, does not cause spurious turn-on of the device. The time interval between forward current zero and zero of the reapplied voltage is defined as the circuit commutated turn-off time t_q. If the reapplied dv/dt occurs after sufficient delay, it becomes identical to off-state dv/dt.

From the mechanism of thyristor operation discussed above, it may be evident that the turn-off time t_q cannot be defined as a constant parameter. In fact, it is influenced by junction temperature T_j and the parameters I_T, V_R, V_{DRM}, reapplied dv/dt, commutation di/dt, and gate bias during the turn-off interval. Various trade-off considerations are possible among these parameters. For example, the GE C158 inverter thyristor has $t_q = 25$ μs in the specification sheet for $T_j = +125°C$, $I_T = 150$ A, $V_R = 50$ V, reapplied $dv/dt = 200$ V/μs, commutation $di/dt = 5$ A/μs, gate bias = 0 V, 100 Ω. This time can be improved to 20 μs with reapplied

dv/dt = 20 V/µs, or it may deteriorate to 40 µs if the reverse voltage is clamped by a bypass diode.

Snubber

It is generally necessary to connect an RC snubber circuit across the power semiconductor device.   The function of the snubber can be summarized as follows:

- It protects the device from supply- and load-side voltage transients.
- reduces the off-state and reapplied dv/dt.
- reduces the magnitude of peak recovery voltage.
- reduces the device's switching loss.

Figure 1.4 shows a thyristor with an unpolarized snubber which is used when the supply voltage is ac.   A polarized snubber with a dc supply voltage is illustrated in Fig. 1.12.   The series inductance L in Fig. 1.4 may be stray inductance or inductance intentionally added to limit di/dt during turn-on of the device.   If a forward voltage V is applied in the off-state, the resulting differential equation is

$$L\frac{di}{dt} + iR + \frac{1}{C} \int i \, dt = V \tag{1.5}$$

which can be solved for v and dv/dt in the underdamped condition as (Ref. 6)

$$v = V\left[1 - \left(\cos \omega t - \frac{\alpha}{\omega} \sin \omega t\right)e^{-\alpha t}\right] \tag{1.6}$$

$$\frac{dv}{dt} = V\left(2\alpha \cos \omega t + \frac{\omega^2 - \alpha^2}{\omega} \sin \omega t\right)e^{-\alpha t} \tag{1.7}$$

where $\omega = \sqrt{(1/LC) - \alpha^2}$ and $\alpha = R/2L$. The device voltage v will overshoot to some extent, depending on the damping factor $\xi = R/2\sqrt{C/L}$.   The capacitor C will eventually charge to the full voltage V.   When the thyristor is turned on, the capacitor will discharge through the device with initial current V/R and this turn-on di/dt will tend to be extremely high.   Fortunately, the device can tolerate this high di/dt if the magnitude of the current is limited to a low value.   Assume that the device is subsequently being turned off by applying a reverse voltage of the

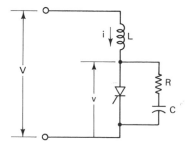

Figure 1.4 Thyristor with snubber circuit.

same magnitude.   The forward current will decay by the slope V/L and reach the peak reverse recovery current I_{RM} as explained in Fig. 1.3.   Then as I_{RM} collapses quickly, the peak recovery voltage contributed by the snubber can be given as

$$V_{RRM} = V\left(1 + \exp\left\{-\frac{\xi}{\sqrt{1 - \xi^2}} \tan^{-1}\left[-\frac{(2\xi - 4\xi^2 x + x)\sqrt{1 - \xi^2}}{1 - 3\xi x - 2\xi^2 + 4\xi^3 x}\right]\right\}\right) \quad (1.8)$$
$$\times \sqrt{1 - 2\xi x + x^2}$$

where $\qquad\qquad\qquad\qquad x = (I_{RM}/V)\ \sqrt{L/C}.$

Again, as the forward voltage is reapplied at the end of turn-off interval, the reapplied dv/dt can be derived as

$$\left(\frac{dv}{dt}\right)_{\text{reapplied}} = 2V\left(2\alpha\ \cos\ \omega t + \frac{\omega^2 - \alpha^2}{\omega}\ \sin\ \omega t\right)e^{-\alpha t} \quad (1.9)$$

A small snubber resistance is desirable for low dv/dt, but it causes a low damping coefficient, which correspondingly increases the overshoot voltage.   A small capacitance value again improves snubber loss but adversely affects the dv/dt value.   The components should be noninductive and the wiring circuit should have minimum leakage inductance.

The snubber parameters can be designed from graphs available in the literature (Ref. 6) and from a handbook based on equations (1.5) – (1.9) and then fine-tuned experimentally.

Power Loss

In normal operation, a thyristor has several types of losses.   These are due to conduction, switching, blocking, and gate drive.   In 60-Hz power-frequency applications, the conduction loss will dominate; on the other hand, the switching loss dominates in high-frequency switching applications.   Figure 1.5 gives information on power dissipation for various duty cycles of a rectangular current wave.   The average power dissipation can be calculated by multiplying the instantaneous voltage and current waves, integrating, and then taking the average value.   In the loss curves, the switching dissipation is neglected but the blocking and gate losses are included.   These curves are typically valid in the frequency range 50 to 400 Hz.

Thermal Impedance

The heat dissipated in the vicinity of a junction flows to the case and then to the ambient through the externally mounted heat sink, causing a rise in the junction temperature.   The maximum junction temperature of a device should be limited because of its adverse effect on leakage current, breakover voltage, turn-off time, thermal stability, and long-term reliability of the device.

Figure 1.6 shows the thermal equivalent circuit of a thyristor mounted on a heat sink.   For sustained power dissipation P at the junction, the junction tem-

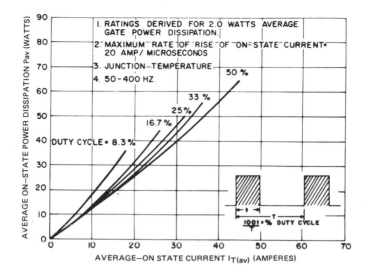

Figure 1.5 Average on-state power dissipation for rectangular current waveform (GE-C147).

perature T_j can be calculated as

$$T_j - T_A = P(\Theta_{JC} + \Theta_{CS} + \Theta_{SA}) \qquad (1.10)$$

where T_A is the ambient temperature and Θ_{JC}, Θ_{CS}, and Θ_{SA} represent the thermal resistance from junction to case, case to sink, and sink to ambient, respectively. From equation (1.10) it is evident that for a limited $T_{j\max}$ (usually 125°C), P can be increased by reducing Θ_{SA}. This means that a more efficient cooling system will increase the power-dissipation capability of a device. An infinite heat sink will result if Θ_{CS} and Θ_{SA} are reduced to zero and the case temperature T_C is locked to the fixed ambient temperature T_A.

In practical operation, the dissipation P is cyclic and therefore the transient RC equivalent circuit shown in Fig. 1.6 should be considered. For pulsed power dissipation, the thermal capacitance effect delays the junction temperature rise, and thus permits heavier loading of the device. The effective impedance of the transient equivalent circuit at the junction is known as the transient thermal impedance. Figure 1.7(a) shows the T_j curve for dissipation of a single pulse of power.

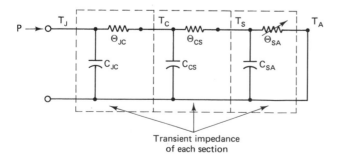

Transient impedance
of each section

Figure 1.6 Transient thermal equivalent circuit of thyristor with heat sink.

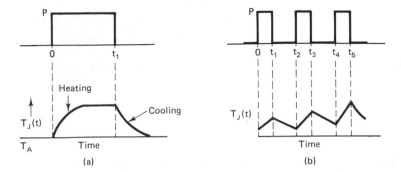

Figure 1.7 Junction temperature rise for pulsed power dissipation.

Taking advantage of the complementary nature of heating and cooling curves, the
following instantaneous equations can be written:

$$T_j(t) = T_A + P\Theta(t) \qquad \text{for } 0 < t < t_1 \tag{1.11}$$

$$T_j(t) = T_A + P[\Theta(t) - \Theta(t - t_1)] \qquad \text{for } t > t_1 \tag{1.12}$$

where $\Theta(t)$ is the transient thermal impedance at time t. Manufacturers usually
provide the transient thermal impedance information for a thyristor with an infinite
heat-sink condition. The effect of finite heat sink can thus be added on, if desired.
Figure 1.7(b) shows the typical junction temperature buildup for three repeated
pulses. $T_j(t)$ at any instant can be calculated from the information in the transient
impedance curve by using the superposition principle.

Current Rating

Based on the criteria of limiting junction temperature as discussed above, Fig. 1.8
shows the average current rating $I_{T(av)}$ versus permissible case temperature T_{Cmax}
for various duty cycles of a rectangular current wave. For example, if the case
temperature is limited to 110°C, the thyristor can carry an average of 20 A with a
duty cycle of 25%. If a better heat sink is provided to limit T_{Cmax} to 102°C, the
current can be increased to 30 A. For the same $T_{Cmax} = 110$°C, a 50% duty cycle
will improve the form factor of the current wave, reducing the peak-to-peak junction
temperature and therefore allowing the current to be increased to 27.5 A. All
the curves converge to $T_{Cmax} = 125$°C, where the current rating is reduced to zero
(i.e., $T_{Cmax} = T_{jmax} = 125$°C. On the other end, the curves terminate in vertical
lines which correspond to a fixed root-mean-square (rms) current rating for the
device. The rms current rating $[I_{T(rms)}]$ is limited to prevent excessive heating of
resistive elements such as joints and leads. Figure 1.8 can be used with Fig. 1.5
to design the heat-sink thermal resistance. Similar sets of figures are available in
specification sheets for sinusoidal current waveforms.

The other important current ratings for a thyristor are peak one-cycle surge
current (I_{TSM}) and I^2t, which are valid during temporary overload or fault condi-

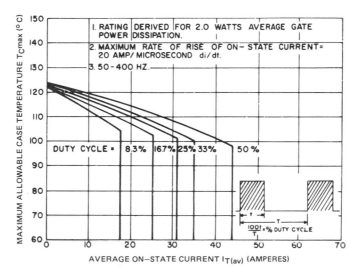

Figure 1.8 Maximum allowable case temperature for rectangular current waveform (GE-C147).

tions.   Under such conditions the junction temperature may temporarily exceed the specified limit without causing any damage.   These nonrecurrent current ratings are used to design protection into the devices.

High-Frequency Current Rating

The discussion on current rating so far has been confined to phase-control-type thyristors, where switching losses could be neglected.   For high-frequency applications such as in a PWM inverter, the switching losses become very significant during both turn-on and turn-off processes.   At high-frequency switching, the device does not fully turn on and the losses tend to be high near the gate-cathode periphery.   Discharge of the snubber capacitor during turn-on worsens the losses. The specification sheet provides information on current ratings for both rectangular and sine-wave current shapes.   For a rectangular current wave, the peak current decreases with higher di/dt, higher switching frequency, or higher duty cycle.   Similar information is also available for a sine current wave.   The switching loss data in high-frequency applications are given in units of watt-seconds per pulse, which can be converted to average loss by knowing the switching frequency for heat-sink design purposes.

Gate Drive

The fundamental gate triggering characteristics of thyristors were explained before. In designing a gate circuit, it is necessary to drive the gate hard with a sharp rise time so that the device turns on quickly with a large initial di/dt.   This consideration is especially important for high-frequency application. Although a short firing pulse is adequate for turning on a device, it is the usual practice in converter design to give pulse-train firing during the desired conduction interval so that the thyristor cannot revert to a blocking state if the load current falls below the holding current accidentally.

Figure 1.9 shows gate trigger characteristics where instantaneous gate voltage has been plotted with instantaneous gate current for various trigger conditions. The locus of possible gate trigger points is bounded by the two boundary lines X and Y as shown in the figure. The possible dc trigger area shown is bounded by a horizontal line known as dc trigger voltage (V_{GT}) and a vertical line known as dc trigger current (I_{GT}) in the specification sheet. In this area, there is a probability that the device will fire, but for reliable firing the trigger point should be outside this area. The parameters V_{GT} and I_{GT} vary with case temperature as shown in the figure. It is the usual practice to draw a load line on the gate characteristics so that it crosses the preferred area and determines the voltage and impedance of the gate drive circuit. Figure 1.9 shows a 20-V 20-Ω load line (curve on log-log graph) on which satisfactory trigger points may lie anywhere within the curves X and Y. With pulse firing, it may be possible to increase the peak gate power dissipation where the peak power varies inversely with the pulse width. For pulse-train firing with a certain duty cycle, both the peak power and average power are to be limited within the specification sheet values. Figure 1.10 shows a practical gate drive circuit for converter application. The output pulse transformer provides electrical isolation between the control circuit and the power circuit. The series diode decouples the transformer from the gate and connects the shunt resistance R_S when gate drive does not exist. The pulse-train control signal from the integrated circuit is amplified in several stages before feeding to the gate.

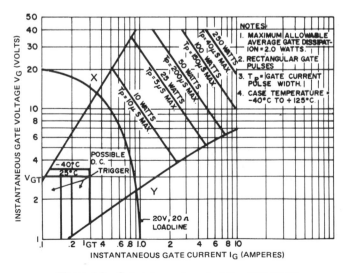

Figure 1.9 Gate trigger characteristics (GE-C147).

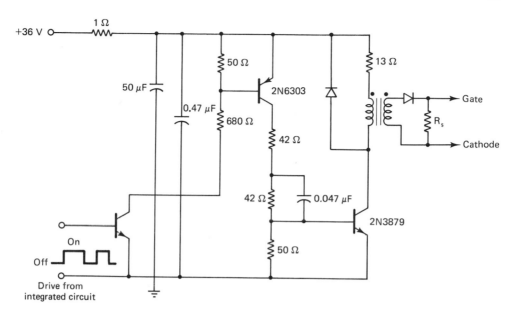

Figure 1.10   Practical gate drive circuit.

1.3 TRIAC

Functionally, a triac can be considered as an integration of a pair of phase-control thyristors connected in inverse-parallel. The circuit symbol of a triac and its volt–ampere characteristic in the first and third quadrants are shown in Fig. 1.11. The structure of a triac is more complex than simply a *P-N-P-N* device in parallel with an *N-P-N-P* device. A triac is normally operated in the following modes:

Mode I + : This is a conventional thyristor mode of operation when the applied voltage at terminal T_2 is positive with respect to T_1 and the device is fired by a positive gate current.

Mode III − : This is a remote gate thyristor operation when the terminal T_1 is positive with respect to T_2 and the device is fired by injecting a negative gate current.

A triac is economical compared to a pair of antiparallel thyristors and the control circuit is simpler. But the device has the following limitations:

- A triac has poor reapplied *dv/dt* capability, which makes it difficult to apply with inductive load.
- The gate circuit sensitivity is somewhat poorer.
- The turn-off time t_q is longer.

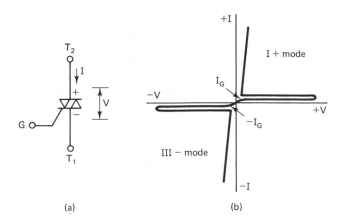

(a)   (b)

Figure 1.11 Triac symbol and volt–ampere characteristics.

Triacs are normally used for phase control with a resistive load (i.e., lamp dimming and heater control types of applications). They are also used commonly in solid-state relays, but to a limited extent for motor control applications. The supply frequency is usually limited to 60 or 400 Hz. The state-of-the-art device voltage and current ratings are much smaller than thyristors.

1.4 ASYMMETRICAL THYRISTOR

In inverter or chopper-type applications, a thyristor is not required to block reverse voltage and it is normally bypassed by a free-wheeling diode. An asymmetrical thyristor (ASCR) is fabricated to have limited reverse voltage capability and, as a result, it permits reduction of turn-on time, turn-off time, and conduction drop. A typical ASCR may have a reverse blocking voltage of 30 V and forward blocking voltage in the range 400 to 2000 V. A 1500-V ASCR may typically have a turn-off time of 10 to 15 μs compared to 20 to 30 μs for a similar thyristor device.

A reverse-conducting thyristor (RCT) is a special-case asymmetrical thyristor where the reverse blocking capability is entirely removed by integrating monolithically an antiparallel diode on the same chip. This reduces the heat-sink size and therefore enhances the mechanical compactness of the converter. The undesirable effect of stray inductance between the thyristor diode loops is also eliminated.

1.5 GATE-TURN-OFF THYRISTOR

A gate-turn-off thyristor (GTO) is a device that can be turned on like a thyristor with a single pulse of gate current, but in addition it has the capability of being turned off by injecting a negative gate current pulse. GTOs have been in existence almost from the beginning of thyristor era, but only recently are these devices being developed with large power-handling capabilities and improved performance, and they are gaining popularity in conversion equipment.

The thyristor-transistor-like behavior of GTOs can be explained by the two-transistor analogy shown in Fig. 1.1. If the transistor T_2 current gain α_2 is designed to be near unity in the on-state, and if negative gate current can completely divert the *P-N-P* collector current I_{C1} out of the gate, the device can be turned off successfully. The turn-off gain, defined as the ratio of anode current prior to turn-off to the negative gate current required for turn-off, is very low, typically between 3 and 5. For example, a 2500-V 600-A GTO will need a peak 150-A negative gate current for turn-off. GTOs are a serious competitor to thyristors and high-powered transistors in inverter and chopper applications.

Switching Characteristics

The switching characteristics of GTOs are somewhat different from those of thyristors and therefore require further explanation. Figure 1.12 shows a GTO chopper circuit with a polarized snubber. The snubber decreases the rate of voltage rise (*dv/dt*) across the GTO and improves its turn-off capability. The GTO is turned on with a positive gate current pulse. Prior to turn-on, the snubber capacitor C_s is charged to the full supply voltage V_d. At turn-on, C_s discharges through the resistor R_s and GTO in series, dumping most of its energy into the resistor R_s. When the GTO is turned off by negative gate current pulse, C_s charges resonantly through the diode D_s and leakage inductances in series, limiting *dv/dt* across the device. The power dissipation of the snubber circuit is approximately given as

$$P_S \simeq \tfrac{1}{2} C_s V_d^2 f \tag{1.13}$$

where f is the operating frequency of the chopper. Figure 1.13 shows a typical gate drive circuit of a GTO. GTO performance is greatly influenced by the design of the gate drive circuit. It consists of three parts: on-gating circuit, off-gating circuit, and bias circuit. In the normally off-condition, it is desirable to put a negative bias voltage in the gate to prevent spurious triggering by *dv/dt*. This is because with the unshorted emitter structure of GTO, the *dv/dt* capability is somewhat poor. The power supply for the gate drive is isolated by transformers. Both the on-gating and bias circuits are supplied from a high-frequency power supply through diode rectifiers. The on-gating signal is optically isolated and turns on transistor T_1, which in turn switches on T_3 to apply positive gate current pulse to

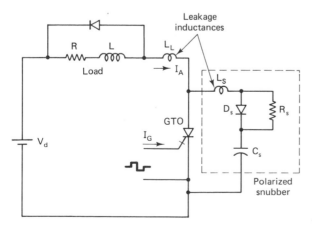

Figure 1.12 GTO chopper circuit with polarized snubber.

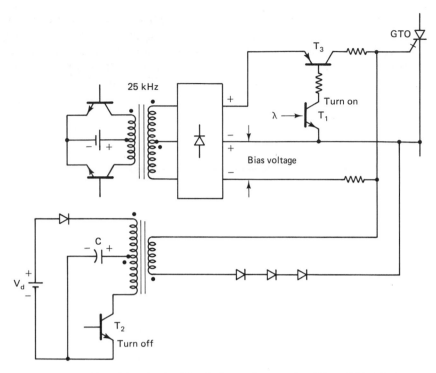

Figure 1.13   Gate drive circuit of GTO. From "Present Condition of High Power GTO Application," by N. Seki in PCI September 1982 Proceedings Intertec Communications, Inc.

the GTO. The off-gating pulse is initiated by turning on transistor T_2. Prior to switching on T_2, the capacitor C resonantly charges to $2V_d$ from the supply voltage V_d. When T_2 is on, the stored energy of the capacitor creates a large negative gate current pulse. The series diodes and transistor T_3 prevent interaction between on-gating and off-gating circuits.

The turn-on behavior of a GTO is essentially similar to a thyristor's. Both the delay time and turn-on time can be decreased by boosting the on-gating current. The turn-off characteristics of a GTO are somewhat different and are shown in Fig. 1.14. As the negative gate current I_{GR} is established, the controllable anode current I_A begins to fall after a time delay. The fall time of I_A is abrupt, typically less than 1μs. As the forward voltage begins to develop and the anode current tends to bypass through the snubber circuit, the leakage inductance L_s creates a spike voltage as shown in Fig. 1.14. A large spike voltage is extremely harmful because current concentration may create localized heating, causing what is known as second breakdown failure. This emphasizes the necessity of minimizing snubber circuit leakage inductance. After the spike voltage, the anode voltage overshoots due to snubber circuit resonance before settling down to normal forward blocking voltage V_d. During the overshoot, the anode circuit shows a displacement current, defined as tail current. The overshoot voltage and tail current can be decreased

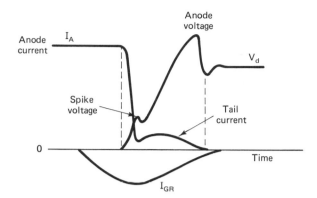

Figure 1.14 Turn-off characteristics of GTO.

by increasing the size of the snubber capacitor, which should be compromised with the snubber loss.   The capacitor size of a GTO snubber is normally several times higher than that of a thyristor.   The controllable current I_A can be increased and the turn-off time can be decreased by increasing the rate of rise of I_{GR}.

Since the GTO power losses are somewhat higher during switching, the GTO circuits are normally restricted to operate typically at or below 1 kHz of switching frequency.   The gate driving circuit losses are somewhat higher than those of the thyristor, but the elimination of forced commutation circuit improves the overall efficiency of the converter.   This also results in less size and cost for the converter and eliminates operational dependency on line-voltage fluctuation.

1.6 POWER TRANSISTOR

Power transistors are finding increasing popularity in low- to medium-power applications, where they compete successfully with thyristors and GTOs.   Of course, in the low-power range, its popularity has been challenged by power MOSFET devices, which are discussed in the next section.   A power transistor has low current gain and requires continuous base drive during on-state conditions but does not require forced commutation circuitry.

Power transistors can be used in high switching frequency, permitting size reduction of electromagnetic components, and can provide current-limit protection by the base drive circuit.   Of course, it cannot withstand reverse voltage and application is therefore limited to dc voltage-fed inverters and choppers.

A Darlington power transistor consists of a cascade connection of two transistor stages and can have a large dc common-emitter current gain (h_{FE}).   Figure 1.15 shows a monolithic Darlington transistor where a bypass diode has been fabricated on the same chip. The shunt resistances R_1 and R_2 reduce the collector leakage current and help to establish bias voltages across base–emitter junctions. The intermediate base lead B_2 can be used for fast turn-off of T_2 transistor.   The base drive power requirement is lower in a Darlington transistor, but it is at the cost of slightly reduced switching frequency and higher conduction drop.

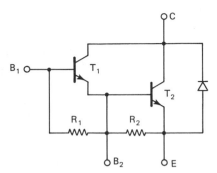

Figure 1.15 Darlington power transistor with built-in bypass diode.

Voltage and Current Ratings

The transistor maximum voltage and current ratings are specified at a case temperature $T_C = 25°C$. The voltage rating $V_{CEO(SUS)}$ is defined as collector-to-emitter breakdown voltage at a relatively high value of collector current with the base circuit open. It is an important parameter for inductive load switching considerations. The collector currents of a device are specified as continuous (dc) current (I_{dc}) and peak current I_{Cmax} (Fig. 1.17). The currents are limited on the basis of maximum junction temperature. The continuous current can be increased on a duty-cycle basis, taking advantage of the thermal capacitance effect, but is limited to the peak current value to prevent fusing the bonding wires within the package. The collector current can be increased further (I_{CSM}) for single nonrepetitive pulse application.

Current Gain

The dc current gain of a transistor is a strong function of collector current, as shown in Fig. 1.16. The parameter is also influenced by the junction temperature and collector voltage. This gain is low at low collector current; it then increases with collector current, reaches a peak value, and then decreases again. Because of the variation of h_{FE}, the base drive has to be designed conservatively either to track with collector current or on a worst-case basis for reliable operation. If, for example, the collector current varies in the range 0 to 150 A, the worst-case h_{FE} = 70 at $T_j = 150°C$. Then, considering an overdrive factor of 1.1, the base current should be 2.36 A. The base overdrive will tend to reduce the turn-on time, but dissipation and storage time will be affected adversely.

Safe Operating Area

During switching, power transistors show a complex phenomenon known as the second breakdown effect. This may be the reason for frequent device failure if the circuit is not designed carefully. While avalanche breakdown is defined as first breakdown, the second breakdown can be defined as breakdown of the junction due to localized heating effects. A transistor can have second breakdown during

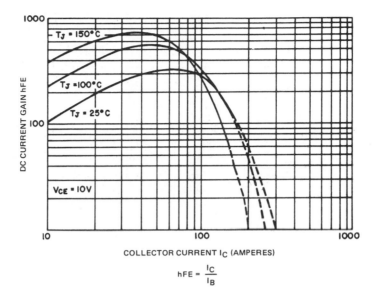

Figure 1.16   Current gain variation with collector current of a Darlington transistor (GE-D67DE).

turn-on (FBSB) as well as during turn-off (RBSB).   In either case, the collector current becomes concentrated in a small area of reverse-biased collector junction where a hot spot is formed and the device fails by thermal runaway, known as second breakdown.   The rise in junction temperature at the hot spot accentuates the current concentration due to the negative temperature coefficient of the drop, and this regeneration effect causes collapse of the collector voltage, destroying the device.   During turn-on the collector current tends to crowd near the base–emitter periphery, which is improved by interdigitated fabrication.

The second breakdown effect tends to be more serious during turn-off of an inductive load when the collector current becomes constricted in a small region due to base–emitter reverse biasing.   For such a breakdown to occur, the device needs only a small amount of energy and cannot be influenced by heat sink design. The manufacturer provides specifications in terms of safe operating areas (SOAs) during turn-on and turn-off, which are shown in Figs. 1.17 and 1.18, respectively. In Fig. 1.17, the collector current is permissible on a continuous basis below the A-B-C-D line.   This current can be exceeded up to I_{Cmax} line if operated on a pulsed basis with no more than a 1% turn-on duty cycle.   The regions A and B are based on the bonding wire limit and junction heating limit, respectively.   Regions C and D, which operate at a high collector voltage, are limited by the second breakdown effect.   The turn-on load line should be shaped by circuit design so that not only is FBSB avoided, but turn-on dissipation is minimized.   Figure 1.18 shows RBSOA under an inductive-load clamped condition where the test conditions are as shown in the diagram.   Again, a practical turn-off load line should be designed so that it remains with this curve and the turn-off dissipation is minimized.

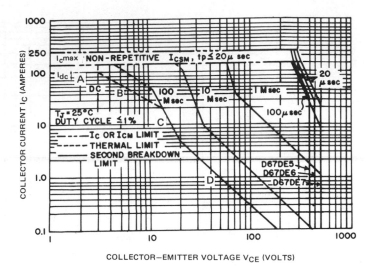

Figure 1.17 Forward bias safe operating area (FBSOA) (GE-D67DE).

Switching Characteristics

From the discussion above, it is obvious that an effective snubber circuit is mandatory, which will reduce the device switching losses and protect the device from second breakdown. Figure 1.19 shows a transistor chopper with a polarized snubber and Fig. 1.20 explains the switching waveforms. The single quadrant chopper has an inductive load which is bypassed by fast recovery diodes D_1 and D_3. The snubber circuit consists of the elements R_1, L_1, R_2, C_2, and D_2. Assume that initially the base current $I_B = 0$ and the transistor is turned off. At this condition, the inductive load current free-wheels entirely through L_1 and D_3, and C_2 remains charged to supply voltage V_d. If base current I_B is established to turn on the

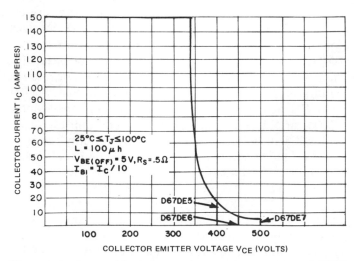

Figure 1.18 Reverse bias safe operating area (RBSOA) (GE-D67DE).

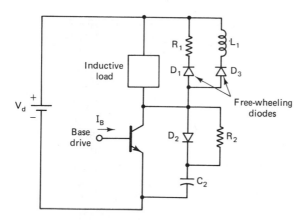

Figure 1.19 Transistor chopper with polarized snubber.

device, the collector voltage will fall to zero after turn-on time $t_{on} = t_d + t_r$, where t_d is the delay time and t_r is the rise time, and the supply voltage will be absorbed across L_1. Without an L_1 circuit and $R_1 = 0$ as in a conventional chopper, the D_1 recovery current will be taken by the transistor with full impressed V_d, causing a high turn-on dissipation and possibly second breakdown failure. Of course, at

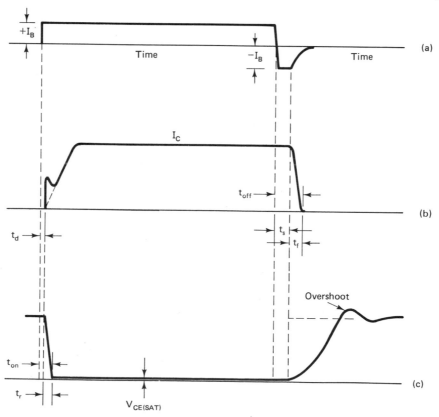

Figure 1.20 Switching characteristics of transistor with snubber: (a) base current; (b) collector current; (c) collector voltage.

turn-on, C_2 discharges initially through the transistor limited by R_2, but the load current builds up in it with a ramp slope V_d/L_1 until the L_1 current is zero. In the on condition, load current I_C flows in the transistor and the C_2 drop is equal to $V_{CE(\text{sat})}$. The turn-off of the device is initiated by injecting a negative base current within the limit of emitter–base avalanche voltage V_{EBO}. During storage time t_s, the collector junction remains forward biased and excess minority carriers in the base sweep out of the base lead. Then during fall time t_f, the collector begins to block until turn-off is complete. As the device turns off, the collector voltage rises gradually because the device current is shunted to C_2 through series D_2. Meanwhile, the load current tends to free-wheel initially through D_1 and R_1 and then through D_3 and L_1 in the steady state. Without C_2, the collector voltage will jump to the supply voltage as soon as the collector current begins to fall, resulting in high dissipation and consequent second breakdown failure. Note that V_{CE} overshoots to some extent as the load current transiently transfers from the R_1, D_1 to the D_3, L_1 circuit. The overshoot voltage should be duly considered in selecting the voltage rating of the device. The practical switching frequency should be limited so that the snubber transient disappears before beginning another cycle.

Base Drive

Figure 1.21 shows a practical base drive circuit for a GE-D66D Darlington transistor which has ratings of $V_{CEO(SUS)} = 500$ V and $I_{Cmax} = 30$ A. The logic signal from

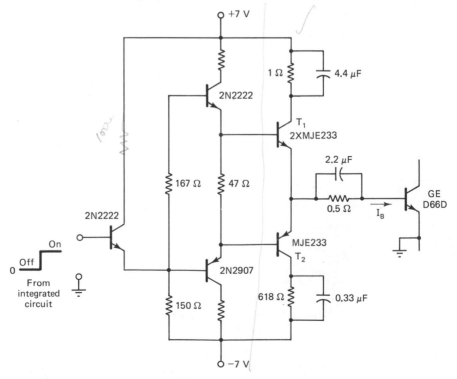

Figure 1.21   Transistor base drive circuit.

an integrated circuit is amplified in several stages and converted to a bipolar signal for base drive. The positive base current is established by switching on T_1, whereas negative current is impressed by switching on T_2. The speed-up capacitors shown provide an overdrive with high dI_B/dt at the leading edges of both positive and negative currents and therefore help to decrease switching times. The base drive power supply and logic signal normally require electrical isolation when used in an inverter circuit. The signal is normally isolated optically and the power supply for each transistor in an inverter is built separately with transformer isolation.

1.7 POWER MOSFET

The MOSFET as power semiconductor devices have shown great promise for low-power (up to a few kWs), high-frequency applications. Originally, the devices employed surface-groove technology, called VMOS, but today planar DMOS structure, shown in Fig. 1.22, is used for high-voltage devices. The circuit symbol of the device is shown in Fig. 1.22(b). Unlike a bipolar transistor, it is a voltage-controlled majority-carrier device. With a positive voltage applied to the gate with respect to the source terminal (NMOS), it induces an N-channel and permits electron current to flow from the source to the drain with applied voltage V_{DS}. Because of SiO_2 layer isolation, the gate circuit impedance is extremely high, typically in the range $10^9 \ \Omega$. This feature permits power MOSFET drive directly from CMOS or TTL logic. The devices have an integrated reverse rectifier which permits free-wheeling current of the same magnitude as that of the main device.

Static Characteristics

The fundamental drain–source static characteristics of a power MOSFET is shown in Fig. 1.23, which also shows collector–emitter characteristics for a conventional

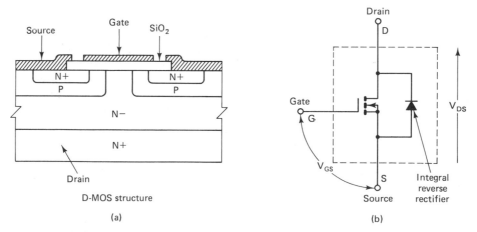

Figure 1.22 Power MOSFET structure and circuit symbol. Courtesy of International Rectifier Corporation.

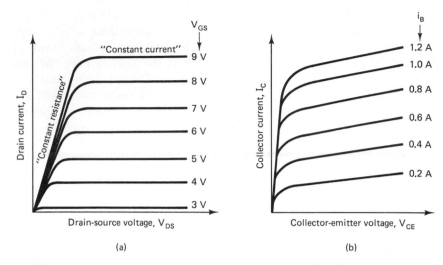

Figure 1.23 Comparison of idealized output characteristics of (a) power MOS-FET, and (b) bipolar power transistor. Courtesy of International Rectifier Corporation.

bipolar transistor for comparison. The gate circuit has a threshold voltage, typically 2 to 4 V below which the drain current is very small. With V_{GS} beyond the threshold value, I_D–V_D characteristics have two distinct regions, a constant-resistance $[R_{DS(\text{on})}]$ region and a constant-current region. The $R_{DS(\text{on})}$ of a MOSFET is a key parameter and determines the conduction voltage drop. For a device, the $R_{DS(\text{on})}$ increases with the voltage rating and has positive temperature coefficient characteristics. For example, the conduction drop of a 30-A 100-V MOSFET is typically 1.5 V, which increases to 3 V at a maximum junction temperature of 150°C. For a 500-V device the drop is 4 V at 10 A and 25°C, which increases to 8 V at 150°C.

Whereas the conduction loss of a high-voltage MOSFET is very high, its switching loss is almost negligible. The device does not have minority-carrier delay time as in a bipolar device, and its switching times are determined essentially by the ability of the drive to charge and discharge a tiny input capacitance C_{ISS}, defined as $C_{ISS} = C_{GS} + C_{GD}$ with C_{DS} shorted, where C_{GS} is the gate-to-source capacitance, C_{GD} the gate-to-drain capacitance, and C_{DS} the drain-to-source capacitance. Although MOSFET can be controlled statically by the voltage source, it is normal practice to drive it by a current source dynamically followed by a voltage source to minimize switching delays. Figure 1.24 shows a typical gate drive circuit of a MOSFET with a simple RC snubber across the device.

Safe Operating Area

As mentioned before, a MOSFET is a majority-carrier device and therefore it has a positive temperature coefficient of resistance and its second breakdown effect is minimal compared to that of a bipolar transistor. If localized and potentially destructive heating occurs within the device, the positive temperature coefficient

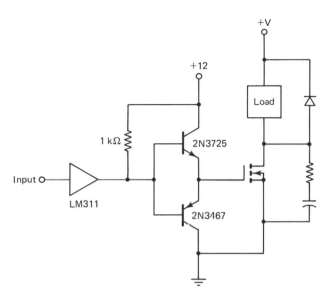

Figure 1.24 MOSFET gate drive circuit with snubber. Courtesy of International Rectifier Corporation.

effect of resistance forces local current concentration to be uniformly distributed across the total area. The maximum safe operating area of a MOSFET is shown in Fig. 1.25. The device has a continuous current rating of 5 A, which can be raised to an absolute maximum value of 10 A on a pulsed basis. Similarly, the maximum drain-to-source voltage V_{DS} is limited to 400 V without causing avalanche breakdown. The safe operating areas of a MOSFET are determined solely by thermal considerations, and the power limit can be calculated from thermal impedance information. As shown in the figure, dc current is limited in the full voltage range by 125 W of power dissipation. This corresponds to a dc junction-to-case thermal resistance $\Theta_{jC} = 1°\text{C/W}$ for $T_{jmax} = 150°\text{C}$ and infinite heat sink with

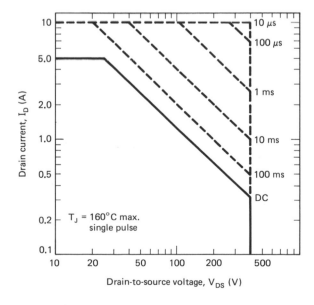

Figure 1.25 Safe operating areas of MOSFET (IR-305). Courtesy of International Rectifier Corporation.

$T_A = 25°C$. Note that there is no secondary slope on SOA curves indicating second breakdown as commonly seen on bipolar SOA curves (Fig. 1.17). The figure also shows SOA curves based on single-pulse operation. For example, at $V_{DS} = 150$ V, a dc current of 0.83 A (125 W) can be increased to 7 A (1050 W) with 1-ms single pulse. The SOA curves can be modified for operation of the device on a duty-cycle basis. Figure 1.26 shows transient thermal impedance curves for a MOSFET on a single-pulse and duty-cycle basis. The junction power dissipation can be calculated from the expression

$$P = \frac{T_{jmax} - 25}{\Theta_{\text{eff}}(t_p, D)} \tag{1.14}$$

where $T_{jmax} = 150°C$, $T_C = T_A = 25°C$, Θ_{eff} is the transient thermal impedance, t_p the pulse duration, and $D = t_p/T$ the duty cycle. For example, a duty cycle of 50% and a pulse width of 1 ms gives $\Theta_{\text{eff}} = 0.8$ (i.e., $P = 156.25$ W).

With inductive load, the drain-to-source voltage transient is coupled to the gate via the drain-to-gate capacitance C_{DG}. To protect the gate from damaging overvoltage, zener diode protection is necessary if the drive circuit impedance is low. The inductive load is normally bypassed by a fast recovery diode, but the stray inductance may cause destructive overvoltage. This overvoltage can be attenuated by a simple RC snubber as shown in Fig. 1.24.

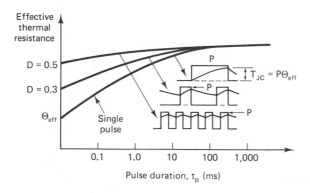

Figure 1.26 Transient thermal impedance curves of MOSFET. Courtesy of International Rectifier Corporation.

1.8 HYBRID DEVICES

A number of hybrid power semiconductor devices have recently received attention, of which only two will be discussed briefly here.

Insulated Gate Transistor

An insulated gate transistor (IGT) is a MOS gate turn on/off bipolar transistor which combines the attributes of the MOSFET, bipolar transistor and thyristor. The equivalent circuit of IGT is shown in Fig. 1.27. The device has the high input impedance of a MOSFET but a low on-state conduction drop similar to that of a bipolar transistor. The switching speed and safe operating area of bipolar transistor

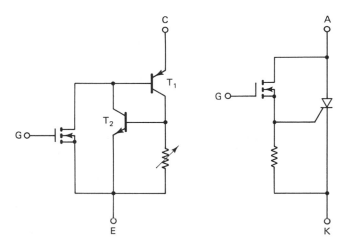

Figure 1.27 Insulated gate transistor (IGT).

Figure 1.28 FET-controlled thyristor.

are retained. The storage time of the bipolar tends to be long because of its incapability to drive negative base current. However, the device has thyristor-like reverse voltage blocking capability.

FET-Controlled Thyristor

A FET-controlled thyristor in which a MOSFET and a thyristor are connected in parallel is shown in Fig. 1.28. The device has the input impedance of a FET but the conduction drop of a thyristor. With the gate voltage exceeding the threshold voltage, current flows in the shunt resistance and switches on the thyristor. The device can be fired optically from an integrated circuit, providing electrical isolation between the control and power circuits.

REFERENCES

1. B. R. Pelly, "Power Semiconductor Devices—A Status Review," *Conf. Rec. IEEE/IAS Int. Sem. Power Conv. Conf.*, pp. 1–19, May 1982.
2. *SCR Manual*, 6th ed., General Electric Company, 1979.
3. H. LeHuy, P. Viarouge, and Y. Jean, "Power Transistor Base Drive Circuits Using Power MOSFETS," *Conf. Rec. IEEE/IAS Annu. Meet.*, pp. 782–793, Oct. 1982.
4. *Semiconductor Data Handbook*, 3rd ed., General Electric Company, 1977.
5. *Transistors—Diodes* (Electronic Data Library), General Electric Company, 1982.
6. W. McMurray, "Optimum Snubbers for Power Semiconductors," *IEEE Trans. Ind. Appl.*, Vol. IA-8, pp. 593–600, Sept.–Oct. 1972.
7. N. Seki, K. Ichikawa, Y. Tsuruta, and K. Matsuzaki, "Present Condition of High Power GTO Applications," *Conf. Rec. Power Conv. Intl.*, pp. 391–403, 1982.

2

AC MACHINES

2.0 INTRODUCTION

Ac machines are the workhorses in adjustable-speed ac drive systems and understanding their characteristics thoroughly is mandatory in designing today's complex drive systems. Traditionally, ac machines have been used in open-loop constant-speed applications where the steady-state characteristics are of principal importance. In closed-loop adjustable-speed drive applications, a consideration of static as well as dynamic behavior is important. The dynamic behavior of an ac machine is considerably more complex than that of dc machine.

In this chapter we study the basic static and dynamic characteristics of induction and synchronous machines. The dynamic $d-q$ model of ac machines is introduced. The performance characteristics are discussed with particular emphasis on variable-speed applications. Further details on ac machine characteristics are presented in subsequent chapters.

2.1 INDUCTION MACHINE

The induction machine, particularly the cage type, is most commonly used in adjustable-speed ac drive systems. Figure 2.1 shows an idealized three-phase two-pole induction motor where each phase in the stator and rotor windings has a concentrated coil. The stator windings are supplied with balanced three-phase ac voltage, which induces current in the short-circuited rotor windings by induction or transformer action.

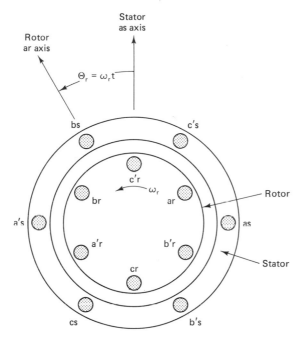

Figure 2.1 Idealized three-phase two-pole induction motor.

Torque Production

Neglecting the effect of harmonics due to nonideal distribution of windings and nonsinusoidal voltage and current waves, it can be shown that the stator establishes a spatially distributed sinusoidal flux density wave in the air gap which rotates at synchronous speed given by

$$N_e = \frac{120 f_e}{P} \tag{2.1}$$

where N_e is the speed in rpm, f_e the stator frequency in hertz, and P the number of poles. If the rotor is initially stationary, its conductors will be subjected to a sweeping magnetic field, inducing rotor current at the same frequency. It is the interaction of air gap flux and rotor magnetomotive force (mmf) which produces torque in the machine, as explained by the waveforms of Fig. 2.2. At the synchronous speed of the machine, the rotor cannot have any induction and therefore torque cannot be produced. At any other speed N_r, the speed differential $N_e - N_r$ creates slip, and correspondingly the per unit slip S is defined as

$$S = \frac{N_e - N_r}{N_e} = \frac{\omega_e - \omega_r}{\omega_e} = \frac{\omega_{sl}}{\omega_e} \tag{2.2}$$

where ω_e is the stator angular frequency, ω_r the rotor electrical speed in angular frequency, and ω_{sl} the slip angular frequency. The air gap flux moving at slip frequency ω_{sl} relative to the rotor induces slip frequency voltage in the rotor, which

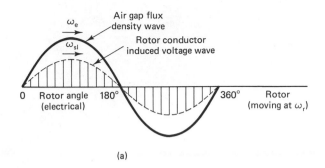

(a)

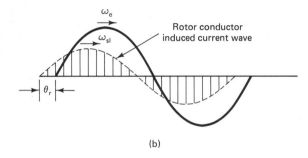

(b)

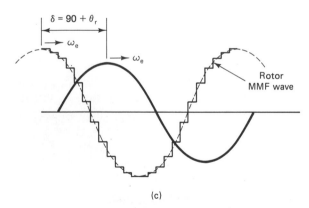

(c)

Figure 2.2 Torque production by interaction of airgap flux and rotor mmf wave. From *Electric Machinery* by Fitzgerald and Kingsley © 1971, McGraw-Hill Book Company.

correspondingly produces slip frequency current in the short-circuited rotor. In Fig. 2.2(a), the sinusoidal air gap flux density wave moving at speed ω_e induces voltage in the rotor conductors, shown by the vertical lines. The resulting rotor current wave lags the voltage wave by the rotor power factor angle θ_r. The stepped rotor mmf wave can be constructed from the current wave, which can be approximated by the dashed line shown in the figure. Since the rotor is moving at speed ω_r and its current wave is moving at speed ω_{sl} relative to the rotor, the rotor mmf wave moves at the same speed as that of the air gap flux wave. The derivation of an exact torque expression is somewhat involved and can be shown as (Ref. 1)

$$T_e = \pi\left(\frac{P}{2}\right)lrB_pF_p \sin \delta \qquad (2.3)$$

where P is the number of poles, l the axial length of the machine, r the machine radius, B_p the peak value of the air gap flux density, F_p the peak value of the rotor mmf, and $\delta = 90 + \theta_r$. Equation (2.3) can also be derived in the form

$$T_e = \frac{3}{2}\left(\frac{P}{2}\right)|\hat{\psi}_m|\,|\hat{I}_r|\sin\delta \qquad (2.4)$$

where $|\hat{\psi}_m|$ is the peak value of the air gap flux linkage per pole and $|\hat{I}_r|$ is the peak value of the rotor current.

Equivalent Circuit

The physical explanation of induction motor operation given so far helps to develop a transformer-like per phase equivalent circuit shown in Fig. 2.3 which is extremely important for steady-state performance analysis. The synchronously rotating air

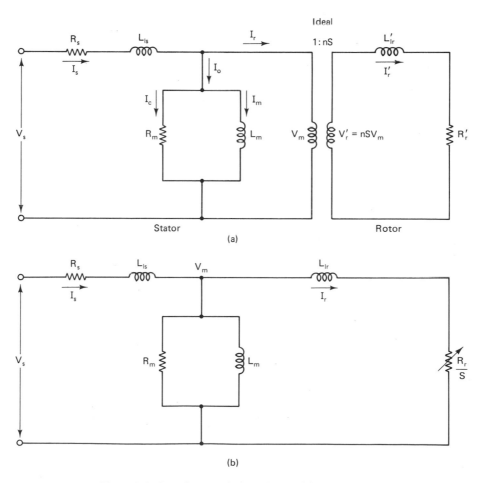

Figure 2.3 Per phase equivalent circuit of induction motor.

$\therefore \dfrac{1}{n} = \dfrac{N_{stator}}{N_{rotor}} \qquad \dfrac{N_{rotor}}{N_{stator}} \approx n$

gap flux wave generates a counter emf V_m, which is then converted to slip voltage $V'_r = nSV_m$ in rotor phase, where n is the rotor-to-stator turns ratio and S = slip [per unit (pu)]. The stator terminal voltage V_s differs from voltage V_m by the drops in resistance R_s and leakage inductance L_{ls}. The no-load excitation current I_0 consists of two components: a core loss component $I_c = V_m/R_m$ and a magnetizing component $I_m = V_m/\omega_e L_m$, where R_m is the equivalent resistance for excitation loss and L_m the magnetizing inductance. The rotor induced voltage V'_r causes rotor current I'_r at slip frequency ω_{sl}, which is limited by rotor resistance R'_r and leakage impedance $\omega_{sl}L'_{lr}$. The stator current I_s consists of excitation component I_0 and the rotor reflected current I_r. Figure 2.3(b) shows the equivalent circuit with respect to the stator, where I_r is given as

$$I_r = nI'_r = \frac{n^2 SV_m}{R'_r + j\omega_{sl}L'_{lr}}$$

$$= \frac{V_m}{\left(\dfrac{R_r}{S}\right) + j\omega_e L_{lr}} \tag{2.5}$$

where the parameters R_r and L_{lr} are referred to the stator. At standstill, $S = 1$ and therefore Fig. 2.3(b) corresponds to the short-circuited transformer equivalent circuit. At synchronous speed (i.e., $S = 0$) the current $I_r = 0$ and the machine takes excitation current I_0 only. At a subsynchronous speed ($0 < S < 1.0$) and with a small value of S, the rotor current I_r is principally influenced by R_r/S ($R_r/S \gg \omega_e L_{lr}$).

The phasor diagram of equivalent circuit is shown in Fig. 2.4. The torque

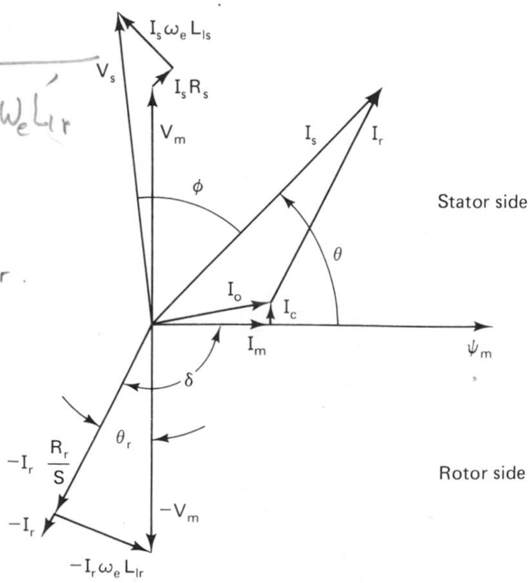

Handwritten margin notes:

$I'_r = \dfrac{V'_r}{R'_r + j\omega_{sl}L'_{lr}}$

$= \dfrac{nSV_m}{R'_r + j\omega_{sl}L'_{lr}}$

$= \dfrac{nV_m}{\dfrac{R'_r}{S} + j\dfrac{\omega_{sl}}{S}L'_{lr}}$

$I_r = nI'_r$

$= \dfrac{n^2 V_m}{\dfrac{R'_r}{S} + j\dfrac{\omega_{sl}}{S}L'_{lr}}$

$= \dfrac{V_m}{\dfrac{1}{n^2}\left(\dfrac{R'_r}{S}\right) + j\dfrac{1}{n^2}\omega_e L_{lr}}$

$= \dfrac{V_m}{\left(\dfrac{R_r}{S}\right) + j\omega_e L_{lr}}$

Figure 2.4 Phasor diagram for equivalent circuit of Fig. 2.3(b).

(Labels: $I_s\omega_e L_{ls}$, V_s, $I_s R_s$, V_m, I_s, I_r, ϕ, Stator side, θ, I_0, I_c, I_m, ψ_m, δ, $-I_r\dfrac{R_r}{S}$, θ_r, $-I_r$, Rotor side, $-V_m$, $-I_r$, $-I_r\omega_e L_{lr}$)

and by substituting $-S_m$, the regeneration breakdown torque is

$$T_{eg} = -\frac{3}{4}\frac{P}{\omega_e}\frac{V_s^2}{\sqrt{R_s^2 + \omega_e^2(L_{ls} + L_{lr})^2} - R_s} \tag{2.22}$$

Note that $|T_{em}| = |T_{eg}|$ if stator resistance R_s is neglected. A further simplification of the equivalent circuit of Fig. 2.5 can be made by neglecting the stator parameters R_s and L_{ls}. This assumption is not unreasonable for an integral-horsepower machine and if the speed is typically above 10%. Then the torque equation (2.18) can be simplified as

$$T_e = 3\left(\frac{P}{2}\right)\left(\frac{V_s}{\omega_e}\right)^2\frac{\omega_{sl}R_r}{R_r^2 + \omega_{sl}^2 L_{lr}^2} \tag{2.23}$$

The equation (2.23) can be shown to be analogous to equation (2.6) by substituting

$$I_r = \frac{SV_s}{\sqrt{R_r^2 + \omega_{sl}^2 L_{lr}^2}} \tag{2.24}$$

$$\cos\theta_r = \frac{R_r}{\sqrt{R_r^2 + \omega_{sl}^2 L_{lr}^2}} \tag{2.25}$$

and recognizing that the air gap flux ψ_m can be given by

$$\psi_m = \frac{V_s}{\omega_e} \tag{2.26}$$

In a low-slip region, equation (2.23) can be approximated as

$$T_e = 3\left(\frac{P}{2}\right)\frac{1}{R_r}\psi_m^2\omega_{sl} \tag{2.27}$$

where $R_r^2 \gg \omega_{sl}^2 L_{lr}^2$. Equation (2.27) indicates that at constant ω_{sl}, $T_e \propto \psi_m^2$ or for constant ψ_m, $T_e \propto \omega_{sl}$.

Variable-Voltage Operation

A simple and economical method of speed control of a cage-type induction motor is to vary the stator voltge at constant frequency. The stator voltage at line frequency can be controlled by phase-angle control of antiparallel thyristors connected in each phase. Figure 2.7 shows the torque–speed curves with variable stator voltage which have been plotted from equation (2.18). A load–torque curve for a fan or blower-type drive ($T_L \propto \omega_r^2$) is also shown in the figure, where the points of intersection define stable points for variable-speed operation. The motors with high slip S_m (i.e., with high rotor resistance under the NEMA Class D category) are normally used in this method of speed control, and this correspondingly causes higher copper loss in the machine. The range of speed control will evidently be diminished if a low-slip machine is used. If, on the other hand, the machine is designed such that $S_m \geq 1$, a constant-torque type of load can be controlled in the

equation (2.4) can be expressed in the form

$$T_e = K\psi_m I_r \sin\delta \tag{2.6}$$

where ψ_m and I_r are the rms values shown in the phasor diagram. If the core loss current component I_c is neglected, equation (2.6) can be simplified as

$$T_e = K'I_m I_r \sin\delta$$
$$= K'I_m I_s \sin\theta \tag{2.7}$$
$$= K'I_m I_a$$

ignoring the effect of the sign. The torque equation (2.7) is analogous to that of a dc machine, where I_m is the magnetizing or flux component of stator current, $I_a = I_s \sin\theta$ = the armature or torque component of the stator current, and K' the torque constant. The current components I_m and I_a are orthogonal or mutually decoupled.

Equivalent-Circuit Analysis

The loss and power expressions of a machine can be summarized as:

$$\text{input power } P_{in} = 3V_s I_s \cos\phi \tag{2.8}$$
$$\text{stator copper loss } P_{ls} = 3I_s^2 R_s \tag{2.9}$$
$$\text{core loss } P_{lc} = 3\frac{V_m^2}{R_m} \tag{2.10}$$
$$\text{power across air gap } P_g = 3I_r^2\frac{R_r}{S} \tag{2.11}$$
$$\text{rotor copper loss } P_{lr} = 3I_r^2 R_r \tag{2.12}$$
$$\text{output power } P_0 = P_g - P_{lr} = 3I_r^2 R_r\frac{1-S}{S} \tag{2.13}$$
$$\text{shaft power } P_{sh} = P_o - P_{FW} \tag{2.14}$$

where P_{FW} is friction and windage loss. Since the output power is the product of developed torque T_e and speed, T_e can be expressed as

$$T_e = \frac{P_o}{\omega_m} = \frac{3}{\omega_m}I_r^2 R_r\frac{1-S}{S} = 3\left(\frac{P}{2}\right)I_r^2\frac{R_r}{S\omega_e} \tag{2.15}$$

where $\omega_m = (2/P)\omega_r$ is the rotor mechanical speed (rad/s). Substituting equation (2.11) in (2.15) yields

$$T_e = \left(\frac{P}{2}\right)\frac{P_g}{\omega_e} \tag{2.16}$$

which indicates that torque can be calculated from the air gap power by knowing

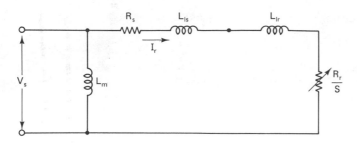

Figure 2.5   Approximate equivalent circuit.

the stator frequency. The power P_g is often defined as torque in synchronous watts.

The equivalent circuit of Fig. 2.3(b) can be simplified to that of Fig. 2.5, where the core-loss resistor R_m has been dropped out and magnetizing inductance L_m has been transferred to the input. The approximation is valid for an integral-horsepower machine where $|(R_s + j\omega_e L_{ls})| \ll \omega_e L_m$. In Fig. 2.5, the current I_r can be solved as

$$I_r = \frac{V_s}{\sqrt{(R_s + R_r/S)^2 + \omega_e^2(L_{ls} + L_{lr})^2}} \tag{2.17}$$

Substituting equation (2.17) in (2.15) yields

$$T_e = 3\left(\frac{P}{2}\right)\frac{R_r}{S\omega_e}\frac{V_s^2}{(R_s + R_r/S)^2 + \omega_e^2(L_{ls} + L_{lr})^2} \tag{2.18}$$

Torque–speed curve.   If the supply voltage and frequency are constants, the torque T_e can be calculated as a function of slip S from equation (2.18). Figure 2.6 shows the torque–speed ($\omega_r/\omega_e = 1 - S$) curve, where the value of the slip is extended beyond the region $0 \le S \le 1.0$. The zones can be defined as plugging ($1.0 \le S \le 2.0$), motoring ($0 \le S \le 1.0$), and regeneration ($S < 0$). In the normal motoring region, $T_e = 0$ at $S = 0$, and as S increases (i.e., speed decreases), T_e increases in a quasi-linear curve until breakdown torque T_{em} is reached. In this region, the stator drop is small and the air gap flux remains approximately constant. Beyond the breakdown torque, T_e decreases with the increase of S. The machine starting torque T_{es} at $S = 1$ can be written from equation (2.18) as

$$T_{es} = 3\left(\frac{P}{2}\right)\frac{R_r}{\omega_e}\frac{V_s^2}{(R_s + R_r)^2 + \omega_e^2(L_{ls} + L_{lr})^2} \tag{2.19}$$

In the plugging region, the rotor rotates in the opposite direction to that of air gap flux, so that $S > 1$. This condition may arise if the stator supply phase sequence is reversed when the rotor is moving or because of an overhauling type of load which drives the rotor in the opposite direction. The torque during plugging appears as braking torque. The energy due to the plugging brake is dissipated

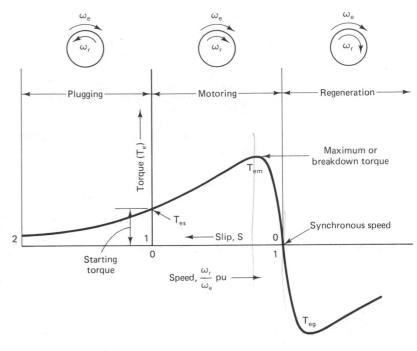

Figure 2.6   Torque–speed curve at constant voltage and frequency.

within the machine and therefore excessive machine heating must be guarded against. In the regeneration region, the rotor moves at supersynchronous speed in the same direction as the air gap flux so that the slip becomes negative, creating negative or regeneration torque. The negative slip corresponds to negative equivalent resistance R_r/S in Fig. 2.5. The positive resistance R_r/S consumes energy during motoring, but the negative R_r/S generates energy and sends it back to the source. The machine therefore operates in an induction generator mode. In a variable-frequency induction motor drive, the stator frequency can be controlled to be lower than the rotor speed ($\omega_e < \omega_r$) to obtain a regenerative braking effect. An induction motor can, of course, continually operate as a generator if its shaft is rotated at supersynchronous speed by a prime mover.

If equation (2.18) is differentiated with respect to slip and equated to zero, then

$$S_m = \pm\frac{R_r}{\sqrt{R_s^2 + \omega_e^2(L_{ls} + L_{lr})^2}} \tag{2.20}$$

where S_m is the slip corresponding to breakdown torque T_{em}. Substituting $+S$ in equation (2.18), the motoring breakdown torque is

$$T_{em} = \frac{3}{4}\frac{P}{\omega_e}\frac{V_s^2}{\sqrt{R_s^2 + \omega_e^2(L_{ls} + L_{lr})^2} + R_s} \tag{2.2}$$

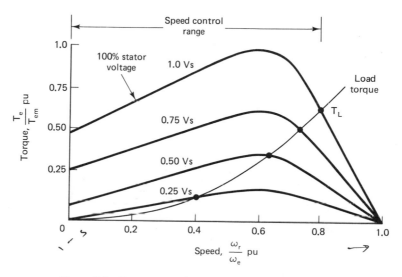

Figure 2.7 Torque–speed curves with variable stator voltage.

full range of speed.   The classical two-phase servomotors and single-phase appliance-type drives operate on a variable-voltage, constant-frequency principle.   In this method of speed control, the developed torque per ampere of stator current is reduced as the stator voltage (i.e., the air gap flux) is reduced.   Therefore, for a constant-load torque, the stator current increases as the speed is reduced, resulting in more copper loss, thereby causing a severe machine heating problem.   For the square-law characteristic of load torque $(T_L \propto \omega_r^2)$, it can be shown (Ref. 4) that the stator current reaches its maximum value which is higher than the full-load stator current at approximately two-thirds of synchronous speed.

Variable-Frequency Operation

If the stator frequency is increased beyond the rated value, the torque–speed curves, derived from equation (2.18) can be plotted as shown in Fig 2.8.   The air gap flux and stator current decrease as frequency increases, and correspondingly the maximum developed torque also decreases.   The maximum torque as a function of slip can be given from equation (2.23)

$$T_{em} = 3\left(\frac{P}{2}\right)\left(\frac{V_s}{\omega_e}\right)^2 \frac{\omega_{slm}R_r}{R_r^2 + \omega_{slm}^2 L_{lr}^2} \tag{2.28}$$

where $\omega_{slm} = R_r/L_{lr}$ is the slip frequency at maximum torque.   The equation shows $T_{em}\omega_e^2$ = constant (i.e., the machine behaves like a dc series motor in variable-frequency operation).

    If an attempt is made to decrease the supply frequency at rated voltage, the air gap flux will saturate, causing excessive stator current.   Therefore, the region below the base frequency ω_b should be accompanied by the corresponding reduction

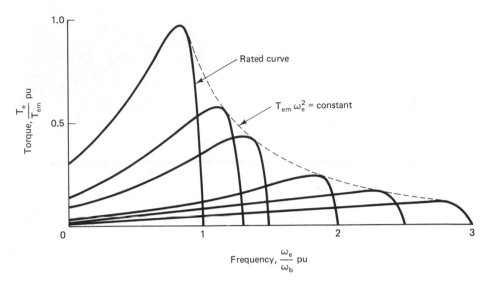

Figure 2.8 Torque–speed curves at variable frequency.

of stator voltage so as to maintain constant air gap flux. Figure 2.9 shows the plot of torque–speed curves where the V_s/ω_e ratio is maintained constant. The maximum torque T_{em} given by equation (2.28) remains approximately valid except in the low-frequency region where the air gap flux is reduced by the stator impedance drop. Therefore, in this region stator drop has to be compensated by an additional voltage boost so as to produce maximum torque.

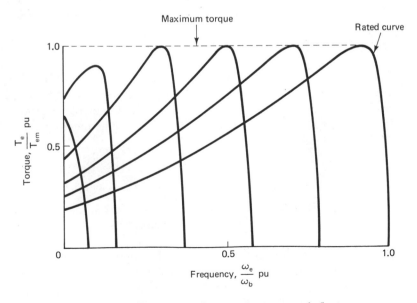

Figure 2.9 Torque–speed curves at constant volts/hertz.

less than voltage.
rated

Since the motor is operated at a constant air gap flux in the constant-torque region, the torque sensitivity per ampere of stator current is high, permitting fast transient response of the drive system. In a variable-voltage, variable-frequency drive system, the machine usually has a low slip characteristic, giving improved efficiency. In spite of the low inherent starting torque for base-frequency operation, the machine can always be started at maximum torque, as indicated in Fig. 2.9. By far the majority of adjustable-speed, industrial ac drives operate with a variable-voltage, variable-frequency power supply.

The different regions of torque–speed curves of a practical drive system with a variable-voltage, variable-frequency supply are shown in Fig. 2.10 and the corresponding voltage–frequency relation is shown in Fig. 2.11. The figure also shows torque, stator current, and slip as a function of frequency. In the constant-torque region, the maximum available torque is shown somewhat lower than the breakdown torque due to the limited inverter current capability.

At the right edge of the constant-torque region, the stator voltage reaches the rated value and then the machine enters the constant-power region. In this region, the air gap flux decreases, but the stator current is maintained constant by increasing the slip. This is equivalent to the field weakening mode of a dc separately excited motor. At the edge of the constant-power region, the breakdown torque T_{em} is reached and then the machine speed can be further increased by increasing the frequency as shown in Fig. 2.8 with the reduction of stator current. Any operating point below the maximum-torque envelope can be obtained by controlling the voltage and/or frequency and will be discussed in Chapter 7.

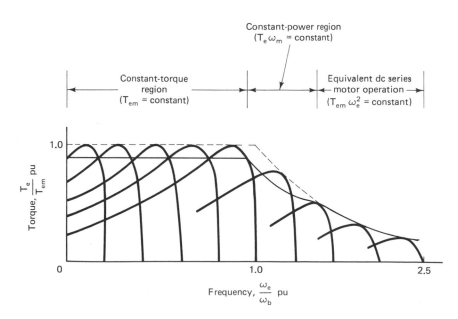

Figure 2.10 Regions of torque–speed curves with variable-voltage, variable-frequency power supply.

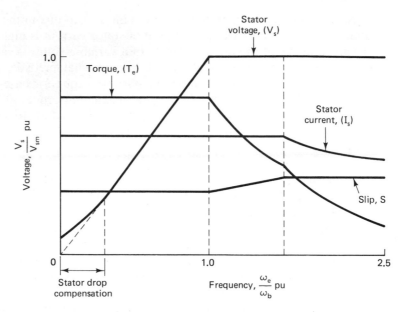

Figure 2.11 Voltage-frequency relation of induction motor.

Variable-Stator Current

Instead of controlling the stator voltage, the stator current can be controlled directly to control the developed torque. With current control, the torque characteristic depends on relative distribution of magnetizing current and rotor current for a fixed stator current magnitude but is independent of stator parameters R_s and L_{ls}. The distribution is affected by the inverse ratio of parallel impedances, which in turn are dependent on frequency and slip.

Neglecting the rotor leakage inductance ($\omega_e L_{lr} \ll R_r/S$) and core loss, the distribution of currents can be given as

$$I_m = \frac{R_r/S}{\sqrt{\omega_e^2 L_m^2 + (R_r/S)^2}}\, I_s \qquad (2.29)$$

$$I_r = I_a = \frac{\omega_e L_m}{\sqrt{\omega_e^2 L_m^2 + (R_r/S)^2}}\, I_s \qquad (2.30)$$

Substituting equations (2.29) and (2.30) in (2.7), the torque expression is

$$T_e = K' I_s^2 \frac{S\omega_e R_r L_m}{R_r^2 + S^2 \omega_e^2 L_m^2} \qquad (2.31)$$

The equation (2.31) gives torque as a function of stator current, frequency, and slip. The motor torque–speed curves at different stator currents but at fixed fre-

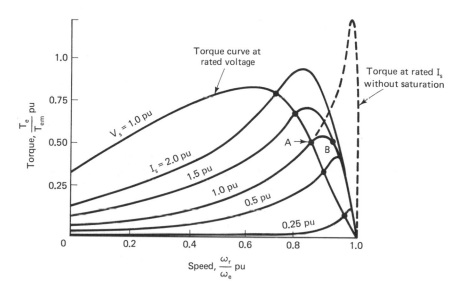

Figure 2.12   Torque–speed curves with variable stator current.

quency are shown in Fig. 2.12.   If, for example, the machine is operated at rated current ($I_s = 1.0$ pu), the starting torque will be very low compared to that of a voltage-fed machine at $V_s = 1.0$ pu.   The reason is that the air gap flux will be very low due to the rotor short-circuiting effect.   As the speed increases (i.e., the slip decreases) the stator voltage rises and as a result the torque increases with higher air gap flux.   If saturation of the machine is neglected, the torque rises to a high value, as shown by the dashed line, and then decreases to zero with a steep slope at synchronous speed.   In a practical machine, however, the saturation will limit the developed torque, as shown by the solid line.   A torque curve with the rated voltage is also shown in Fig. 2.12, where the part with negative slope can be considered to have stable operation with the rated air gap flux.   This curve inter-sccts $I_s = 1.0$ pu torque curve at point A.   The machine can be operated either at A or B for the same torque demand.   Because of lower slip at point B, the rotor current will be lower and the gap flux will be somewhat higher, causing partial saturation.   This will result in higher core loss and harmonic torque pulsation, which will be discussd later.   The stator copper loss is the same at A and B, but the rotor copper loss is somewhat higher at A.   Overall, operation at A is more desirable.   However, since A is in the unstable region of the torque curve, closed loop operation of the machine is mandatory.   The torque can be varied by varying the stator current and slip so that the air gap flux remains constant (i.e., the locus is on the negative slope of the equivalent voltage-generated torque curve).   The various operation points in the torque–frequency plane (as in Fig. 2.10) may be established by a variable-current, variable-frequency power supply.   Current-fed drives are discussed further in Chapter 7.

Harmonic Effects

In adjustable-speed drives, the machines are fed by converters which contain harmonics at the output. These harmonics have the following harmful effects: (1) heating and (2) torque pulsation.

Heating. With a voltage-fed drive, the nonsinusoidal stator voltage can be resolved into fundamental and harmonic components by Fourier analysis. For a symmetrical waveform, only the odd harmonics will be present. Again, the triplen harmonics, which are cophasal, cannot cause any current in delta or star load without a neutral. The Fourier series of phase voltages for lower-order harmonics can be given as

$$v_{as} = V_{1m} \sin \omega_e t + V_{5m} \sin 5\omega_e t + V_{7m} \sin 7\omega_e t + \cdots \qquad (2.32)$$

$$v_{bs} = V_{1m} \sin (\omega_e t - 120°) + V_{5m} \sin 5(\omega_e t - 120°)$$
$$+ V_{7m} \sin 7(\omega_e t - 120°) + \cdots \qquad (2.33)$$

$$v_{cs} = V_{1m} \sin (\omega_e t + 120°) + V_{5m} \sin 5(\omega_e t + 120°)$$
$$+ V_{7m} \sin 7(\omega_e t + 120°) + \cdots \qquad (2.34)$$

Equations (2.33) and (2.34) can be simplified as

$$v_{bs} = V_{1m} \sin (\omega_e t - 120°)$$
$$+ V_{5m} \sin (5\omega_e t + 120°) + V_{7m} \sin (7\omega_e t - 120°) + \cdots \quad (2.35)$$

$$v_{cs} = V_{1m} \sin (\omega_e t + 120°) + V_{5m} \sin (5\omega_e t - 120°)$$
$$+ V_{7m} \sin (7\omega_e t + 120°) + \cdots \qquad (2.36)$$

For each harmonic component, the machine can be approximately represented by a constant-parameter linear equivalent circuit and the resultant current can be calculated by the superposition principle. The per phase equivalent circuit of Fig. 2.3(b) can be converted to a harmonic equivalent circuit as shown in Fig. 2.13, where the core-loss resistor R_m has been omitted. In this figure, n is the order of the harmonic and S_n is the slip at the nth harmonic. Equations (2.32), (2.35), and (2.36) show that the 5th harmonic voltage V_{5m} has a negative phase

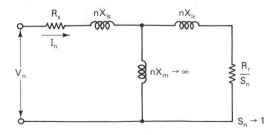

Figure 2.13 Harmonic per phase equivalent circuit.

sequence and therefore the corresponding magnetic fields rotate in the backward direction at speed $5\omega_e$. The same equations indicate that the 7th harmonic magnetic field rotates in the forward direction at speed $7\omega_e$. Since the rotor speed is related to fundamental frequency only, the rotor appears practically stationary with respect to a harmonic field (i.e., $S_n \simeq 1.0$). Mathematically, the slip at the nth harmonic can be given as

$$S_n = \frac{n\omega_e \mp \omega_r}{n\omega_e} \tag{2.37}$$

where negative and positive signs relate to forward- and backward-rotating fields, respectively. Substituting $\omega_r/\omega_e = 1 - S$ in equation (2.37) and simplifying yields

$$S_n = \frac{(n \mp 1) \pm S_1}{n} \simeq 1.0 \tag{2.38}$$

where S_1 is the fundamental-frequency slip. For example, in equation (2.38) if S_1 varies from 0 to 1, S_5 will vary from 1.2 to 1.0 and S_7 will vary from 0.857 to 1.0. The equivalent circuit of Fig. 2.13 then approximates a passive equivalent circuit where harmonic currents can be calculated by knowing the harmonic voltages. This also means that the harmonic currents are not influenced by a fundamental-frequency operating condition (i.e., these are independent of torque and speed of the machine). Assuming that $nX_m \to \infty$ and $(nX_{ls} + nX_{lr}) \gg (R_s + R_r)$,

$$I_n = \frac{V_n}{n(X_{ls} + X_{lr})} \tag{2.39}$$

and the corresponding rms harmonic current can be given as

$$I_h = \sqrt{I_5^2 + I_7^2 + I_{11}^2 + I_{13}^2 + \cdots} \tag{2.40}$$

$$= \sqrt{\sum_{n=5,7\ldots} I_n^2} \tag{2.41}$$

where I_5, I_7, and so on, are the rms harmonic components of the current. The total stator and rotor copper losses can then be calculated as

$$P_{ls} = 3(I_{sl}^2 + I_h^2)R_s \tag{2.42}$$

$$P_{lr} = 3(I_{rl}^2 + I_h^2)R_r \tag{2.43}$$

where I_{sl} and I_{rl} are the fundamental rms currents. The harmonics also increase the core loss, but its magnitude is small compared to the copper loss.

So far it has been assumed that the equivalent-circuit parameters are constant, but this assumption is hardly true. Both the stator and rotor resistances increase with temperature. The skin effect at fundamental-frequency operation can be neglected, but at harmonic frequency the effect becomes dominant, especially for rotor resistance. All the inductances in equivalent circuit vary with saturation (Ref. 5).

Torque pulsation.   Pulsating torques are produced by the interaction of air gap flux and rotor mmf waves of different harmonic order. The general torque expression as a function of air gap flux, rotor current, and phase angle between the air gap flux and the rotor current is given in equation (2.6). For the fundamental-frequency components or at any other, such as harmonic frequency, the angle δ remains constant and therefore only unidirectional torque is produced. A harmonic component of the air gap flux induces rotor current at the same frequency $(S_n \simeq 1)$, and therefore torque is produced in the same direction as the rotating air gap flux. The 7th harmonic torque, for example, adds to the fundamental torque, but the 5th harmonic torque opposes it.

The torque pulsation is produced when the angle δ varies with time. This occurs when ψ_m of one frequency interacts I_r of another frequency, modulating δ at a rate which is the difference between speeds of the corresponding rotating

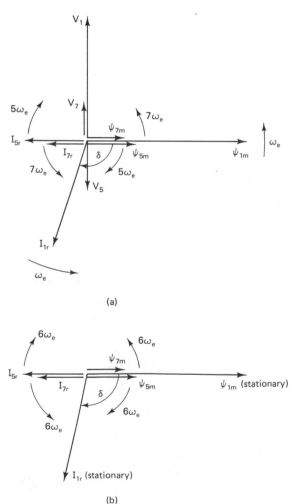

(a)

(b)

Figure 2.14 Phasor diagrams explaining sixth harmonic torque.

phasors. The pulsating torques can be calculated by superposing the flux and rotor current phasors of various frequencies in a single diagram, as shown in Fig. 2.14(a). In the figure, the effects of only the fundamental, 5th, and 7th harmonic voltages are considered and the flux phasors are assumed to be cophasal at an instant $t = 0$. Each harmonic voltage will cause the corresponding flux and rotor current components as shown. The equivalent circuit resistances for the 5th and 7th harmonics are neglected, and therefore harmonic currents will lag the respective flux components by $180°$. The fundamental and 7th harmonic phasors rotate in the anticlockwise direction at speeds ω_e and $7\omega_e$, respectively, whereas the 5th harmonic phasors rotate in the clockwise direction at speed $5\omega_e$. Figure 2.14(b) has been constructed from Fig. 2.14(a) by giving the whole diagram a clockwise rotation at ω_e to make the fundamental phasors stationary. From the diagram it can be seen that the 6th harmonic torque is contributed by the interaction of fundamental flux with the 5th and 7th harmonic currents, and of fundamental current with the 5th and 7th harmonic fluxes. Mathematically, the 6th harmonic torque expression can be written as

$$T_{e6} = K[\psi_{1m}I_{7r} \sin (\pi - 6\omega_e t) + \psi_{7m}I_{1r} \sin (\delta + 6\omega_e t)$$

$$+ \psi_{1m}I_{5r} \sin (\pi + 6\omega_e t) + \psi_{5m}I_{1r} \sin (\delta - 6\omega_e t)] \tag{2.44}$$

$$- K[\psi_{1m}(I_{7r} - I_{5r}) \sin 6\omega_e t + I_{1r}(\psi_{7m} + \psi_{5m}) \cos 6\omega_e t] \tag{2.45}$$

where $\delta \simeq 90°$. Since the harmonic flux components ψ_{7m} and ψ_{5m} are very small, the contribution of the second term can be neglected.

The pulsating torque will tend to cause jitter in the machine speed, but the effect of high-frequency components will be smoothened due to system mechanical inertia. The speed jitter may be aggravated if the pulsating frequency is low or if the system mechanical inertia is small. The pulsating torque frequency may be near the mechanical resonance of the drive system, and this may result in severe shaft vibration, causing fatigue, wearing of gear teeth, and unsatisfactory performance in the feedback control system.

Dynamic Model

So far we have considered the per phase equivalent circuit of the machine, which is valid only in steady-state operation. In an adjustable-speed drive system, the machine normally constitutes an element within a feedback loop, and therefore its dynamic behavior has to be taken into consideration. The dynamic performance of an ac machine is somewhat complex because of the coupling effect between the stator and rotor phases, where the coupling coefficients vary with rotor position. Therefore, the machine model can be described by differential equations with time-varying coefficients.

If the power supply is balanced three-phase, as is usually true when fed by a converter, the two-axis or d–q theory is normally used for dynamic modeling. In this theory, the time-varying parameters are eliminated and the variables and parameters are expressed in orthogonal or mutually decoupled direct (d) and quadrature (q) axis. The d–q dynamic model of a machine can be expressed in either a stationary or a rotating reference frame. In a stationary reference frame, the reference d^s and q^s axes are fixed on the stator, whereas in a rotating reference frame these are rotating. The rotating frame may either be fixed on the rotor or move at synchronous speed. The advantage in a synchronously rotating frame model is that with sinusoidal supply the variables appear as dc quantities in steady-state conditions.

Axes transformation.   Let us consider the stator supply voltages only and derive transformation relations between as–bs–cs axes and d^s–q^s axes, where both are in the stationary reference frame shown in Fig. 2.15. The other quantities, such as current and flux, can be transformed in a similar manner. The angle θ is arbitrary between the two sets of axes. The phase voltages in terms of d^s and q^s voltages can be written in matrix form as

$$\begin{bmatrix} v_{as} \\ v_{bs} \\ v_{cs} \end{bmatrix} = \begin{bmatrix} \cos \theta & \sin \theta & 1 \\ \cos (\theta - 120°) & \sin (\theta - 120°) & 1 \\ \cos (\theta + 120°) & \sin (\theta + 120°) & 1 \end{bmatrix} \begin{bmatrix} v_{qs}^s \\ v_{ds}^s \\ v_{0s}^s \end{bmatrix} \qquad (2.46)$$

and the corresponding inverse relation is

$$\begin{bmatrix} v_{qs}^s \\ v_{ds}^s \\ v_{0s}^s \end{bmatrix} = \frac{2}{3} \begin{bmatrix} \cos \theta & \cos (\theta - 120°) & \cos (\theta + 120°) \\ \sin \theta & \sin (\theta - 120°) & \sin (\theta + 120°) \\ 0.5 & 0.5 & 0.5 \end{bmatrix} \begin{bmatrix} v_{as} \\ v_{bs} \\ v_{cs} \end{bmatrix} \qquad (2.47)$$

where v_{0s}^s is the zero-sequence component. For balanced three-phase condition, the zero-sequence component does not exist. It has been considered only to yield the unique transformation relations.

It is convenient to set $\theta = 0$, so that the q^s-axis is coincident with the as-axis. Also ignoring the zero-sequence component, the transformation relations can be

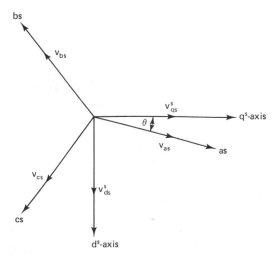

Figure 2.15 Stationary as–bs–cs to d^s–q^s axes transformation. (See Fig. 2.1)

simplified as

$$v_{as} = v_{qs}^s \tag{2.48}$$

$$v_{bs} = -\frac{1}{2}v_{qs}^s - \frac{\sqrt{3}}{2}v_{ds}^s \tag{2.49}$$

$$v_{cs} = -\frac{1}{2}v_{qs}^s + \frac{\sqrt{3}}{2}v_{ds}^s \tag{2.50}$$

and

$$v_{qs}^s = \frac{2}{3}v_{as} - \frac{1}{3}v_{bs} - \frac{1}{3}v_{cs} = v_{as} \tag{2.51}$$

$$v_{ds}^s = -\frac{1}{\sqrt{3}}v_{bs} + \frac{1}{\sqrt{3}}v_{cs} \tag{2.52}$$

The voltages in the stationary $d^s - q^s$ frame can be converted to the synchronously rotating $d^e - q^e$ frame with the help of Fig. 2.16 as follows:

$$v_{qs} = v_{qs}^s \cos \omega_e t - v_{ds}^s \sin \omega_e t \tag{2.53}$$

$$v_{ds} = v_{qs}^s \sin \omega_e t + v_{ds}^s \cos \omega_e t \tag{2.54}$$

For convenience the superscript e has been dropped from the synchronously rotating frame parameters.   Equations (2.53) and (2.54) can be inverted to define relations of stationary frame variables in terms of rotating frame variables as follows:

$$v_{qs}^s = v_{qs} \cos \omega_e t + v_{ds} \sin \omega_e t \tag{2.55}$$

$$v_{ds}^s = -v_{qs} \sin \omega_e t + v_{ds} \cos \omega_e t \tag{2.56}$$

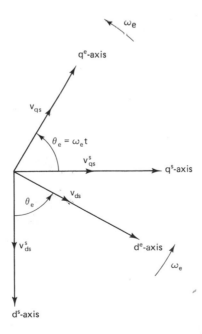

Figure 2.16 Stationary d^s-q^s axes to synchronously rotating d^e-q^e axes transformation.

Assume, for example, that the phase voltages are balanced and sinusoidal. Then

$$v_{as} = V_{sm} \cos \omega_e t \tag{2.57}$$

$$v_{bs} = V_{sm} \cos (\omega_e t - 120°) \tag{2.58}$$

$$v_{cs} = V_{sm} \cos (\omega_e t + 120°) \tag{2.59}$$

Substituting these in equations (2.51) and (2.52) yields

$$v_{qs}^s = V_{sm} \cos \omega_e t \tag{2.60}$$

$$v_{ds}^s = - V_{sm} \sin \omega_e t \tag{2.61}$$

Again, substituting those in equations (2.53) and (2.54), we have

$$v_{qs} = V_{sm} = \hat{V}_m \tag{2.62}$$

$$v_{ds} = 0 \tag{2.63}$$

These relations verify that the sinusoidal variables appear as dc quantities in a synchronously rotating reference frame.

Synchronously rotating frame model. It is possible to express the three-phase stator equations of the machine in stationary coordinates *as, bs,* and *cs.* The stator equation in vector form can be given as

$$\overline{v}_s^s = R_s \overline{i}_s^s + \frac{d\overline{\psi}_s^s}{dt} \tag{2.64}$$

where $\overline{v}_s^s$, $\overline{i}_s^s$ and $\overline{\psi}_s^s$ are instantaneous voltage, current, and flux linkage vectors, respectively, in the stationary frame. Each vector can be expressed in terms of component unit vectors in the form

$$\overline{X}^s = X_{as} \overline{U}_{as} + X_{bs} \overline{U}_{bs} + X_{cs} \overline{U}_{cs} \tag{2.65}$$

If the coordinate axes rotate at synchronous speed ω_e, equation (2.64) can be written in the form

$$\overline{v}_s = R_s \overline{i}_s + \frac{d\overline{\psi}_s}{dt} + \overline{\omega}_e \times \overline{\psi}_s \tag{2.66}$$

where the vectors are the same as in equation (2.64), but an additional term $\overline{\omega}_e \times \overline{\psi}_s$ has been added. The cross product $\overline{\omega}_e \times \overline{\psi}_s$ is defined as speed voltage due to rotation of the reference frame.

After performing the cross product, equation (2.66) can be written in terms of component d^e and q^e voltages in a synchronously rotating frame as

$$v_{qs} = R_s i_{qs} + \frac{d\psi_{qs}}{dt} + \omega_e \psi_{ds} \tag{2.67}$$

$$v_{ds} = R_s i_{ds} + \frac{d\psi_{ds}}{dt} - \omega_e \psi_{qs} \tag{2.68}$$

The substitution of $\omega_e = 0$ in the equations above will result in stator equations in the stationary d^s-q^s frame.

If the rotor is not moving, the rotor equations for a doubly fed machine will be similar to equations (2.67) and (2.68):

$$v_{ds} = R_s i_{ds} + \frac{d\psi_{ds}}{dt} - \omega_e \psi_{qs} \qquad (2.69)$$

$$v_{dr} = R_r i_{dr} + \frac{d\psi_{dr}}{dt} - \omega_e \psi_{qr} \qquad (2.70)$$

where all the variables and parameters are referred to the stator. Since the rotor moves at speed ω_r, the $d-q$ axes fixed on the rotor move at a speed $\omega_e - \omega_r$ relative to the synchronously rotating reference frame. Therefore, equations (2.69) and (2.70) should be modified as

$$v_{qr} = R_r i_{qr} + \frac{d\psi_{qr}}{dt} + (\omega_e - \omega_r)\psi_{dr} \qquad (2.71)$$

$$v_{dr} = R_r i_{qr} + \frac{d\psi_{dr}}{dt} - (\omega_e - \omega_r)\psi_{qr} \qquad (2.72)$$

Figure 2.17 shows the d^e-q^e equivalent circuits of the machine in a synchronously

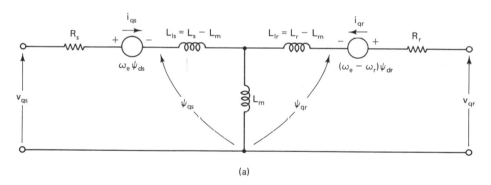

(a)

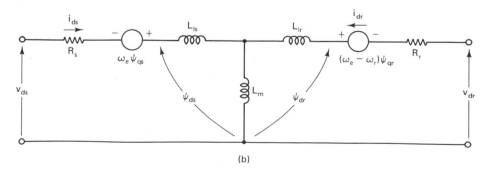

(b)

Figure 2.17 $D-Q$ equivalent circuits at synchronously rotating reference frame: (a) q^e-axis circuit; (b) d^e-axis circuit.

rotating reference frame. The flux linkage expressions in terms of the currents can be written from Fig. 2.17 as

$$\psi_{qs} = L_{ls}i_{qs} + L_m(i_{qs} + i_{qr}) \tag{2.73}$$

$$\psi_{qr} = L_{lr}i_{qr} + L_m(i_{qs} + i_{qr}) \tag{2.74}$$

$$\psi_{ds} = L_{ls}i_{ds} + L_m(i_{ds} + i_{dr}) \tag{2.75}$$

$$\psi_{dr} = L_{lr}i_{dr} + L_m(i_{ds} + i_{dr}) \tag{2.76}$$

Combining the expressions above with equations (2.67), (2.68), (2.71), and (2.72), the model of electrical dynamics in terms of voltages and currents can be given in matrix form as

$$\begin{bmatrix} v_{qs} \\ v_{ds} \\ v_{qr} \\ v_{dr} \end{bmatrix} = \begin{bmatrix} R_s + SL_s & \omega_e L_s & SL_m & \omega_e L_m \\ -\omega_e L_s & R_s + SL_s & -\omega_e L_m & SL_m \\ SL_m & (\omega_e - \omega_r)L_m & R_r + SL_r & (\omega_e - \omega_r)L_r \\ -(\omega_e - \omega_r)L_m & SL_m & -(\omega_e - \omega_r)L_r & R_r + SL_r \end{bmatrix} \begin{bmatrix} i_{qs} \\ i_{ds} \\ i_{qr} \\ i_{dr} \end{bmatrix} \tag{2.77}$$

where S is the Laplace operator. For a singly fed machine, the voltages v_{qr} and v_{dr} should be assumed as zero. If the speed ω_r is considered as constant, then knowing the inputs v_{qs}, v_{ds}, and ω_e, the currents i_{qs}, i_{ds}, i_{qr}, and i_{dr} can be solved from equation (2.77). The voltages v_{qs} and v_{ds} can correspond to nonsinusoidal waves as is true in a converter-fed machine. For a current-fed machine, the quantities i_{qs}, i_{ds}, and ω_e are independent. Then the dependent variables v_{qs}, v_{ds}, i_{qr}, and i_{dr} can be solved from equation (2.77). If only steady-state solution of equation (2.77) is desired, all the S-related terms should be zero. In the steady state, all variables in a synchronously rotating reference frame appear as dc quantities with sinusoidal excitation, as mentioned before.

The speed signal ω_r in equation (2.77) cannot normally be treated as a constant. It can be related to the torques as

$$T_e - T_L = J\frac{d\omega_m}{dt} = \frac{2}{P}J\frac{d\omega_r}{dt} \tag{2.78}$$

where T_L is the load torque and J the system inertia.

Developed torque. The development of torque by the interaction of air gap flux and rotor mmf was discussed at the beginning of this chapter. Here it will be expressed in more general form, relating the d–q components of variables. From equation (2.4) the torque can be given in general vector form as

$$T_e = \frac{3}{2}\left(\frac{P}{2}\right)\hat{\psi}_m \times \hat{I}_r \tag{2.79}$$

or in terms of d^e–q^e components as

$$T_e = \frac{3}{2}\left(\frac{P}{2}\right)(\psi_{dm}i_{qr} - \psi_{qm}i_{dr}) \tag{2.80}$$

Substituting the relations between fluxes and currents, several other forms of T_e are

$$T_e = \frac{3}{2}\left(\frac{P}{2}\right)(\psi_{dm}i_{qs} - \psi_{qm}i_{ds}) \qquad (2.81)$$

$$= \frac{3}{2}\left(\frac{P}{2}\right)(\psi_{ds}i_{qs} - \psi_{qs}i_{ds}) \qquad (2.82)$$

$$= \frac{3}{2}\left(\frac{P}{2}\right)L_m(i_{qs}i_{dr} - i_{ds}i_{qr}) \qquad (2.83)$$

Equations (2.77), (2.78), and (2.83) give the complete model of the electro-mechanical dynamics of an induction machine. The composite system equation is of fifth order and nonlinearity in the model is very evident. The equation can easily be written in state-space form. The electromechanical model with the transformation equations can be simulated on a computer to study the transient- and steady-state performances.

Stationary frame model.   The dynamic machine model in the stationary reference frame can be derived by substituting $\omega_e = 0$ in equation (2.77). In this form, it is known as the Stanley (Ref. 3) equation in d^s–q^s notation, and the corresponding equivalent circuits are shown in Fig. 2.18. As mentioned before,

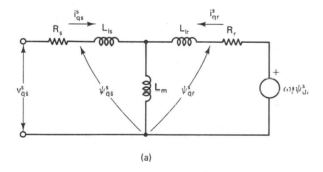

(a)

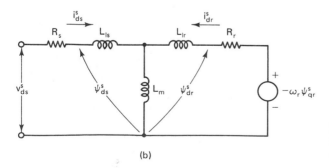

(b)

Figure 2.18 D–Q equivalent circuits in stationary reference frame: (a) q^s-axis circuit; (b) d^s-axis circuit.

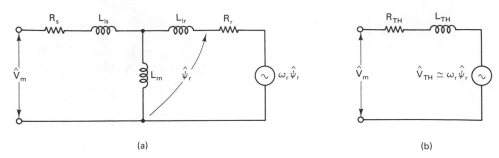

Figure 2.19 Derivation of per phase equivalent circuit with counter emf.

in the stationary reference frame the variables appear as sine waves in the steady state with sinusoidal inputs. It should be noted that torque equations (2.79) to (2.83) also remain valid with the corresponding stationary reference frame variables. The d^s–q^s equivalent circuits in Fig. 2.18 can be combined in steady state to form the per phase equivalent circuit with counter emf as shown in Fig. 2.19(a), where the phasors $\hat{V}_m$ and $\hat{\psi}_r$ are related by

$$|\hat{V}_m| = \sqrt{v_{qs}^{s2} + v_{ds}^{s2}}$$

$$|\hat{\psi}_r| = \sqrt{\psi_{qr}^{s2} + \psi_{dr}^{s2}}$$

where $|\hat{V}_m|$ and $|\hat{\psi}_r|$ are the peak values of the phasors. The simplified Thévenin equivalent circuit of Fig. 2.19(b) can be derived from Fig. 2.19(a), where $R_{TH} \simeq R_s + R_r$, $L_{TH} \simeq L_{ls} + L_{lr}$, and $\hat{V}_{TH} \simeq \omega_r\hat{\psi}_r$ assuming that L_m is large. This circuit can represent a simplified per phase transient equivalent circuit of the machine and is frequently used for converter analysis.

2.2 SYNCHRONOUS MACHINE

A synchronous machine is a serious competitor for an induction machine in adjustable-speed drive systems. The two classes of machines are analogous in many aspects, and much of the discussion in the preceding section holds true for synchronous machines. Therefore, only the salient features of a synchronous machine will be reviewed here.

A synchronous machine, as the name indicates, must rotate at synchronous speed (i.e., the speed is uniquely related to the supply frequency). Figure 2.20 shows an idealized three-phase, two-pole synchronous machine. The stator winding of the synchronous machine is identical to that of the induction machine, but the rotor has a winding that carries dc current and produces flux in the rotor. The machine shown is characterized as salient pole because of the nonuniform air gap around the rotor which contributes to asymmetrical magnetic reluctance in the d and q axes. This is in contrast to a machine with a cylindrical rotor structure having a uniform air gap, defined as a nonsalient-pole machine. Although salient-pole synchronous machines are common, high-speed machines used as turbo-generators in power stations are nonsalient-pole types. In addition to field winding,

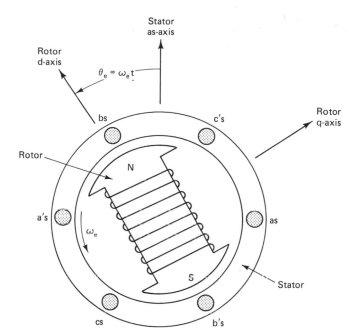

Rotor
d-axis

Stator
as-axis

$\theta_e = \omega_e t$

bs

c's

Rotor
q-axis

Rotor

N

a's

ω_e

as

S

Stator

cs

b's

Figure 2.20   Idealized three-phase two-pole synchronous machine (salient pole).

the rotor usually contains amortisseur windings, which are like short-circuited squirrel-cage bars in an induction motor.

The mechanism of torque production in a synchronous machine is the same as that of an induction motor, and therefore the previous discussions on torque production are valid here. Since the rotor always moves at synchronous speed in steady state (i.e., the slip is zero), there is no induction to the rotor and therefore the rotor mmf is supplied exclusively by the field winding.

Equivalent Circuit

A simple per phase steady-state equivalent circuit for a nonsalient-pole synchronous machine can be derived from the same physical considerations as those for an induction motor, as shown in Fig. 2.21. Figure 2.21(a) shows the transformer-like coupled equivalent circuit linking the stator and the moving rotor winding. The rotor is supplied by a field current I_f due to the supply voltage v_f. The rotor section can be substituted in terms of the stator by a current source I_f' of frequency ω_e, as shown in Fig. 2.21(b), where n is the ratio relating the rms magnitude of I_f' to the magnitude of the dc field current I_f. At steady-state operation, the power transferred to the rotor is zero, and all the power across the air gap is converted to mechanical power. Neglecting the core-loss resistor R_m, Fig. 2.21(b) can be drawn in the form of Fig. 2.21(c) using Thévenin's theorem, where $V_f = \omega_e L_m n I_f$ is the Thévenin voltage and $\omega_e L_m$ is the Thévenin impedance. The voltage V_f is defined as excitation emf and is directly proportional to the field current I_f. The sum of leakage reactance $\omega_e L_{ls}$ and magnetizing reactance $\omega_e L_m$ is known as the synchronous reactance (i.e., $X_s = \omega_e L_s = \omega_e L_{ls} + \omega_e L_m$) and the total impedance

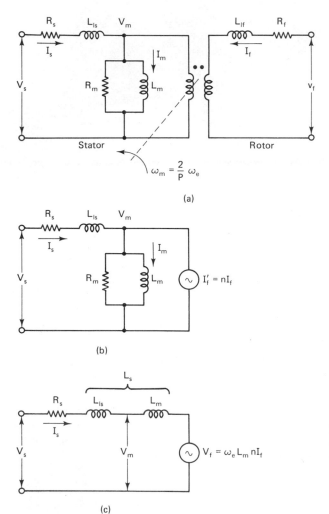

(a)

(b)

(c)

Figure 2.21 Development of per phase equivalent circuit (nonsalient-pole machine). From *Electric Machines* by R. Slemon and A. Straughten © 1980, Addison-Wesley Publishing Company, Reading, Ma. Reprinted with permission.

$Z_s = R_s + jX_s$ is known as the synchronous impedance. The synchronous impedance can be measured from the open-circuit and short-circuit characteristics of the machine.

Unlike an induction machine, a synchronous machine can operate at any desired power factor: leading, lagging, or unity. The power factor can be controlled by the magnitude of the field excitation. At a given frequency ω_e, the air gap voltage V_m tends to balance the supply voltage V_s; therefore, the air gap flux or the corresponding magnetizing current I_m in Fig. 2.21(b) tends to be constant. The current I_m is contributed by field component I_f and the reactive current component of I_s. If the machine is overexcited, the lagging reactive current can be supplied to the output (i.e., it runs at a leading power factor). On the other hand, if the machine is underexcited, it takes lagging current from the stator to supplement the excitation.

Figure 2.22 shows the phasor diagrams for the equivalent circuit of Fig. 2.21(c) under both motoring and generating conditions. The motoring mode is shown with overexcitation (i.e., the machine operating at leading power factor. The angle δ between V_s and V_f is generally known as the power angle of the synchronous machine, and it is negative in the motoring mode but positive in the generating mode. The phasor diagram in the generating mode is shown with underexcitation (i.e., when it consumes lagging reactive current).

Torque characteristics.   For simplicity, let us neglect the stator resistance R_s. Then from Fig. 2.22(a) the phase relations are

$$\bar{I}_s = \frac{\overline{V}_s \angle 0 - \overline{V}_f \angle -\delta}{jX_s}$$

$$= \frac{\overline{V}_s}{X_s} \angle -90° - \frac{\overline{V}_f}{X_s} \angle -(\delta + 90°) \tag{2.84}$$

or

$$I_s \cos \phi = \frac{V_s}{X_s} \cos (-90°) - \frac{V_f}{X_s} \cos (-\delta - 90°)$$

$$= -\frac{V_f}{X_s} \cos (\delta + 90°) \tag{2.85}$$

The power input to the machine is

$$P_{in} = 3V_s I_s \cos \phi \tag{2.86}$$

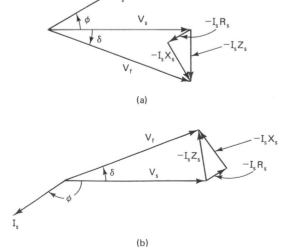

(a)

(b)

Figure 2.22 Phasor diagrams of nonsalient-pole machine: (a) motoring mode; (b) generation mode.

Substituting equation (2.85) in (2.86) yields

$$P_{in} = \frac{3V_s V_f}{X_s} \sin \delta \qquad (2.87)$$

If the losses are ignored, the power P_{in} is also delivered to the shaft,

$$P_{in} = P_s = \frac{2}{P} \omega_e T_e \qquad (2.88)$$

Combining equations (2.87) and (2.88) gives

$$T_e = 3\left(\frac{P}{2}\right) \frac{V_s}{\omega_e} \frac{V_f}{X_s} \sin \delta \qquad (2.89)$$

Equation (2.89) gives the torque–power angle characteristics of the machine, which is plotted in Fig. 2.23 for both the motoring and generating modes. The developed torque is zero at $\delta = 0$ and becomes maximum at $\delta = \pm 90°$. The stability considerations dictate that the machine should be operated with δ angle within $\pm 90°$. At a fixed frequency and supply voltage, the torque curve is proportional to the field excitation current as shown in the figure, where the saturation effect is neglected. Or for a fixed δ angle and field excitation, the torque remains unchanged if the supply voltage-to-frequency ratio remains unchanged.

Equations (2.86) and (2.88) can be combined to write the torque expression in the form

$$T_e = 3\left(\frac{P}{2}\right) \frac{V_s}{\omega_e} I_s \cos \phi$$

$$= 3\left(\frac{P}{2}\right) \psi_m I_s \cos \phi$$

$$= K' I_m I_a \qquad (2.90)$$

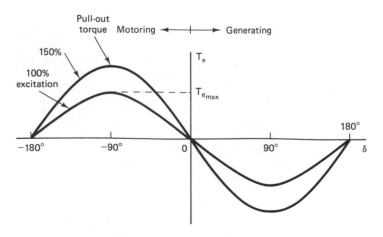

Figure 2.23 Torque–power angle curves of a nonsalient-pole machine.

where ψ_m is the air gap flux, I_m the magnetizing or flux component of current, and I_a the active or torque component of current.   The torque equation (2.90) is identical in form to that of equation (2.7).

V-curves.   The characteristics of a synchronous machine at constant supply voltage and frequency but at variable excitation are given by the V-curves shown in Fig. 2.24.   The curves can be derived from the phasor diagram, Fig. 2.22.   At no-load condition, the excitation current I_f can be adjusted so that the machine operates at unity power factor (i.e., the stator current I_s is minimum).   The input power factor can be made lagging or leading by decreasing or increasing the excitation, respectively.   The resulting I_s current curve is defined as the V-curve. The V-curves at half load and full load are also shown in Figure 2.24.

Salient-Pole Machine Characteristics

So far we have discussed the characteristics of a cylindrical rotor machine.   The characteristics of a salient-pole machine differ from those of a cylindrical rotor machine because of the nonuniform air gap reluctances in the d^e and q^e axes.   The resulting asymmetry in the direct and quadrature axes magnetizing reactances causes the corresponding synchronous reactances to be unsymmetrical (i.e., $X_{ds} \neq X_{qs}$). Figure 2.25 shows phasor diagrams of a salient-pole machine for the motoring and generating modes.   The phasor diagrams of flux linkages are added in the figures. Again, for simplicity the stator resistance has been dropped.   The excitation or speed emf V_f, which is perpendicular to field flux ψ_f, is shown aligned with the q^e axes, whereas ψ_f is aligned with the d^e axes.   The phase voltage V_s and phase current I_s are resolved into corresponding d^e and q^e components and a voltage phasor diagram is drawn with the corresponding reactive drops.   In a flux linkage

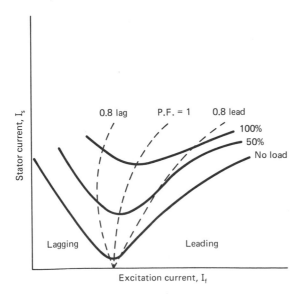

Figure 2.24 Synchronous motor V-curves. From *Electric Machines* by R. Slemon and A. Straughten © 1980, Addison-Wesley Publishing Company, Reading, Ma. Reprinted with permission.

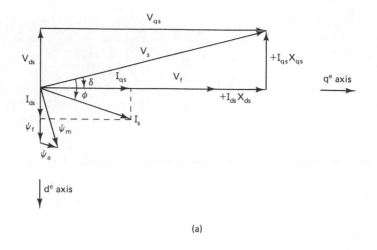

(a)

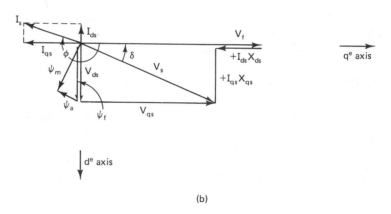

(b)

Figure 2.25 Phasor diagrams of salient-pole machine ($X_{ds} \neq X_{qs}$): (a) motoring mode; (b) generating mode.

phasor diagram, the armature reaction flux ψ_a aids the field flux to result in the air gap flux ψ_m as shown. The motoring mode phasor diagram, which is drawn for lagging power factor, $\psi_m > \psi_f$, whereas in the generating mode, $\psi_m < \psi_f$, because it is operating at leading power factor. It should be noted that a d^e–q^e axes phasor diagram can also be drawn for the cylindrical rotor machine where $X_{ds} = X_{qs}$.

From the phasor diagram of Fig. 2.25(a) we can write

$$I_s \cos \phi = I_{qs} \cos \delta - I_{ds} \sin \delta \qquad (2.91)$$

Substituting equation (2.91) in (2.86), the input power P_{in} can be given as

$$P_{\text{in}} = 3V_s(I_{qs} \cos \delta - I_{ds} \sin \delta) \qquad (2.92)$$

Again, from the phasor diagram we can write

$$I_{ds} = \frac{V_s \cos \delta - V_f}{X_{ds}} \qquad (2.93)$$

$$I_{qs} = \frac{V_s \sin \delta}{X_{qs}} \qquad (2.94)$$

Substituting equations (2.93) and (2.94) in (2.92) yields

$$P_{in} = 3 \frac{V_s V_f}{X_{ds}} \sin \delta + 3V_s^2 \frac{X_{ds} - X_{qs}}{2 X_{ds} X_{qs}} \sin 2\delta \qquad (2.95)$$

Again, combining equations (2.88) and (2.95), we have

$$T_e = 3\left(\frac{P}{2}\right) \frac{1}{\omega_e} \left(\frac{V_s V_f}{X_{ds}} \sin \delta + V_s^2 \frac{X_{ds} - X_{qs}}{2 X_{ds} X_{qs}} \sin 2\delta \right) \qquad (2.96)$$

Equation (2.96) gives the torque–power angle characteristics of a salient-pole synchronous machine. The first component of the equation gives the torque of the cylindrical rotor machine and is identical to equation (2.89) except that X_s is replaced by X_{ds}. The second component is defined as the reluctance torque and arises only due to the rotor saliency (i.e., $X_{ds} \neq X_{qs}$). This component is contributed by the rotor, which tends to align with the position of minimum reluctance and is not influenced by the field excitation. The torque–power angle curves are plotted in Fig. 2.26 for different excitations and for both motoring and generating condi-

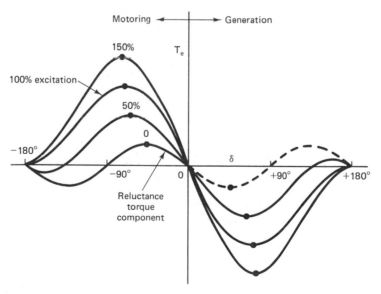

Figure 2.26   Torque–power angle curves of a salient-pole machine $(X_{ds} > X_{qs})$. From *Electric Machines* by R. Slemon and A. Straughten © 1980, Addison-Wesley Publishing Company, Reading, Ma. Reprinted with permission.

tions. The steady-state stability limit corresponds to the maximum points and are indicated by the dots. The reluctance torque component is the lowest curve where stability limit is reached at $\delta = \pm 45°$. At a certain excitation, the cylindrical rotor torque is added to it, and at higher excitation the effect of the salient torque appears less evident. It may be evident from equation (2.95) that if V_s/ω_e is maintained constant (i.e., the supply voltage is changed proportional to the frequency at a fixed excitation and δ angle), the developed torque remains constant.

Variable-Frequency Operation

The speed of a synchronous machine is uniquely related to frequency, as mentioned before. The torque–speed characteristics of a machine with a variable-voltage, variable-frequency power supply can now be drawn with the torque expressions discussed so far. Figure 2.27 shows such characteristics, assuming that the machine is nonsalient and its terminal power factor is maintained at unity. The machine has two regions of operation: a constant-torque region and a constant-power region separated by the base frequency ω_b. In the constant-torque region, the ratio V_s/ω_e is maintained constant as in the induction motor, which makes the magnetizing current I_m constant. Let us consider a point A in the constant-torque region and assume that the frequency remains constant. From equation (2.90), the torque can be increased in a vertical line by increasing the stator current $I_s(I_s = I_a)$ until the maximum torque is reached at point B. With an increase in torque, the angle δ will increase, as is evident from Fig. 2.22(a). If the frequency is set to a higher value, the voltage is to be increased proportionately to reach the same maximum torque, but the current I_s remains the same as before. The voltage-to-frequency ratio has a linear relation except at low frequency, where an additional voltage boost is required to compensate for the stator drop. At the right edge of the constant-torque region, the full supply voltage is reached and the machine enters into the constant-power or field weakening region. Here the torque is reduced because for the same stator current I_s, the magnetizing current I_m is reduced with

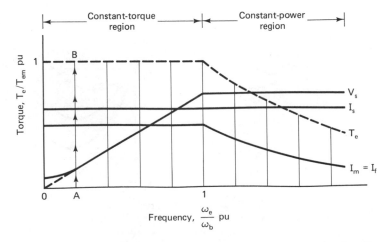

Figure 2.27   Torque–frequency curves of synchronous motor ($\cos\phi = 1$).

the reduced V_s/ω_e ratio.   As the frequency ω_e is increased, the torque is reduced in the form of a rectangular hyperbola, so that the output power remains constant. The field current i_f can be adjusted at an operating point to maintain the desired power factor at the machine terminal.

Dynamic Model

A nonsalient-pole synchronous machine without amortisseur windings can be represented by the simplified per phase transient equivalent circuit shown in Fig. 2.28. The voltage V_m is the air gap voltage, which is also defined as the voltage behind the transient impedance $R_s + j\omega_e L_{1s}$.   The air gap flux is assumed to remain constant during transient operation, and therefore impedance beyond the voltage V_m does not appear in the circuit.

The more comprehensive dynamic performance of a salient-pole synchronous machine can be studied by synchronously rotating d^e–q^e frame model, known as Park's (Ref. 9) equation.   This is the only practical model of a synchronous machine because of the presence of rotating field winding and saliency of rotor poles.   The d^e–q^e model can be derived following the same procedure as in an induction machine.   Including the effect of the d^e and q^e components of amortisseur windings, the machine electromechanical dynamics can be represented by the following equations:

$$
\begin{bmatrix} v_{qs} \\ v_{ds} \\ 0 \\ 0 \\ V_{fr} \end{bmatrix} = \begin{bmatrix} R_s + SL_{qs} & \omega_e L_{ds} & SL_{qm} & \omega_e L_{dm} & 0 \\ -\omega_e L_{qs} & R_s + SL_{ds} & -\omega_e L_{qm} & SL_{dm} & \omega_e L_{dm} \\ SL_{qm} & 0 & R_{qr} + SL_{qr} & 0 & 0 \\ 0 & SL_{dm} & 0 & R_{dr} + SL_{dr} & \omega_e L_{dm} \\ 0 & SL_{dm} & 0 & SL_{dm} & R_{fr} + S(L_{lfr} + L_{dm}) \end{bmatrix} \times \begin{bmatrix} i_{qs} \\ i_{ds} \\ i_{qr} \\ i_{dr} \\ I_{fr} \end{bmatrix} \quad (2.97)
$$

$$
T_e = \frac{3}{2}\left(\frac{P}{2}\right)(\psi_{ds} i_{qs} - \psi_{qs} i_{ds})
$$
$$
= T_L + \frac{2}{P} J \frac{d\omega_r}{dt} \quad (2.98)
$$

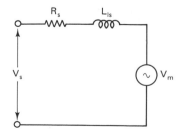

Figure 2.28   Per phase transient equivalent circuit of synchronous machine.

where the field circuit parameters V_{fr}, I_{fr}, R_{fr}, and L_{lfr} are referred to the stator circuit and all other symbols are given in standard notation. Equation (2.97) can be represented with the equivalent circuits shown in Fig. 2.29. All other equations relating to coordinate and phase transformations remain the same as in induction motor. It may be interesting to note that if saliency and field coupling effects are ignored and the induction motor is operated at synchronous speed ($\omega_e = \omega_r$), the d^e–q^e equivalent circuits of the induction and synchronous machines becomes identical.

Synchronous Machine Classification

Synchronous machines can be classified into several types. The wound-field machine which has been emphasized so far is generally large to justify the additional power supply for the field circuit. The dc excitation current for the rotor can be

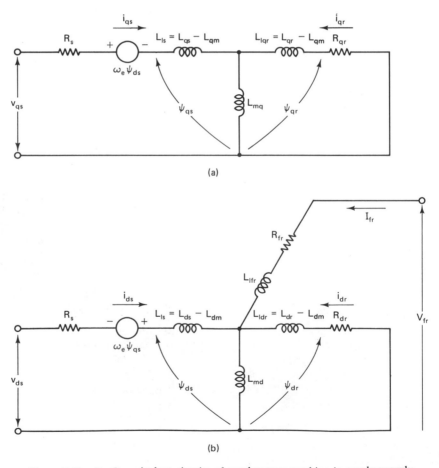

Figure 2.29 D–Q equivalent circuits of synchronous machine in synchronously rotating reference frame: (a) q^e-axis circuit; (b) d^e-axis circuit.

supplied by several methods, such as static excitation, where the ac power is converted to dc by a static rectifier and fed to the rotor by slip rings and brushes; or a shaft-mounted ac excitor may generate the excitation power and feed to the field circuit through a rotating diode rectifier. The ac excitor may be replaced by the induction generator where the stator may be fed by variable voltage three-phase ac supply. In an inductor homopolar or clawpole type of machine, the field winding can be transferred to the stator, thus making the rotor construction simple with the elimination of commutators and brushes. But the machine size becomes larger with larger field mmf requirements, which contributes to higher cost and a reduction in efficiency. Such machines have been considered for applications such as flywheel energy storage and linear motor propulsion.

Several more synchronous motor types which are frequently used for adjustable-speed drives are reviewed briefly here. The analysis of synchronous machines that has been given so far in general holds true for all types of machines.

Synchronous reluctance motor. The idealized structure of a reluctance motor is the same as that of the salient-pole synchronous machine shown in Fig. 2.20, except that the rotor does not have a field winding. The stator has a three-phase symmetrical winding, which creates a rotating magnetic field in the air gap, and reluctance torque is developed because the induced magnetic field in the rotor will have a tendency to cause the rotor to align with the stator field at a minimum reluctance position. The developed torque of the reluctance machine as derived in equation (2.96) can be given as

$$T_e = 3\left(\frac{P}{2}\right)\frac{1}{\omega_e} V_s^2 \frac{X_{ds} - X_{qs}}{2X_{ds}X_{qs}} \sin 2\delta \qquad (2.99)$$

Equation (2.99) has been plotted in Fig. 2.26, indicating that the stability limit is reached at $\delta = \pm 45°$. A reluctance machine is designed with flux barriers in the rotor to strengthen the saliency effect (i.e., $X_{ds} \gg X_{qs}$), so that the available maximum torque is high.

The inherent simplicity of the reluctance machine has made it traditionally very popular in low-horsepower fiber-spinning-mill types of applications, where a number of machines are supplied from a common source so that these operate in exact synchronism. The reluctance machine has no controllable rotor excitation, and since the entire excitation is supplied from the stator side, it must operate at a poor lagging power factor. The typical power factor may be 0.65, but power factors as high as 0.75 have been reported. Again, as with other synchronous machines, the reluctance motor has no inherent starting torque. The machine can be started by a squirrel-cage winding and then pulled into synchronism by the reluctance torque.

Permanent-magnet motor. Permanent-magnet synchronous machines have received considerable attention in recent years for variable-speed drives in the low- to medium-horsepower range. Again, the stator of a permanent-magnet machine has the conventional winding of a three-phase machine. The rotor magnetic field excitation is provided by permanent magnets instead of discrete winding carrying the dc current. The absence of field copper loss helps to improve the machine efficiency. Figure 2.30 shows the cross-sectional view of a permanent-magnet machine, where part (a) shows the configuration with surface or peripheral mounting and part (b) shows interior or buried mounting. An additional squirrel-cage winding may be provided for 60-Hz startup operation. A surface magnet machine

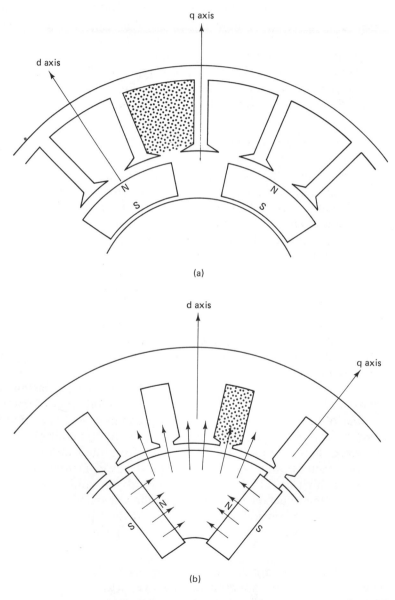

Figure 2.30 Cross-sectional views of permanent-magnet machines: (a) surface magnet machine; (b) interior magnet machine.

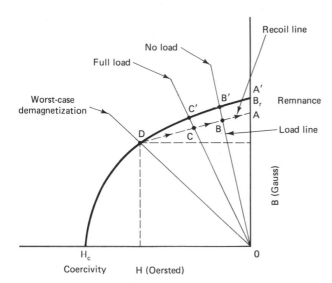

Figure 2.31 Permanent-magnet machine operating points on $B-H$ curve.

can be considered a nonsalient-pole type and since air gap is generally large its armature reaction effect on pole flux is very weak.   The interior magnet machine has salient poles and therefore the total torque is contributed by excitation torque as well as reluctance torque.   Its armature reaction effect cannot be ignored.   The quadrature axis reactance of this type of machine is higher than direct axis reactance $(L_{qs} > L_{ds})$.

Figure 2.31 shows the operation of the machine at different portions of the demagnetization curve of the $B-H$ loop. The maximum flux B_r corresponding to point A' will be available initially if the magnet is short-circuited with steel keepers. When the magnet is installed in the machine, operating point B' will correspond to the no-load line, where the flux density is reduced to some extent due to the air gap effect.   With load current flowing in the stator winding, the d^e-axis armature reaction effect will further reduce the air gap flux density.   A load line corresponding to worst-case demagnetization, which may be due to starting, pull-in or other transient conditions, is also shown in Fig. 2.31.   Once the operating point reaches D and the demagnetization effect is removed, the magnet will recover along the recoil line, which has approximately the same slope as the original $B-H$ curve near $H = 0$.   Subsequently, the stable operating point will be determined by the intersection of the load line and the recoil line.   The magnet is therefore permanently demagnetized at no-load operation, corresponding to the vertical distance between A' and A.   The worst-case demagnetization point is therefore vitally important for machine performance and should be closely controlled.   Alternatively, if the material of the permanent magnet is selected to have a straight-line demagnetization curve, the recoil line will always trace back along the demagnetization curve irrespective of the worst-case demagnetization point (i.e., the demagnetization effect will be negligible).

Figure 2.32 shows the characteristics of several possible permanent-magnet materials.   Ferrite material is commonly used in permanent-magnet machines.

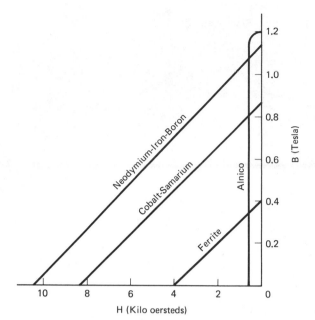

Figure 2.32 Permanent-magnet characteristics (1 Tesla = 10 kilogauss).

The material is low in cost and has excellent linearity in demagnetization, but the remnance is low, making the machine somewhat bulky. Alnico has the highest remnance but the coercivity is small, with a very nonlinear demagnetization curve. The Neodymium-Iron-Boron magnet has high remnance and good coercivity but shows some temperature sensitivity. The cobalt-samarium (rare earth) material has by far the best combination of characteristics, but the material is very expensive and is used only in specialized applications, where the size and weight reduction can be justified at the expense of higher cost.

Since permanent-magnet machines operate at constant excitation, the flexibility of field control to adjust the machine terminal power factor is lost. If the field excitation is designed so that the machine operates at unity power factor at full load, then at light load the power factor will be excessively leading, causing poor efficiency.

REFERENCES

1. A. E. Fitzgerald, C. Kingsley, and S. D. Umans, *Electric Machinery*, McGraw-Hill, New York, 1983.

2. G. R. Slemon and A. Straughen, *Electric Machines*, Addison-Wesley, Reading, Mass., 1980.

3. H. C. Stanley, "An Analysis of the Induction Machine," *AIEE Trans.*, Vol. 57, pp. 751–759, 1938.

4. D. A. Paice, "Induction Motor Speed Control by Stator Voltage Control," *IEEE Trans. Power Appar. Syst.*, Vol. PAS-87, pp. 585–590, Feb. 1968.

5. V. B. Honsinger, "Induction Motors Operating from Converters," *Conf. Rec. IEEE/ IAS Annu. Meet.*, pp. 1276–1285, Oct. 1980.

6. B. K. Bose, "Adjustable Speed AC Drives—A Technology Status Review," *Proc. IEEE*, Vol. 70, pp. 116–135, Feb. 1982.

7. S. D. T. Robertson and K. M. Hebber, "Torque Pulsations in Induction Motors with Inverter Drives," *IEEE Trans. Ind. Appl.*, Vol. IA-7, pp. 318–323, Mar.–Apr. 1971.

8. A. B. Plunkett and D. L. Plette, "Inverter–Induction Motor Drive for Transit Cars," *IEEE Trans. Ind. Appl.*, Vol. IA-73, pp. 26–37, Jan.–Feb. 1977.

9. R. H. Park, "Two-Reaction Theory of Synchronous Machines—I," *AIEE Trans.*, Vol. 48, pp. 716–721, June 1920.

3

PHASE-CONTROLLED
CONVERTERS
AND CYCLOCONVERTERS

3.0 INTRODUCTION

The function of a converter generally is to convert ac to dc, which is defined as rectification, or to convert from dc to ac, which is defined as inversion. A cyclo-converter is a frequency changer that converts ac from one input frequency directly to the output of another frequency. The class of converters discussed in this chapter uses the phase control principle and has evolved over the past several decades. A converter circuit can be considered as a matrix of power semiconductor switches which fabricates the output voltage from the segments of the input voltage and, as a result, generates rich harmonics to both input and output lines. A converter can be viewed as a switching-mode power amplifier with high gain because of its low signal power requirement. Low power loss in the switching devices makes the converter efficiency high, typically in the vicinity of 98%. The switching-mode operation of the converter makes a power electronic system discrete time and nonlinear, which will be discussed later.

The application of phase-controlled converters may include the following:

- Electrochemical processes, such as electroplating, anodizing, metal refining, and hydrogen production.
- Dc motor speed control
- HVDC conversion
- Dc supply for inverters
- Dc–ac conversion from fuel cells, solar cells, and so on

68

- Magnet power supply, such as machine excitation, fusion reactor supply, and so on

In this chapter we review different types of converter circuits and analyze their performance. The control principles and modeling of converters are then discussed. Cycloconverters are discussed in the last section. The ac line voltages are assumed to be balanced and sinusoidal, and drops in the semiconductor switches are neglected.

3.1 SINGLE-PHASE CONVERTER (*Rectifier*).

Phase-controlled converters may have many possible circuit configurations. In this section the fundamentals of the single-phase bridge converter are studied and then extrapolated to other types of converters. Figure 3.1 shows the circuit of a single-phase bridge converter, which consists of four thyristors arranged to form the two legs. A single-phase ac supply is connected to the center of the legs through a transformer. The transformer is optional but can be used for electrical isolation and voltage-level change. The dc output can be connected either to a passive resistance–inductance load or to an active load with counter emf. The dc load current can flow only in one direction, but the load voltage can be of either polarity. In a bridge converter, diagonally opposite pairs of thyristors Q_1, Q_2 and Q_3, Q_4 conduct in sequence for half-cycle intervals. The circuit can be considered as series cascading of a center-tap positive converter and a center-tap negative converter. The current i_d sourcing in the positive converter will sink in the negative converter, making the dashed-line connection redundant. The operation of the component converters is identical except that they operate with a 180° phase shift. Figure 3.2 shows waveforms for the positive center-tap converter where the load inductance is assumed to be sufficiently large to maintain continuous conduction. The thyristor Q_1 or Q_3 can be fired at any instant during the half-cycle when its anode voltage is positive, and once fired the load voltage follows the profile of ac voltage as shown. The conduction of the thyristor will continue beyond the 180° angle (i.e., when the anode voltage is negative), until the next thyristor is fired.

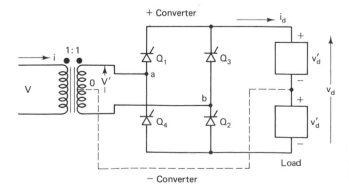

Figure 3.1   Single-phase bridge converter.

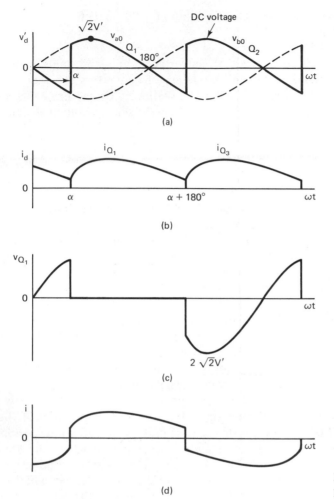

Figure 3.2 Waveforms of center-tap positive converter ($\alpha = 60°$).

At angle $\alpha + 180°$, when thyristor Q_3 is fired, a reverse voltage is impressed across the outgoing device, turning it off, and then the conduction will be undertaken by the incoming device. This process of current transfer is known as natural or line commutation. The two devices will conduct in a symmetrical manner and the average dc voltage can be given as

$$V'_d = \frac{1}{\pi} \int_{\alpha}^{\alpha+\pi} \sqrt{2}\, V' \sin \omega t \, d\omega t = V'_{d0} \cos \alpha \tag{3.1}$$

where $\sqrt{2}\, V'$ is the peak voltage and $V'_{d0} = 2\sqrt{2}/\pi V'$. The voltage V'_d can be controlled by controlling the firing angle α. At $\alpha = 0$, $V'_d = V'_{d0}$ and the circuit operates as a full-wave diode rectifier. The α angle can be retarded to the maximum value of 90° for the rectifier mode of operation when $V'_{d0} = 0$. This condition requires practically infinite load inductance to maintain continuous conduction.

    The angle α can be retarded beyond 90° and continuous conduction can be maintained if the load contains a counter emf of opposite polarity. In this mode of operation, the converter operates as an inverter and power from the dc source is pumped back to the ac line. Figure 3.3 shows waveforms for $\alpha = 135°$. Ideally, α angle can be extended up to 180°, but in practice it should be limited to a few degrees below 180°. The reason for this limit angle is that reverse voltage must be impressed for a definite time across the outgoing thyristor for successful line commutation. A converter that can operate both as a rectifier and as an inverter is defined as two-quadrant converter. This is in contrast to the one-quadrant converter, which operates in the rectification mode only. A two-quadrant converter, for example, is desirable for dc motor speed control, where α can be varied between 0 and 90° in the motoring mode, but regenerative braking is possible by reversing the counter emf and simultaneously retarding the firing angle beyond 90°.

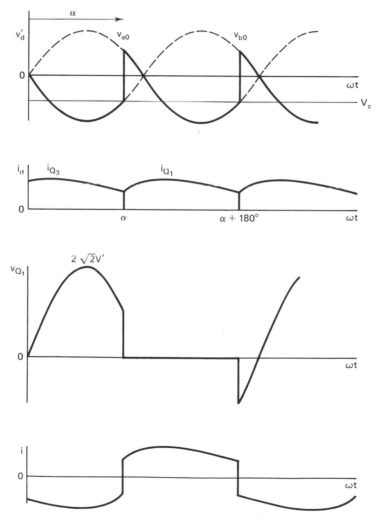

Figure 3.3   Waveforms of center-tap positive converter ($\alpha = 135°$).

Load Voltage and Harmonics

When the positive and negative converters are superimposed to constitute the full bridge converter, all the waveforms remain the same except that the load voltage magnitude becomes double. The instantaneous load circuit equation of bridge can be written as

$$v_d = R_d i_d + L_d \frac{di_d}{dt} + V_c \tag{3.2}$$

where R_d, L_d, and V_c are the resistance, inductance, and counter emf of the load, respectively. Equation (3.2) can be averaged to derive the dc load circuit equation as

$$V_d = \frac{1}{\pi} \int_\alpha^{\alpha+\pi} \left(R_d i_d + L_d \frac{di_d}{dt} + V_c \right) d\omega t$$

$$= I_d R_d + V_c$$

or

$$I_d = \frac{V_d - V_c}{R_d} \tag{3.3}$$

The average power delivered to the load is given by

$$P_0 = \frac{1}{\pi} \int_\alpha^{\alpha+\pi} v_d i_d \, d\omega t = I_0^2 R_d + V_c I_d \tag{3.4}$$

where I_0 is the rms load current. The load voltage v_d contains harmonics and can be given by the general Fourier expression

$$v_d = V_d + \sum_{n=1}^{\infty} (a_n \cos n\omega t + b_n \sin n\omega t) \tag{3.5}$$

where

$$a_n = \frac{2}{\pi} \int_\alpha^{\alpha+\pi} v_d \cos n\omega t \, d\omega t$$

$$b_n = \frac{2}{\pi} \int_\alpha^{\alpha+\pi} v_d \sin n\omega t \, d\omega t$$

The unsymmetrical nature of the v_d waveform indicates that it contains even harmonics and its fundamental frequency is twice that of the ac source. The harmonics can be attenuated by an inductance, capacitance, or inductance–capacitance filter. For a dc machine load, the armature or field inductance may provide adequate filtering. For a linear passive load, harmonic current components can be calculated corresponding to each voltage harmonic and then the superposition principle can be applied.

The line current of a phase-controlled converter as shown in Fig. 3.2(d) contains harmonics which can also be analyzed by Fourier series. The harmonics flowing in the utility system are undesirable, because of unnecessary loading of the power equipment and interference to neighboring telephone lines. The line voltage may also be distorted to some extent, depending on source impedance. It can be shown that the line current will approach a square wave if the load is assumed to be perfectly filtered. Figure 3.4 shows the line voltage and current waves under this condition.

Distortion Factor

The degree of line current distortion can be determined by the distortion factor. It is defined as

$$DF = \frac{\text{rms value of fundamental current}}{\text{rms value of total current}} \tag{3.6}$$

$$= \frac{I_1}{\sqrt{I_1^2 + \sum_{n=2}^{\infty} I_n^2}}$$

For square-wave current, as shown in Fig. 3.4,

$$DF = \frac{(4/\pi)\,(1/\sqrt{2})I_d}{I_d} = \frac{2\sqrt{2}}{\pi}$$

Displacement Factor

The fundamental line current of a phase-controlled converter always lags the line voltage by a displacement angle ϕ (i.e., the line must supply lagging reactive power demanded by the converter). The displacement factor can be defined as

$$\text{Dis.F.} = \frac{\text{average input power}}{\text{rms fundamental voltage} \times \text{rms fundamental current}} \tag{3.7}$$

$$= \frac{P_1}{VI_1} = \frac{VI_1 \cos \phi}{VI_1} = \cos \phi$$

In Fig. 3.4, i_1 is the fundamental line current and therefore ϕ is the true displacement angle. It may be noted here that the displacement angle always equals the firing angle α for a perfectly filtered load. Again, assuming the converter as lossless, the input power must balance the output power.

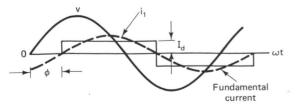

Fundamental current

Figure 3.4   Line voltage and current waves.

Power Factor

The input power factor can be defined as

$$\text{PF} = \frac{\text{average input power}}{\text{rms supply voltage} \times \text{rms supply current}} \qquad (3.8)$$

$$= \frac{P_1}{V\sqrt{I_1^2 + \sum_{n=2}^{\infty} I_n^2}}$$

Substituting equations (3.6) and (3.7) in (3.8) yields

$$\text{PF} = \frac{VI_1 \cos \phi}{V\sqrt{I_1^2 + \sum_{n=2}^{\infty} I_n^2}} = \cos \phi \; \frac{I_1}{\sqrt{I_1^2 + \sum_{n=2}^{\infty} I_n^2}}$$

$$= \text{Dis.F.} \times \text{DF} \qquad (3.9)$$

For a sine current wave DF = 1 and therefore the power factor is the same as the displacement factor.* For a distorted wave, DF is less than unity and therefore PF < Dis.F. Note that for a diode bridge converter, Dis.F. = 1 and PF = DF, which is always less than unity.

Semibridge Converter

If a free-wheeling diode is connected across the load of a bridge converter, it will not be able to sustain any negative voltage and therefore its inversion capability will be lost. A more economical one-quadrant operation is possible using the semibridge converters shown in Fig. 3.5, which use two thyristors and two diodes. Figure 3.6 shows waveforms for the circuit of Fig. 3.5(a), assuming a perfectly

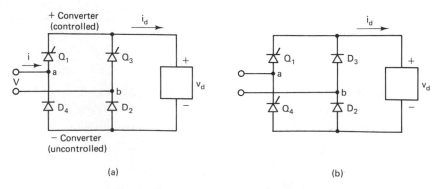

Figure 3.5 Single-phase semibridge converter: (a) uncontrolled negative converter; (b) uncontrolled leg.

*The Dis.F. is loosely defined in the literature as PF. In the remainder of the book, the same convention will be followed.

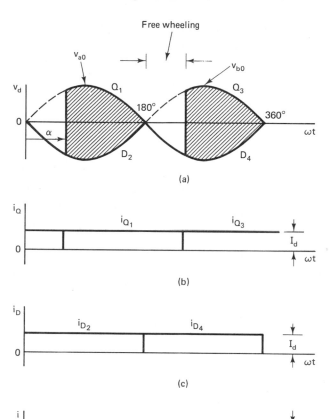

Figure 3.6 Semibridge converter waves for Fig. 3.5(a).

filtered load.   The semibridge circuit can be considered as series cascading of a positive controlled converter and a negative uncontrolled converter, and therefore the waveforms of Fig. 3.6 can be drawn by the superposition principle.   The firing angle of positive converter can be controlled in the range $0 \le \alpha \le 180°$, but a diode of the negative converter will conduct whenever its cathode is more negative than the other.   The resulting load voltage wave is shown by the hatched areas. The current i_{Q1} becomes phase shifted with respect to current i_{D2} and there will be free-wheeling between the series elements every half-cycle, whenever the load voltage tends to be negative.   During the free-wheeling interval, the line does not contribute any current as shown.   The dc load voltage V_d can be given as

$$V_d = V_{d0}(1 + \cos \alpha) \tag{3.10}$$

where $V_{d0} = \dfrac{\sqrt{2}V}{\pi}$.   The dc current I_d is

$$I_d = \frac{V_d - V_c}{R_d} \tag{3.11}$$

The semibridge output voltage as a function of α angle is compared with that of the bridge in Fig. 3.7. Besides being economical, the semibridge converter has the advantage that the load harmonic magnitudes are less because of the suppression of negative voltage. This also results in some improvement of the line-side displacement factor.

Discontinuous Conduction

The converter circuits discussed so far were assumed to operate in continuous conduction where the load voltage is fabricated by the segments of the supply voltage wave. In practice, a converter may also operate in the discontinuous conduction mode, which is more likely with a counter emf load. A two-quadrant converter may have discontinuous conduction both in rectification and in inversion, but we will study the former case only.

Figure 3.8 shows the waveforms for a bridge converter under discontinuous conduction with R_d, L_d, and counter emf load. The thyristor pair Q_1, Q_2 can be fired to conduct when the supply voltage v_{ab} is positive and exceeds the counter emf. The load current i_d will grow up to angle α_2, but will continue up to angle θ due to the load inductance effect. Then v_d will equal V_c until the Q_3, Q_4 pair is fired symmetrically at α angle in the next half-cycle. Obviously, the range of α can be given as $\alpha_1 \leq \alpha \leq \alpha_2$, where $\alpha_1 = \sin^{-1}(V_c/\sqrt{2}V)$ and $\alpha_2 = \pi - \alpha_1$: The conduction may change from discontinuous to continuous if α or V_c is low, or the load inductance L_d is high. The equation during conduction can be given as

$$L_d \frac{di_d}{dt} + R_d i_d = \sqrt{2}\, V \sin \omega t - V_c \tag{3.12}$$

which can be solved for i_d as

$$i_d = A e^{-R_d t / L_d} + \frac{\sqrt{2}\, V}{|Z|} \sin(\omega t - \phi) - \frac{\sqrt{2}\, V}{R_d} \tag{3.13}$$

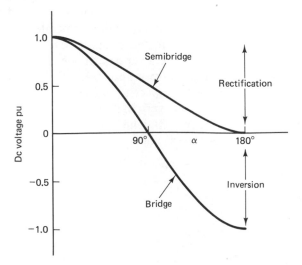

Figure 3.7 Bridge and semibridge output characteristics.

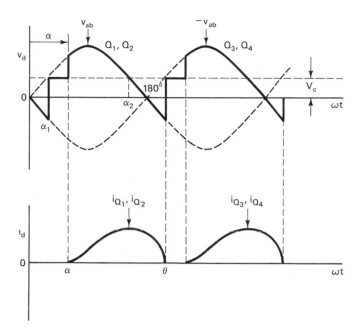

Figure 3.8 Load voltage and current waves of bridge converter under discontinuous conduction.

where

$$|Z| = \sqrt{R_d^2 + \omega^2 L_d^2} \quad \text{and} \quad \phi = \tan^{-1}\frac{\omega L_d}{R_d}$$

Noting that $i_d = 0$ at $\omega t = \alpha$, the equation (3.13) can be written as

$$i_d = \frac{\sqrt{2}\,V}{R_d}\left\{\frac{R_d}{|Z|}\sin(\omega t - \phi) - m\right.$$

$$\left. + \left[m - \frac{R_d}{|Z|}\sin(\alpha - \phi)\right]e^{-(R_d/\omega L_d)(\omega t - \alpha)}\right\} \quad (3.14)$$

where $m = V_c/\sqrt{2}V$ is the counter emf coefficient. Equation (3.14) is valid in the range $\alpha \le \omega t \le \theta$ for discontinuous conduction. From the waveform, $i_d = 0$ again at $\omega t = \theta$. Substituting this condition in equation (3.14) yields

$$e^{R_d\theta/\omega L_d}\frac{\cos\phi\,\sin(\theta - \phi) - m}{\cos\phi\,\sin(\alpha - \phi) - m} = e^{R_d\alpha/\omega L_d} \quad (3.15)$$

This is a transcendental equation relating the parameters α, m and $\omega L_d/R_d$ and can be solved by a computer program. Figure 3.9 gives a plot showing the relations between the parameters. The curves are bounded on the lower side by the α-limit curves, where $\alpha_1 = \sin^{-1}(V_c/\sqrt{2}V)$ and $\alpha_2 = \pi - \alpha_1$, and on the upper left side by the continuous conduction boundary $\alpha = \theta - \pi$. The load dc and rms currents can be calculated from equation (3.14) with the help of Fig. 3.9. However, since the inductance cannot sustain any average voltage, the dc load current I_d can be

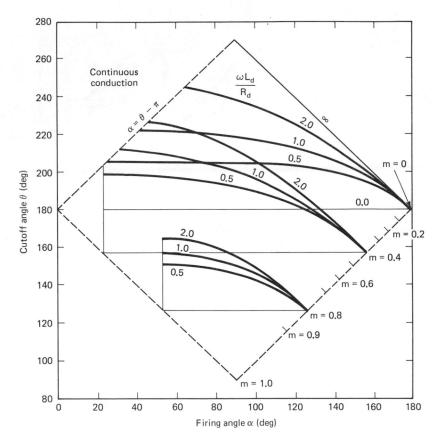

Figure 3.9 Cutoff angle as a function of firing angle for discontinuous conduction. From *Engineering Electronics* J. D. Ryder © 1967, McGraw-Hill Book Company.

calculated as

$$I_d = \frac{1}{\pi} \int_\alpha^\theta \frac{\sqrt{2}\,V \sin \omega t - V_c}{R_d} \, d\omega t = \frac{\sqrt{2}\,V}{\pi R_d} [\cos \alpha - \cos \theta - m(\theta - \alpha)] \quad (3.16)$$

The discontinuous conduction case can be extrapolated to reach the continuous conduction. Writing equation (3.13) for $\omega t = \alpha$ and for $\omega t = \pi + \alpha$ and equating yields

$$A = \frac{\sqrt{2}\,V}{|Z|} \frac{\sin(\alpha - \phi)}{(e^{-R_d\pi/\omega L_d} - 1)e^{R_d\alpha/\omega L_d}} \quad (3.17)$$

and therefore the i_d expression for continuous conduction is given as

$$i_d = \frac{\sqrt{2}\,V}{R_d} \left[\cos \phi \sin (\omega t - \phi) - m \right.$$
$$\left. - \frac{2 \cos \phi \sin (\alpha - \phi)}{1 - e^{-R_d\pi/\omega L_d}} \times e^{-(R_d/\omega L_d)(\omega t - \alpha)} \right] \quad (3.18)$$

The dc load current I_d can be derived from equation (3.18) as

$$I_d = \frac{1}{\pi} \int_\alpha^{\alpha+\pi} i_d \, d\omega t = \frac{\sqrt{2}\,V}{\pi R_d}(2\cos\alpha - m\pi) \tag{3.19}$$

which can be shown to be identical with equation (3.3).

3.2 THREE-PHASE HALF-WAVE CONVERTER

For a load power requirement of several kilowatts or more, it is desirable to use a three-phase converter. A polyphase converter, in general, not only imposes balanced loading on the utility system, but provides considerable improvement in load voltage and line current harmonics, as will be shown later. As a result, the harmonic filtering requirement becomes a nominal problem.

Among all types of polyphase converters, the three-phase bridge converter as shown in Fig. 3.10 is most commonly used. It consists of six thyristors arranged in the form of three legs, the center points of which are connected to three-phase ac power supply. The transformer connection is optional but can be used for voltage-level shifting and isolation. A three-phase bridge converter can be constructed by series cascading of a three-phase half-wave positive converter and a three-phase half-wave negative converter, which are shown in Fig. 3.11. The two component converters operate in an identical manner except with a phase shift of 60°. Therefore, we will study the operation of the positive converter only.

Figure 3.12 shows the waveforms of the positive converter. Again, continuous conduction and a load with perfect filtering have been assumed. The three

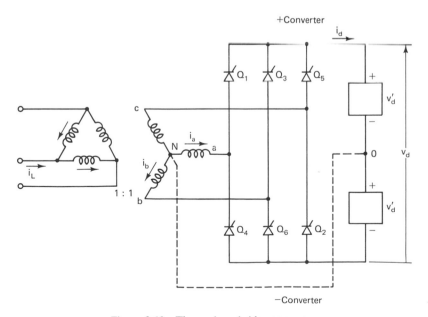

Figure 3.10   Three-phase bridge converter.

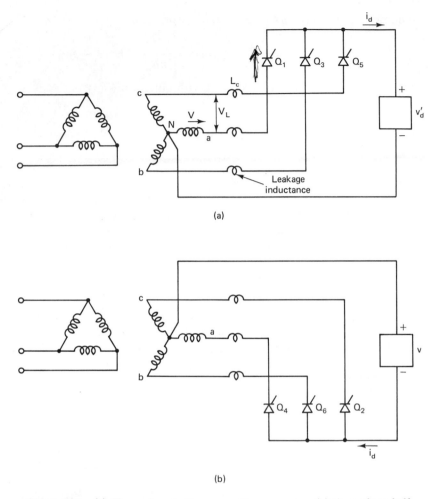

Figure 3.11 (a) Three-phase half-wave positive converter; (b) three-phase half-wave negative converter.

thyristors Q_1, Q_3, and Q_5 conduct symmetrically, each for 120° through the load and provide a common return to the transformer neutral point N. A thyristor can be fired to conduct when its anode voltage is positive with respect to cathode (i.e., with respect to the load voltage) and conduction will continue until the subsequent thyristor is fired after a 120° angle. The commutation from outgoing to incoming thyristor occurs naturally by a segment of negative line voltage as shown in Fig. 3.12(c). The firing angle α is defined from the crossover point of phase voltages, which is the earliest point when a thyristor can assume conduction. At $\alpha = 0$, the thyristors can be considered to be operating as diodes. The dc load voltage V'_d can be derived as

$$V'_d = \frac{3}{2\pi} \int_{(\pi/6)+\alpha}^{[(\pi/6)+\alpha]+2\pi/3} \sqrt{2}\, V \sin \omega t \, d\omega t = V'_{d0} \cos \alpha \qquad (3.20)$$

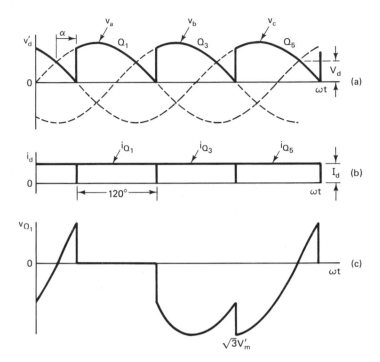

Figure 3.12 Waveforms of three-phase half-wave converter in rectification mode ($\alpha = 30°$).

where $V'_{d0} = (3\sqrt{3}/\sqrt{2})(V/\pi) = 0.675V_L$ and V_L is the line rms voltage. The dc voltage V'_d can be varied by controlling the firing angle α. Up to $\alpha = 30°$, the instantaneous load voltage is always positive, and therefore continuous conduction is assured if the load does not contain any counter emf. For $\alpha > 30°$, the instantaneous load voltage will be negative for part of the cycle and the possibility of discontinuous conduction arises. In the following analysis, we will ignore discontinuous conduction and assume a load with perfect filtering. At $\alpha = 90°$, it can be shown that the positive and negative areas of v'_d wave will balance, giving $V'_d = 0$. The angle α can be controlled beyond 90°, giving inverter operation as discussed before, provided that an energy source exists in the load. Figure 3.13 shows waveforms for inverter operation at $\alpha = 150°$. The angle β, known as the advance angle, is important for line commutation because the reverse voltage shown by the shaded area is impressed across the outgoing thyristor. Typically, the β angle is limited to 10 to 15°.

An examination of load voltage wave indicates that it contains triplen harmonic frequencies (i.e., 3rd, 6th, 9th, etc.). The increase in pulse number from two to three has increased the dc voltage at the output.

In a three-phase converter, each thyristor conducts for one-third of the cycle and therefore carries the average current of $I_d/3$ and rms current of $I_d/\sqrt{3}$. A 120° unidirectional current pulse/cycle in the transformer secondary is objectionable because it may cause dc saturation in the core. This problem may be avoided by providing a zigzag connection in the secondary. Although the circuit of Fig. 3.11(a)

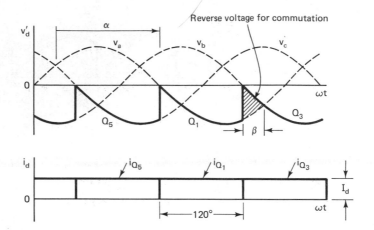

Figure 3.13 Waveforms of three-phase half-wave converter in inversion mode ($\alpha = 150°$).

is not used in practice, its analysis is important because the circuit is a basic functional element in all polyphase converters and cycloconverters.

Effect of Transformer Leakage Inductance

So far we have neglected the source leakage inductance and assumed that commutation (i.e., transfer of current from the outgoing to the incoming thyristor) occurs instantaneously. Such a conduction is practically impossible, and a finite amount of source leakage inductance will permit current transfer to occur only gradually, as shown in Fig. 3.14. During commutation overlap angle μ as shown, the line-to-line voltage is shorted and the supply volt-seconds area is absorbed by the two leakage L_c in series until the current transfer is completed. During this period, the load voltage dwells at intermediate level between the two phase voltages. Considering commutation from Q_1 to Q_3 in Fig. 3.11(a), we can write the

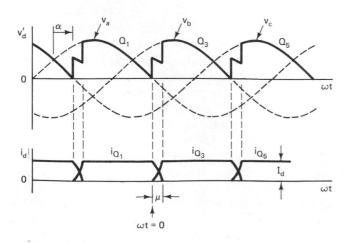

Figure 3.14 Rectifier operation showing commutation overlap (positive converter).

following equations:

$$v_a = L_c \frac{di_{Q_1}}{dt} + v'_d \tag{3.21}$$

$$v_b = L_c \frac{di_{Q_3}}{dt} + v'_d \tag{3.22}$$

During commutation, the load current I_d can be assumed as constant. Therefore,

$$i_{Q_1} + i_{Q_3} = I_d \tag{3.23}$$

That is,

$$\frac{di_{Q_1}}{dt} + \frac{di_{Q_3}}{dt} = 0 \tag{3.24}$$

Combining equations (3.21), (3.22), and (3.24) yields

$$v'_d = \frac{v_a + v_b}{2} \tag{3.25}$$

(i.e., the load voltage is the mean of two phase voltages). From Fig. 3.14 it is evident that every 120° interval some volt-second area is lost during commutation and, as a result, the dc voltage V'_d will be reduced. Combining equations (3.21) and (3.25) yields

$$\frac{di_{Q_1}}{dt} = -\frac{1}{2L_c}(v_b - v_a) \tag{3.26}$$

or

$$i_{Q_1} = -\frac{1}{2L_c}\int (v_b - v_a)\, d\omega t \tag{3.27}$$

Substituting the line-to-line voltage $v_{ba} = v_b - v_a = \sqrt{3}\sqrt{2}\, V \sin(\omega t + \alpha)$ in equation (3.27) and solving gives us

$$i_{Q_1} = \frac{\sqrt{6}\, V}{2\omega L_c} \cos(\omega t + \alpha) + A \tag{3.28}$$

Assuming that $\omega t = 0$ at the beginning of commutation, where $i_{Q_1} = I_d$, the constant A can be evaluated as

$$A = I_d - \frac{\sqrt{6}\, V}{2\omega L_c} \cos\alpha \tag{3.29}$$

Substituting this in equation (3.28) yields

$$i_{Q_1} = I_d - \frac{\sqrt{6}\, V}{2\omega L_c}[\cos\alpha - \cos(\omega t + \alpha)] \tag{3.30}$$

Substituting equation (3.30) in (3.23), we have

$$i_{Q_3} = \frac{\sqrt{6}\,V}{2\omega L_c}\,[\cos\alpha - \cos(\omega t + \alpha)] \tag{3.31}$$

Again, substituting $i_{Q_3} = I_d$ at $\omega t = \mu$ in equation (3.31) gives

$$\cos\alpha - \cos(\mu + \alpha) = \frac{2\omega L_c I_d}{\sqrt{6}\,V} \tag{3.32}$$

Therefore, the commutation angle μ can be expressed as

$$\mu = \cos^{-1}\left(\cos\alpha - \frac{2\omega L_c I_d}{\sqrt{6}\,V}\right) - \alpha \tag{3.33}$$

Equation (3.33) shows that the overlap angle will increase if L_c or I_d increases or as α deviates from the middle of half-cycle. The mean dc voltage loss due to commutation notch can be given as

$$\begin{aligned}
V_x &= \frac{3}{2\pi}\int_0^\mu \frac{1}{2}(v_b - v_a)\,d\omega t \\
&= \frac{3}{4\pi}\int_0^\mu \sqrt{3}\,\sqrt{2}\,V\sin(\omega t + \alpha)\,d\omega t \\
&= -\frac{3\sqrt{3}\,\sqrt{2}\,V}{4\pi}\,[\cos(\mu + \alpha) - \cos\alpha]
\end{aligned} \tag{3.34}$$

Substituting equation (3.32) in (3.34) gives

$$V_x = L_c I_d\,\frac{3\omega}{2\pi} = 3L_c I_d f \tag{3.35}$$

where f is the supply frequency in hertz. Therefore, with loading the dc voltage V_{d1} can be given as

$$V_{d1} = V_d' - V_x = \frac{3\sqrt{3}}{\sqrt{2}}\frac{V}{\pi}\cos\alpha - 3L_c I_d f \tag{3.36}$$

Equation (3.36) indicates that the load voltage is reduced linearly with dc current and the Thévenin resistance of the converter is given as $R_{Th} = 3L_c f$.

Figure 3.15 shows inverter operation of a three-phase half-wave positive converter, where the commutation notch makes the dc voltage more negative. The overlap angle has more significance in inverter operation since it determines how far α angle can be increased (i.e., the minimum advance limit angle β for safe commutation. In the figure,

$$\beta = \mu + \gamma \tag{3.37}$$

where γ is the turn-off angle as shown. Substituting $\alpha = 180° - (\mu + \gamma)$ in

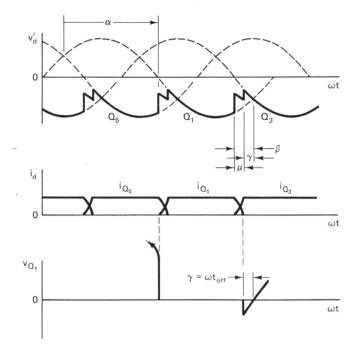

Figure 3.15 Inverter operation showing commutation overlap (positive converter).

equation (3.32) yields

$$\cos \gamma - \cos (\mu + \gamma) = \frac{2\omega L_c I_d}{\sqrt{6}\,V} \qquad (3.38)$$

The thyristors require minimum turn-off time t_{off} for successful commutation, which correspondingly determines the turn-off angle $\gamma = \omega t_{off}$.

3.3 THREE-PHASE BRIDGE CONVERTER   { half-wave converter }

Three-phase bridge converter operation can be analyzed by superposition of waveforms of a positive half-wave converter and a negative half-wave converter. Figure 3.16 shows waveforms of the bridge for firing angle $\alpha = 45°$. The negative converter consisting of thyristors Q_4, Q_6, and Q_2 is fired symmetrically at 120° intervals like the positive converter except that it is phase shifted by 60° as shown. The load voltage v_d, which is enclosed within the envelopes of the component converter waves, has the six-pulse wave shape shown in part (b). A thyristor in both the positive converter and the negative converter must conduct simultaneously to complete the load circuit. The dc load voltage V_d is twice that of a half-wave converter and can be given from equation (3.20) as

$$V_d = 2V'_d = 1.35V_L \cos \alpha \qquad (3.39)$$

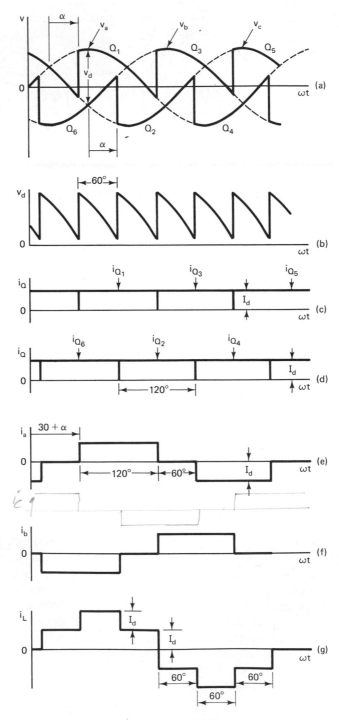

Figure 3.16 Waveforms of three-phase bridge converter in rectification mode ($\alpha = 45°$).

The α angle can be symmetrically controlled for both the component converters in the sequence $Q_1 - Q_2 - Q_3 - Q_4 - Q_5 - Q_6$ to regulate the dc voltage V_d. The Fourier analysis of the v_d wave indicates that it contains harmonics of the order $6n$, where $n = 1, 2, 3$, and so on. The waveform with increasing pulse number is easier to filter and a nominal value of inductance will cause a smooth i_d wave. The phase currents i_a and i_b can be constructed by the superposition of thyristor currents and have the characteristic six-step waveform, which contains harmonics of the order $6n \pm 1$ (i.e., 5th, 7th, 11th, 13th, etc.). If no input transformer is used, i_a and i_b will constitute the line current waves. With a delta–star transformer of unity turns ratio, the input line current wave i_L can be constructed by superposition of the i_a and i_b waves as shown. The transformer does not have a dc saturation problem because of mmf balancing and the order of harmonics of input current is the same as that of the i_a or i_b wave. If the firing angle is retarded further so that $90° < \alpha < 180°$, it can be shown that the converter will operate in the inverting mode, as explained in Fig. 3.17. The general two-quadrant characteristic shown in Fig. 3.7 is also valid for the three-phase bridge converter.

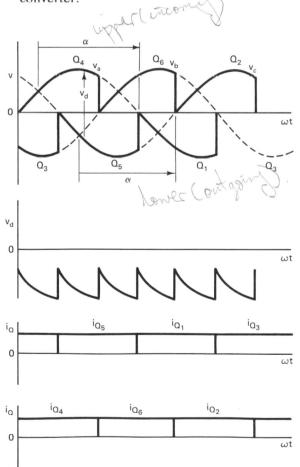

Figure 3.17 Waveforms of bridge converter in inverting mode ($\alpha = 150°$).

Harmonics and Displacement Factor

Assuming that the converter does not have an input transformer, the currents i_a, i_b, and so on, will directly constitute the input line currents. The line current contains only odd harmonics, which can be expressed in the form

$$i_a = \sum_{n=1,3,5,\ldots} a_n \cos n\omega t + b_n \sin n\omega t \tag{3.40}$$

where

$$a_n = \frac{2}{\pi} \int_{\pi/6+\alpha}^{(\pi/6+\alpha)+2\pi/3} I_d \cos n\omega t \, d\omega t$$

and

$$b_n = \frac{2}{\pi} \int_{\pi/6+\alpha}^{(\pi/6+\alpha)+2\pi/3} I_d \sin n\omega t \, d\omega t$$

Evaluating a_n and b_n and substituting in equation (3.40) gives

$$i_a = \frac{2\sqrt{3}}{\pi} I_d \left[\sin (\omega t - \alpha) - \frac{1}{5} \sin 5(\omega t - \alpha) \right.$$

$$- \frac{1}{7} \sin 7(\omega t - \alpha) + \frac{1}{11} \sin 11(\omega t - \alpha)$$

$$\left. + \frac{1}{13} \sin 13(\omega t - \alpha) \cdots \right] \tag{3.41}$$

The i_a wave and its fundamental component have been plotted in the correct phase position with supply phase voltage wave v_a in Fig. 3.18. Figure 3.19 shows the corresponding active and reactive current relations of the fundamental current in both the rectification and inversion modes. The input displacement angle ϕ is equal to the firing angle α, and the active and reactive current components can be given as

$$I_P = r \cos \alpha \tag{3.42}$$

$$I_Q = r \sin \alpha \tag{3.43}$$

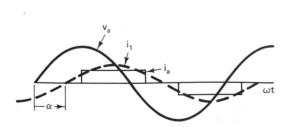

Figure 3.18 Phase relation of input voltage and current waves.

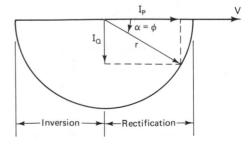

Figure 3.19 Input line active and reactive current characteristics of two-quadrant converter (I_d = constant).

where $r = (2\sqrt{3}/\sqrt{2}\pi)\,I_d$ is the rms value of the fundamental current.   For a firing angle in the range $90 < \alpha < 180$ (i.e., in the inverting mode), the active current I_P becomes negative but the reactive current I_Q remains lagging.

Commutation Overlap

So far the voltage and current waveforms have been considered ideal and the commutation overlap effect has been neglected.   Figure 3.20 shows typical wave-forms with overlap angle μ.   The positive converter and negative converter operate independently and therefore the volt-second area loss per commutation remains the same as that of a half-wave converter.   Since the number of commutations are twice per cycle, the dc voltage loss will also be twice; that is, from equations (3.36) and (3.39),

$$V_{dl} = V_d - 2V_x \tag{3.44}$$

Again, as mentioned earlier, the overlap angle is particularly important in the inverting mode, where the safe minimum γ angle must be maintained under the worst load condition.

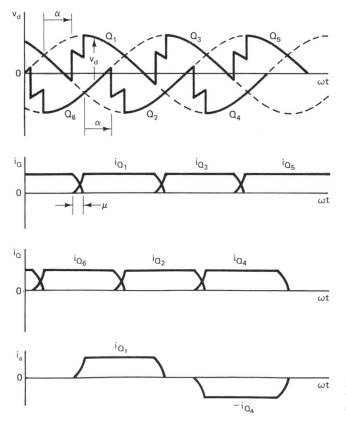

Figure 3.20   Three-phase bridge con-verter waves showing commutation overlap effect.

Design Example

A 2300-V 60-Hz three-phase power supply is connected to a bridge converter through a delta–star transformer. The converter supplies a dc load current of 90A at a voltage varying from $+500$ to -500 V. Design the converter.

Assume that the commutating inductance per phase $L_c = 50$ μH. Then

$$V_x = 3fL_cI_d = 3 \times 60 \times 50 \times 10^{-6} \times 90 = 0.81 \text{ V}$$

Assume a thyristor conduction drop of 1.5 V. Then

$$V_d = 1.35V_L \cos \alpha = 500 + 2 \times 0.81 + 1.5 \times 2$$

$$= 504.62 \text{ V}$$

Select the transformer turns ratio 10:1 so that $V_L = (2300/10)\sqrt{3} = 398.4$ V. Therefore,

$$\cos \alpha = \frac{504.62}{1.35 \times 398.4} = 0.94 \quad \text{or} \quad \alpha = 19.9°$$

The thyristors withstand a peak value of V_L in the reverse direction. Allow a 50% voltage margin for line voltage fluctuation and snubber overshoot. Therefore, the voltage rating is

$$V_{DRM} \simeq V_{RRM} \simeq 398.4 \sqrt{2} \times 1.5 = 845 \text{ V}$$

The thyristor average and rms currents are

$$I_{av} = \frac{90}{3} = 30 \text{ A} \qquad I_{rms} = \frac{90}{\sqrt{3}} = 51.96 \text{ A}$$

Select thyristor type GE-C147T, which has ratings of 900 V and 63 A rms.

In the inverting mode, $V_d = -500 + 1.62 + 3 = 495.38$ V, which gives $\cos \alpha = -0.921$. Therefore,

$$\cos (\alpha + \mu) = \cos \alpha - \frac{2\omega L_c I_d}{\sqrt{6} V}$$

$$= -0.921 - \frac{2 \times 314 \times 50 \times 10^{-6} \times 90 \times \sqrt{3}}{\sqrt{6} \times 398.4}$$

$$= -0.926$$

(i.e., $\alpha + \mu = 157.8°$). Hence $\gamma = 180 - (\alpha + \mu) = 22.21°$ or

$$t_{off} = \frac{22.21 \times 10^3}{314 \times 57.3} = 1.23 \text{ ms}$$

which is adequate. Use Figs. 1.5 and 1.8 with a 33% duty cycle for $I_{av} = 30$ A:

$$T_{C\max} = 104°C$$

$$P_{av} = 43 \text{ W}$$

The thermal resistance $\Theta_{CA} = \Theta_{CS} + \Theta_{SA} = (T_{C\max} - T_A)/P_{av}$. Assuming that $T_A = 25°C$ and $\Theta_{CS} = 0.075°C/W$,

$$\Theta_{SA} = \frac{104 - 25}{43} - 0.075 = 1.76°C/W$$

The transformer currents are

$$I_d \sqrt{\frac{2}{3}} = 90 \sqrt{\frac{2}{3}} = 73.48 \text{ A rms in secondary}$$

and

$$\frac{I_d}{n} \sqrt{\frac{2}{3}} = \frac{90}{10} \sqrt{\frac{2}{3}} = 7.35 \text{ A rms in primary}$$

The transformer VA rating

$$= 3VI_{\text{ph}} = \frac{3 \times 398.4}{\sqrt{3}} \times 73.48$$

$$= 50.7 \text{ kVA}$$

The line rms fundamental current from Fig. 3.16(g) is

$$I_{Lf} = \frac{3}{\pi} \frac{2I_d}{n} \frac{1}{\sqrt{2}} = \frac{3 \times \sqrt{2} \times 90}{\pi \times 10} = 12.16 \text{ A}$$

and the rms current is

$$I_L = \frac{\sqrt{2} \, I_d}{n} = \frac{\sqrt{2} \times 90}{10} = 12.73 \text{ A}$$

In the rectification mode,

$$\text{Dis.F.} = \cos \alpha = 0.94$$

$$DF = \frac{12.16}{12.73} = 0.96$$

and

$$PF = 0.94 \times 0.96 = 0.90$$

The fundamental input power $P_{\text{in}} = \sqrt{3} \times 2300 \times 12.16 \times 0.94 = 45.53$ kW and output power $P_d = V_d I_d = 500 \times 90 = 45$ kW. Therefore, the efficiency $\eta = 45/45.53 = 98.8\%$, where the losses due to conduction drop only has been considered.

3.4 THREE-PHASE SEMIBRIDGE CONVERTER

The three-phase semibridge converter shown in Fig. 3.21 consists of three thyristors and three diodes and can be constructed by series cascading of a positive controlled converter and a negative uncontrolled converter. The waveforms at $\alpha = 45°$ are shown in Fig. 3.22. The negative converter follows the negative phase voltage profile with $\alpha = 0$ and the dc load voltage is controlled by the α angle of the positive converter. By superposition, the dc voltage is given as

$$V_d = 0.675 V_L (1 + \cos \alpha) \qquad (3.45)$$

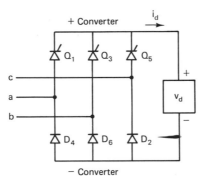

Figure 3.21 Three-phase semibridge converter.

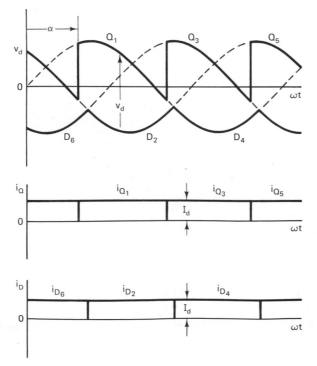

Figure 3.22 Waveforms of three-phase semibridge converter ($\alpha = 45°$).

At $\alpha = 0$, $V_d = V_{d0} = 1.35V_L$ (i.e., the full diode bridge operation occurs). If the overlapping effect is considered under loading condition, the dc voltage can be given as

$$V_d = 0.675V_L(1 + \cos\alpha) - 2V_x \qquad (3.46)$$

For $0 < \alpha < 60°$, v_d is always positive, but as α exceeds 60°, part of the v_d wave will tend to go negative. But the load cannot sustain any negative voltage and therefore free-wheeling between the series elements will occur. Figure 3.23 shows waveforms of the converter at $\alpha = 120°$, which indicates free-wheeling between Q_1 and D_4.

The line current i_a, consisting of i_{Q_1} and i_{D_4} pulses in correct phase relation with the phase voltage v_a, is shown in Fig. 3.24. The i_{D_4} pulse remains in phase with the voltage, and the contribution to the P and Q components of the current is

$$I'_P = 0.5r \qquad (3.47)$$

$$I'_Q = 0 \qquad (3.48)$$

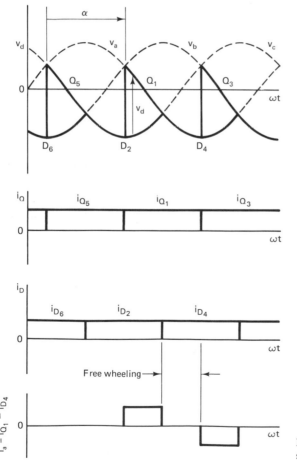

Figure 3.23 Waveforms of three-phase semibridge converter ($\alpha = 120°$).

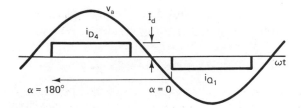

Figure 3.24 Phase relation of input voltage and current for a semibridge converter.

where $r = (2\sqrt{3}/\sqrt{2}\,\pi)\,I_d$. The position of i_{Q_1} varies with α angle and its contribution to the P and Q components is

$$I_P'' = 0.5r\cos\alpha \tag{3.49}$$

$$I_Q'' = 0.5r\sin\alpha \tag{3.50}$$

so that the total active and reactive current in the line is

$$I_P = I_P' + I_P'' = 0.5r(1 + \cos\alpha) \tag{3.51}$$

$$I_Q = I_Q' + I_Q'' = 0.5r\sin\alpha \tag{3.52}$$

The fundamental line current can be given as

$$I_1 = \sqrt{I_P^2 + I_Q^2} = \frac{r}{\sqrt{2}}\sqrt{1 + \cos\alpha} \tag{3.53}$$

and the displacement factor is

$$\cos\phi = \frac{I_P}{I_1} = \frac{1}{\sqrt{2}}\sqrt{1 + \cos\alpha} = \cos\frac{\alpha}{2} \tag{3.54}$$

which indicates an improvement in the displacement factor over that of the bridge converter. The input line active and reactive current characteristics are plotted in Fig. 3.25, which also shows the bridge characteristics for comparison.

Sometimes the load of a semibridge converter is bypassed by an extra free-wheeling diode. In such a case, the free-wheeling operation is always taken by this diode, relieving the outer legs. This has the advantage that the converter can be blocked by the suppression of gate firing for an overload or fault.

The semibridge converter is more economical than the full bridge converter for one-quadrant operation. But its load harmonic contents are higher because of its three-pulse output characteristics and it therefore requires a larger filter.

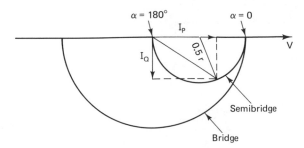

Figure 3.25 Input line active and reactive current characteristics of semibridge converter.

3.5 FOUR-LEGGED BRIDGE CONVERTER

The poor displacement factor of a bridge converter at a reduced output voltage can be improved and at the same time two-quadrant characteristics can be maintained using the four-legged bridge converter shown in Fig. 3.26.   The circuit can be considered as a hybrid connection of a bridge converter and a half-wave converter.   The thyristors Q_7 and Q_8 in the fourth leg can either be blocked to give bridge operation or selectively fired at angle δ from the instant its anode voltage becomes positive to give half-wave converter characteristics.   It can be shown that in the rectification mode, the optimum displacement factor can be obtained by maintaining $\delta = 0$ (i.e., permitting Q_7 and Q_8 to operate as diodes). Figure 3.27 shows the waveforms of the converter at $\alpha = 60°$ and $\delta = 0$.   The thyristors in the positive converter conduct symmetrically, but as the cathode voltage tends to go negative, thyristor Q_7 conducts and locks point G at the neutral potential.

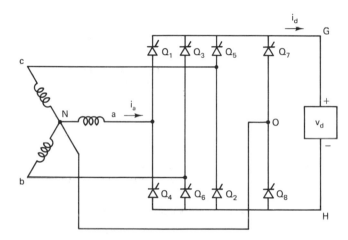

Figure 3.26 Four-legged bridge converter.

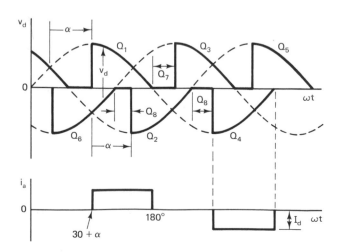

Figure 3.27 Waveforms of four-legged bridge at $\alpha = 60°$ and $\delta = 0°$.

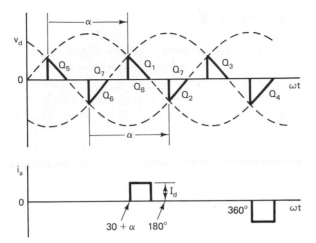

Figure 3.28 Waveforms of four-legged bridge at $\alpha = 120°$ and $\delta = 0°$.

Similarly, for the negative converter, as the common anode potential tends to go positive, thyristor Q_8 conducts, locking point H at the neutral potential. This mode of operation permits the line current i_a to flow when the phase voltage is at the same polarity, thus improving the line displacement factor and load harmonic ripple. It can be shown that for α angle range $0 \le \alpha \le 30°$, the fourth leg does not conduct and the operation is identical to that of a bridge converter. Figure 3.28 shows waveforms for $\alpha = 120°$ and $\delta = 0$. The rectifier operation is obtained in the α angle range $0 \le \alpha \le 150°$, at the end of which both output voltage and current fall to zero.

The best inverter operation of the circuit can be obtained by two sequential control modes. In the first mode, α is maintained at 150° and δ angle is controlled in the range $0 \le \delta \le 120°$ to regulate the output voltage. Figure 3.29 shows waveforms at $\alpha = 150°$ and $\delta = 60°$. This mode of operation maintains v_d at

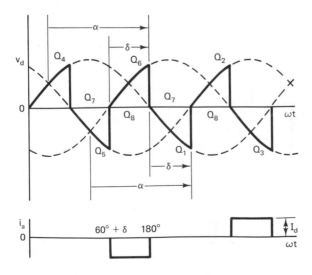

Figure 3.29 Waveforms of four-legged bridge at $\alpha = 150°$ and $\delta = 60°$.

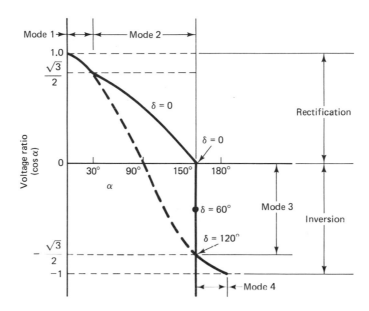

Figure 3.30   Dc voltage characteristics of four-legged bridge converter.

negative polarity and the line current i_a is of opposite polarity to the phase voltage, as shown for the best displacement angle. In the next control mode, firing of Q_7 and Q_8 is inhibited and conventional bridge operation is obtained by controlling α in the range $150° \le \alpha \le 180°$. Figure 3.30 summarizes the converter characteristics at all the control modes.

Although the four-legged bridge has the advantages of improved displacement factor and load harmonic ripple, the harmonic distortion factor of the line current tends to be poorer. In addition, the supply requires a neutral connection which carries triplen harmonic currents. With a delta–star transformer, the harmonics create circulating current in the primary.

3.6 DUAL-BRIDGE CONVERTER

Two bridge converters can be connected in antiparallel to constitute a dual bridge converter, as shown in Fig. 3.31. Here the positive converter supplies the positive load current and the negative load current is taken up by the negative bridge, thus giving four-quadrant characteristics of the circuit. Dual converters are popularly used in thyristor Leonard speed control of dc motors, where reversible and regenerative operations can be obtained. The circuit is used for phase-controlled cycloconversion and can be used to supply voltage-fed inverters where regeneration is desired. A dual converter can be used with an intergroup reactor (IGR) between the component converters, and the load can be connected at the center tap of the reactor. In this case, the circuit can be controlled so as to have circulating current between the positive bridge and the negative bridge (i.e., one operates as a rectifier

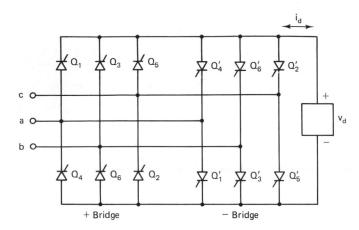

Figure 3.31 Dual-bridge converter.

and the other as an inverter with equal output voltage). The converter without IGR must be controlled to operate without circulating current, which otherwise will cause a line-to-line short circuit.

3.7 SIX-PULSE CENTER-TAP CONVERTER

Two three-phase half-wave converters with a phase shift of 180° can be connected in parallel through an IGR to constitute a six-pulse center-tap converter, as shown in Fig. 3.32. The circuit is popularly used in low-voltage, high-current electro-chemical-type applications. The dc load voltage v_d is the same as that of a half-wave converter, and each converter contributes 50% of the load current i_d. The component converters operate independently, with each thyristor conducting for 120°, and the IGR absorbs the instantaneous potential difference between the common cathodes G and H. As a result, the load voltage v_d has a six-stepped

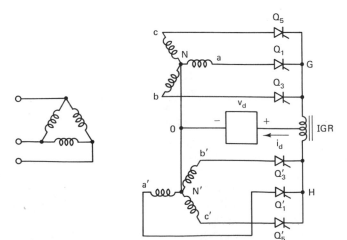

Figure 3.32 Six-pulse center-tap converter with inter-group reactor.

wave, as in a bridge converter.   Note that if the circuit is used without an IGR, a conventional six-pulse star-connected converter operation will result, where each thyristor and transformer secondary winding will conduct symmetrically for 60° per cycle.   Such a circuit operation, although satisfactory, is not desirable due to poor utilization of the transformer and thyristors.   One demerit of the center-tap circuit is that the load current must not fall below a minimum limit, the peak magnetizing current of the IGR.   If this happens, the circuit reverts to the operation of a six-pulse star converter and the load voltage rises by 15% at $\alpha = 0°$ operation.   At light load, the back-and-forth transition between the two modes may cause a stability problem with closed-loop α angle control.

3.8 TWELVE-PULSE SERIES BRIDGE CONVERTER

If the converter current or voltage rating is high so that a single thyristor device is not adequate, multiple devices may be connected in parallel or in series.   The parallel connection of devices is particularly difficult because of the matching problem in static and switching conditions.   Instead, parallel or series operation of converters with phase-shifting transformers is particularly advantageous because of harmonic reduction on the load and source side, although the additional cost of the transformer is involved.   An example of phase-shifted parallel operation was given in Fig. 3.32.   An example of phase-shifted series operation of bridge converters is given in Fig. 3.33.   A single bridge gives six-pulse operation, but

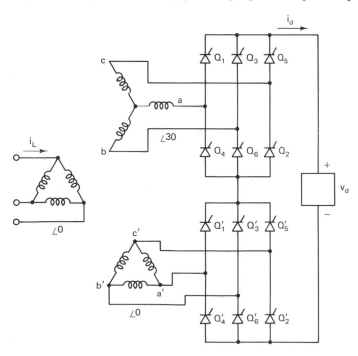

Figure 3.33   12-pulse series connected bridge converter.

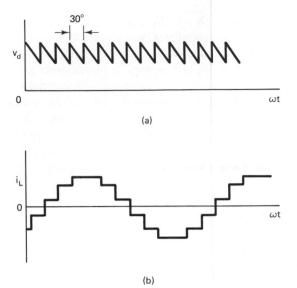

Figure 3.34 Waveforms of 12-pulse series bridge converter: (a) load voltage wave; (b) line current wave (12-stepped).

series connection of two bridges with transformer secondaries at 30° phase shift gives 12-pulse operation. The load voltage and line current waves for this circuit are shown in Fig. 3.34. It can be shown that the order of load voltage harmonics are 12th, 26th, 36th, and so on, and the corresponding order of line current harmonics are 11th, 13th, 23rd, 25th, and so on. Instead of being connected in series, the bridges can also be connected in parallel through an intergroup reactor to give 12-pulse operation. For high-power converter applications, such as HVDC and large dc motor drives, operation with increasing pulse number is very desirable.

Sequential Control

The double-bridge converter shown in Fig. 3.33 is normally controlled symmetrically or concurrently for two-quadrant operation, that is, the consecutive thyristors are fired at a 30° interval in the entire range of α between 0 and 180°. With such a concurrent firing angle control, the line displacement factor deteriorates with reduction of load voltage, as shown in Fig. 3.19. The displacement factor can be considerably improved if one converter is phase controlled while the other remains at full advance (α = 0) or at full retard (α = π) for rectifier or inverter operation, respectively. This means that in the rectification mode, one bridge operates as a diode rectifier while the other bridge is phase controlled. It can easily be seen that a bridge converter operating in such a sequential control mode acts as a semibridge converter, and therefore the improvement of line displacement factor is obvious. Figure 3.35 shows the characteristics in sequential control in both the rectification and inversion modes. The concurrent control characteristics are also superimposed in the same figure for comparison. Although sequential control provides improvement of line displacement factor, the dc output ripple deteriorates (i.e., a 12-pulse converter operates in the six-pulse mode).

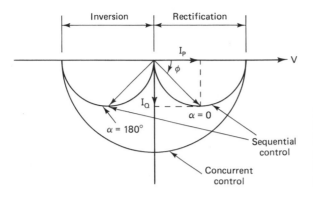

Figure 3.35 Characteristics of series bridge converter with sequential control. From *Thyristor Phase-Controlled Converters and Cycloconverters* by B. R. Pelly © 1971, John Wiley and Sons, Inc.

3.9 *CONTROL OF CONVERTER*

The function of the controller is to control the firing angle of a converter symmetrically in response to a demand of dc voltage or current. The controller usually incorporates the following functions:

- Synchronizing
- Firing angle control
- Advance limit control
- Retard limit control

The synchronizing circuit helps to establish symmetrical firing angle control to all the thyristors with respect to the fixed angular position of the ac voltage wave. The firing control circuit alters the firing angle α in response to a variable input control voltage. The advance and retard limit controls restrict the α angle within safe limits. The advance limit angle can be established as early as $\alpha = 0$, but the retard limit angle should provide a sufficient margin so that a minimum turn-off angle γ is maintained for successful commutation.

Linear Firing Angle Control

Figure 3.36 explains a simple linear firing angle control method for a single-phase bridge converter. The line supply voltage v_{ab} is stepped down through a transformer and converted to a square wave through a zero-crossing detector. A sawtooth wave of twice the supply frequency can be generated as shown in part (c) such that it remains synchronized with the square wave. The sawtooth wave starts with initial voltage A at zero angle and linearly decreases to zero at angle $180°$ and restarts again. The control voltage V_c is compared with the sawtooth wave and the firing angle is generated at the crossover point by the linear relation

$$\alpha = -\frac{\pi}{A} V_c + \pi \qquad (3.55)$$

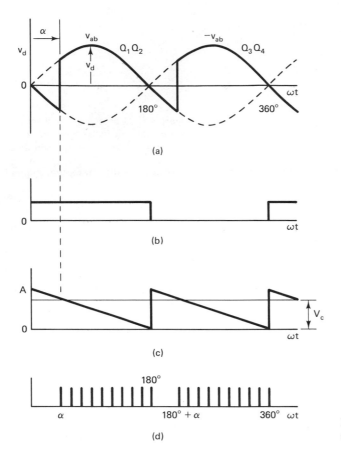

(a)

(b)

(c)

(d)

Figure 3.36 Linear firing angle control scheme of a single-phase bridge converter.

The long firing logic pulse of interval $\pi - \alpha$ is steered to the respective thyristor through a steering circuit. As the control voltage V_c is increased, α decreases, giving a higher dc voltage until α approaches zero at $V_c = A$. On the other extreme, α approaches π at $V_c = 0$ in the absence of any retard limit control. The relation between control voltage V_c and output V_d can be derived by substituting equation (3.55) in $V_d = V_{d0} \cos \alpha$ as

$$\cos^{-1} \frac{V_d}{V_{d0}} = -\frac{\pi}{A} V_c + \pi \tag{3.56}$$

Because of this nonlinear transfer characteristic, the linear firing angle control principle is seldom used.

Cosine Wave-Crossing Method

A popularly used control method where linearity of transfer characteristics is achieved is known as the cosine wave-crossing method. Figure 3.37 explains this method for a single-phase bridge converter. The supply voltage sine wave is phase shifted by 90° to generate the cosine wave and it is phase inverted every half-cycle as

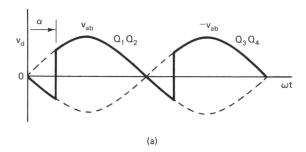

(a)

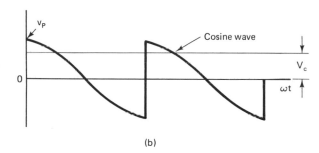

(b)

(c)

Figure 3.37 Cosine wave-crossing control method of single-phase bridge converter.

shown in Fig. 3.37(b). The firing angle is generated by the crossover point of control voltage V_c and cosine wave as

$$\cos \alpha = \frac{V_c}{V_P} \tag{3.57}$$

where V_P is the peak value of the cosine wave. Substituting equation (3.57) in $V_d = V_{d0} \cos \alpha$ yields

$$V_d = \frac{V_{d0}}{V_P} V_c = K V_c \tag{3.58}$$

indicating a linear relation between output and input with a gain factor K. Note that if the cosine wave magnitude varies with the fluctuation of supply voltage, the gain K remains unaltered. It should be noted, however, that equation (3.58) is valid only for continuous conduction. At light load, especially with counter emf, such as in a dc motor, the conduction may become discontinuous. In such a case, the gain becomes nonlinear and depends on the load parameters. Figure 3.38 shows the cosine wave-crossing control method for a three-phase bridge converter and Figs. 3.39 and 3.40 explain its operation. The derivation of firing logic signals for thyristor Q_1 is shown only, but the principle can easily be extended to other thyristors. The line voltage v_{ac} is the reference wave in which the angle 0 to 180°

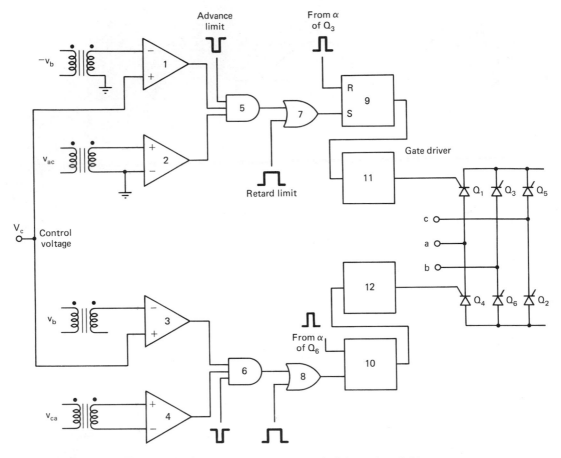

Figure 3.38   Cosine wave-crossing control of three-phase bridge converter.

corresponds to the firing angle range of Q_1.   The phase voltage $-v_b$ leads v_{ac} by 90° and constitutes the cosine reference wave for thyristor Q_1.   The phase and line voltages are stepped down through transformers and connected to the comparators as shown.   The comparator 1, which compares the control voltage V_c with phase voltage $-v_b$, transitions to logic 1 at firing angle α.   The output of comparators 1 and 2 are logically ANDed to trigger flip-flop 9 at the leading edge, which in

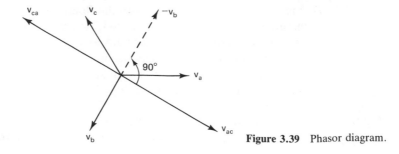

Figure 3.39   Phasor diagram.

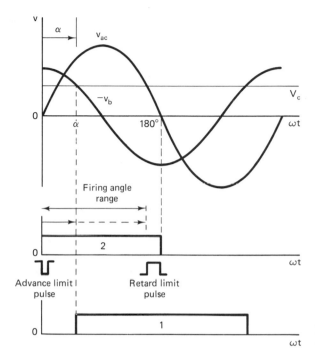

Figure 3.40   Waveforms explaining control circuit operation of thyristor Q_1.

turn couples a pulse train (not shown) to the gate of Q_1. The flip-flop is reset at firing of Q_3, thus limiting the gate pulse duration to 120°. The firing angle of Q_1 can be advanced or retarded by increasing or decreasing, respectively, the magnitude of V_c. The advance limit notch is coupled to AND gate 5 and the retard limit pulse is coupled to the OR gate 7 as shown in Fig. 3.38.

Phase-Locked-Loop Principle

The cosine wave-crossing method described above derives the cosine reference waves directly from the power supply phase voltages. The harmonics generated by the converter flow through the line source impedance and distort the line voltage. Similar distortion and transients can be introduced in the system by the power supply itself or by converters operating in parallel. A large filter to eliminate the distortion may not be satisfactory because of the phase shift, which is sensitive to supply frequency variation. A method to eliminate this problem is to digitally synthesize the cosine wave using the phase-locked-loop (PLL) technique. Figure 3.41 shows the block diagram of a PLL frequency synthesizer, and Fig. 3.42 explains the principle of digital synthesis of a biased cosine wave (Ref. 3) for a single-phase bridge converter. The frequency synthesizer generates the 30.72-kHz clock from the 60-Hz line frequency. The output frequency $f_0 = Nf$ can be programmed by selecting the frequency-divider ratio N so that f_0 tracks the supply frequency f within a definite locking range. The PLL is essentially a digital feedback system where the reference frequency f^* and feedback frequency f are compared in the phase frequency detector and an analog error signal proportional to the phase

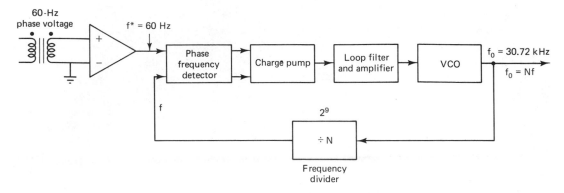

Figure 3.41 Phase-locked-loop (PLL) frequency synthesizer.

difference is generated at the input of a loop filter. The amplified error signal drives a voltage-controlled oscillator (VCO) to generate the desired output frequency. If the output wave tends to fall back in phase (or frequency), the error voltage builds up to correct the VCO output such that the input and feedback waves lock together with a small phase error. In Fig. 3.42, the 30.72-kHz output

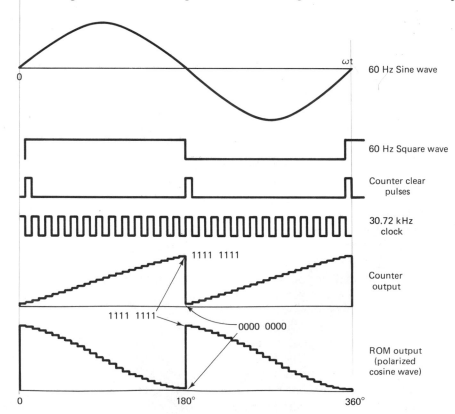

Figure 3.42 Digital synthesis of biased cosine wave for a single-phase bridge converter.

clock, which has frequency and phase synchronization with the ac phase voltage, generates the biased cosine wave through a counter and ROM look-up table. The counter is synchronized to the 60-Hz wave and retrieves the ROM output repetitively every half-cycle. The control voltage can be compared with ROM output either digitally or by an analog method by converting the ROM output through a digital-to-analog (D/A) converter.

An analog phase-locked oscillator control method for a three-phase bridge converter is shown in Fig. 3.43. The method removes the direct dependence of the controller from the line voltage waves, and therefore it is particularly attractive for a soft ac line supply. The converter current control is illustrated by a feedback method where the error signal drives a voltage-controlled oscillator through a proportional-integral (PI) controller. The VCO output drives a six-stage ring counter and the thyristor firing logic signals are derived by OR coupling the consecutive stages as shown. At steady state, the loop error is zero and the PI controller locks the output voltage such that the VCO operates at exactly six times the supply frequency. As the load changes or the command current I_d^* is changed, the VCO frequency will tend to drift, but a change in α angle will compensate I_d so that the system stabilizes at a new α angle.

3.10 MODELING OF CONVERTER

In steady-state condition under continuous conduction, a converter can be looked upon as an amplifier with linear gain characteristics. But in transient operation, it is a discrete-time system where the control voltage is sampled periodically and amplified at the output. Figure 3.44 explains this sampling nature for a three-phase bridge converter. Initially, at steady state the control voltage is sampled every 60° (i.e., at a time interval of $T = 2.78$ ms at a 60-Hz supply frequency). At the instant t_1, the control voltage steps up to a higher value but the effect reflects

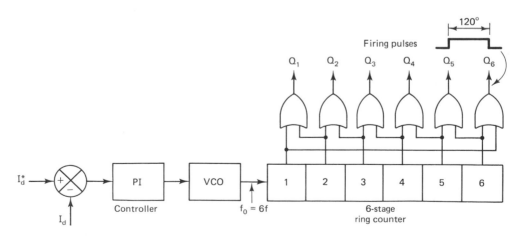

Figure 3.43 Phase-locked oscillator current control of three-phase bridge converter.

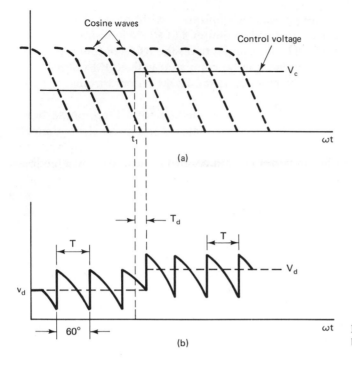

(a)

(b)

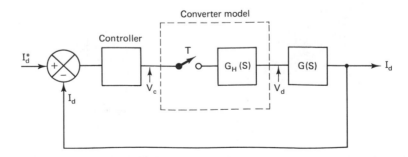

Figure 3.44 Transient response of bridge converter.

to the output after a delay time t_d.   The sampling time is altered during the transient but later resumes its steady-state value of 2.78 ms.   The delay time may vary between 0 and 2.78 ms depending on the instant of step change in the control voltage.   The nature of transient operation suggests that the converter can be represented approximately by a sampler with a zero-order-hold model (Ref. 4) as shown in Fig. 3.45 with a closed-loop current control system.   The term $G_H(S)$ is the transfer function of a sample-and-hold circuit and is given as

$$G_H(S) = K\frac{1 - e^{-TS}}{S} \tag{3.59}$$

where K is the converter gain, T the sampling period, and S the Laplace operator. The term $G(S)$ is the transfer function of the load connected to the converter.   The

Figure 3.45   Closed-loop current control system with converter model.

theoretical model can be simplified if the system bandwidth is small compared to the sampling frequency (i.e., 360 Hz). The approximate transfer function of the converter can be given as

$$\frac{V_d(S)}{V_c(S)} \simeq Ke^{-0.5TS} \qquad (3.60)$$

where $0.5T$ is the equivalent transport lag, which corresponds to half of the sampling interval. This transport lag can be considered as a mean between the extreme values of T_d, as explained in Fig. 3.44.

3.11 CYCLOCONVERTERS

A cycloconverter is a frequency changer that converts ac power at one input frequency to output power at a different frequency with one stage conversion. The phase-control line commutation principle discussed so far can be used for cyclo-conversion. A number of different possible frequency conversion schemes are described in Fig. 3.46. Figure 3.46(a) is the standard cycloconverter that will be discussed in this section. Here, if the output frequency f_o is less than input frequency f_i, it is known as step-down cycloconverter, whereas if $f_o > f_i$, it is known as step-up cycloconverter. In Fig. 3.46(b), ac is converted to ac with the medium of dc, and a rectifier and inverter are used in the two-step conversion process. This type of conversion is discussed in Chapters 4 and 5. Figure 3.46(c) illustrates a high-frequency link cycloconversion principle wherein step-up and step-down cycloconverters are cascaded for conversion from input frequency f_i to output frequency f_o through the medium of high frequency f_H. This type of frequency conversion, although somewhat complex and expensive, has the advantage that the resonant high-frequency link can be used for line commutation of both the step-up and step-down cycloconverters. This permits a wide range of frequency

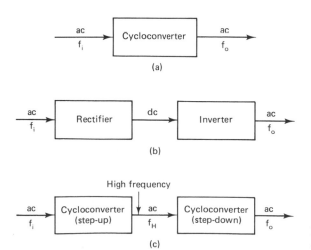

Figure 3.46 Different schemes of frequency conversion.

ratio f_o/f_i with smooth input and output waveforms and a programmable input power factor (Ref. 6).

The phase-controlled cycloconverters can provide a variable-voltage, variable-frequency power supply for driving ac machines. Historically, the first known ac drive used a thyratron cycloconverter to drive a 400-hp synchronous motor in Logan power station (Ref. 7). Because of cost and complexity, the cycloconverters are generally favored for high-horsepower drives. In addition to general speed control of induction and synchronous motors, the application of cycloconverters may include the following:

- Gearless cement and ball mill drives
- Static Scherbius ac drives
- Variable-speed, constant-frequency (VSCF) power generation for shipboard or aircraft power supply
- Static VAR generation
- Induction heating

Single Phase—Single Phase Circuit

The fundamental principle of cycloconversion can be explained with the help of a single phase-to-single phase circuit, shown in Fig. 3.47. A positive and a negative center-tap converter are connected in parallel so that the voltage and current of either polarity can be supplied to the load. The waveforms are drawn assuming a resistive load. In Fig. 3.47(b), an integral half-cycle output wave is fabricated with fundamental frequency $f_o = (1/n)f_i$, where n is the number of input half-cycles per half-cycle of the output. The firing angle can be modulated to control the voltage as well as its harmonic content, as shown in part (c). Instead of step-down conversion, step-up conversion is also possible, as indicated in part (d). The thyristors can be switched alternately between the positive and negative envelopes at a high frequency so that the output becomes carrier-frequency modulated. Of course, in such a mode of operation the thyristors require forced commutation, which is discussed in Chapter 4.

Three-Phase Half-Wave Circuit

The single-phase cycloconverter described above is seldom used in practice. A practical and commonly used cycloconverter uses the three-phase half-wave configuration shown in Fig. 3.48, which is also known as 18-thyristor, three-pulse cycloconverter. The circuit consists of three identical half-wave antiparallel phase groups and is shown with a wye-connected ac machine load. The load neutral can be connected to supply neutral if available. With the neutral connected, the phase groups operate independently. Each phase group functions as a dual converter with four-quadrant capability and therefore the load can sustain varying voltage and current of either polarity. The firing angle of each phase group is modulated sinusoidally but with a 120° phase shift so as to fabricate a mean sinusoidal output

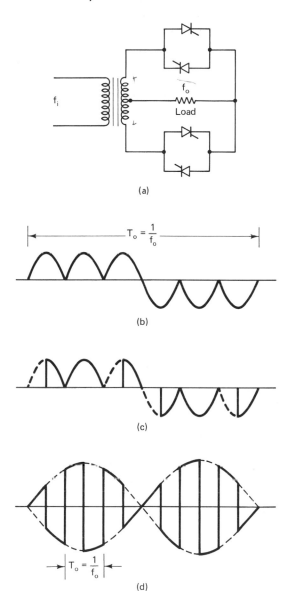

Figure 3.47 (a) single phase-to-single phase cycloconversion principle; (b) integral half-cycle output; (c) waveform with firing angle modulation; (d) waveform for step-up cycloconversion.

voltage, as shown in Fig. 3.49. The output frequency and depth of modulation can be varied to generate a variable-frequency, variable-voltage power supply for an ac motor. The fabricated output voltage contains a complex harmonic pattern which may become adequately filtered by the machine inductance. Any arbitrary power factor load can also be supplied by a cycloconverter. Figure 3.49 shows the phase voltage and current waves in the motoring condition, where the current lags the voltage by angle ϕ. The positive half-cycle of current flows through the positive converter, whereas the negative converter takes the negative half-cycle of current. A component converter operates in the rectification mode if the voltage

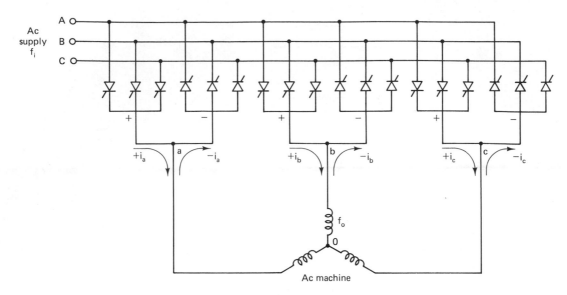

Figure 3.48   18-thyristor half-wave cycloconverter.

and current are of same polarity but in the inversion mode if these are of opposite polarity.   Both of the component converters in a phase group can be controlled simultaneously to fabricate the mean output voltage.   This will permit the bidirectional phase current to flow freely through either converter.   There will, of course, be an instantaneous potential difference between the outputs of two converters, which will tend to cause a short-circuit circulating current.   This can be prevented either by blocking the nonconducting converter or allowing a limited amount of circulating current to flow through an intergroup reactor (IGR).   Figure 3.50 shows the relation between firing angles of positive and negative converters for the same output voltage under continuous conduction.   The output voltage of

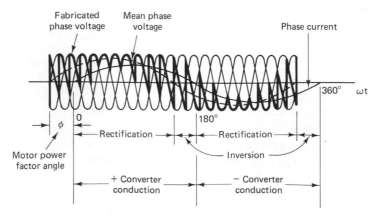

Figure 3.49   Phase voltage and current waves of 18-thyristor half-wave cycloconverter.

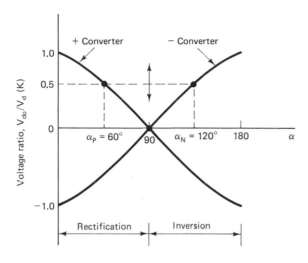

Figure 3.50 Output voltage to firing angle relation of cycloconverter. From *Thyristor Phase-Controlled Converters and Cycloconverters* by B. R. Pelly © 1971, John Wiley and Sons, Inc.

a converter is given by the general expressions

$$V_d = V_{d0} \cos \alpha \tag{3.61}$$

$$V_{d0} = \sqrt{2}\, V \frac{p}{\pi} \sin \frac{\pi}{p} \tag{3.62}$$

where p is the pulse number.   If the output voltage is positive, say for example, the voltage ratio $K = +0.5$, the positive converter operates as a rectifier with firing angle $\alpha_P = 60°$, and the negative converter operates as an inverter with firing angle $\alpha_N = 120°$, so that $V_{d0} \cos \alpha_P = -V_{d0} \cos \alpha_N$. If, instead, the voltage ratio is $K = -0.5$, the negative converter operates as rectifier with $\alpha_N = 60°$ and the positive converter operates as an inverter with $\alpha_P = 120°$.   In cycloconverter operation, the value of K is modulated between $+1.0$ and -1.0, maintaining the relation

$$\alpha_P + \alpha_N = 180° \tag{3.63}$$

The output voltage of the cycloconverter can be given in the form

$$v_0 = \sqrt{2}\, V_0 \sin \omega_o t = V_{d0} \cos \alpha_P = -V_{d0} \sin \alpha_N = m_f V_{d0} \sin \omega_o t \tag{3.64}$$

where V_0 is the rms output voltage and $m_f = \sqrt{2}\, V_0/V_{d0}$ is the modulation factor. The modulation factor is varied between zero and 1, and correspondingly the firing angles are modulated by the relations

$$\alpha_P = \cos^{-1}(m_f \sin \omega_0 t) \tag{3.65}$$

$$\alpha_N = 180° - \alpha_P \tag{3.66}$$

Figure 3.49 illustrates the voltage wave with $m_f = 1$ (i.e., $\alpha_P = 0$ and $\alpha_N = 180°$ at the peak positive voltage).   The output voltage is zero at $m_f = 0$ when $\alpha_P = \alpha_N = 90°$.

Three-Phase Bridge Circuit

Although many different cycloconverter circuit configurations are possible, only one more practical circuit used for large drive applications will be described here. This is the 36-thyristor, six-pulse bridge circuit shown in Fig. 3.51. Each phase group of the cycloconverter consists of a dual-bridge converter with IGR and the load is shown as wye connected with isolated windings. If isolation is not possible in the load, individual phase groups can be isolated at the input using transformers. With a bridge connection, the output voltage magnitude is double, or for the same output voltage demand the supply voltage is half that of an 18-thyristor circuit. This means that the thyristor voltage rating is half that of the 18-thyristor circuit. Since each thyristor carries the same rms current, the total VA rating of the cycloconverter remains the same. The six-pulse operation permits smooth output voltage fabrication, but the circuit with the control is quite complex and expensive.

Output Voltage and Frequency Range

As explained before, the sinusoidal output voltage can be controlled smoothly between zero and the maximum value by controlling the modulation factor m_f between zero and 1. The reduction of voltage causes harmonic deterioration of the voltage wave, which should be evident from Fig. 3.49. The fundamental voltage can, of course, be increased by saturating the cycloconverter, and ultimately the maximum voltage may be obtained in the square-wave mode. In fact, the third or triplen harmonics are sometimes mixed with the fundamental in an isolated neutral system, which permits an increase in fundamental voltage without the penalty of triplen harmonics circulating currents. In the square-wave mode of operation, although the harmonic contents are higher, the control becomes simple and the input displacement factor is somewhat improved. The VSCF system requires sinusoidal operation; the square-wave mode is often preferred for a machine drive system because of advantages cited above and because of inherent filtering in the load.

For line-commutated operation of a cycloconverter, the output frequency should be less than the input frequency. Ideally, the lowest frequency may be zero (i.e., dc operation is possible). Zero-frequency operation is desirable for ac servo and slip recovery drives, which will be discussed later. As the output frequency increases, the harmonic quality of the wave deteriorates and for this reason the output frequency is usually limited to one-third of the supply frequency. The output frequency can be increased above the supply frequency by using forced or load commutation. For a cycloconverter-fed synchronous machine drive, line commutation can be used in the step-down frequency range, but the frequency may be increased to the step-up range using load commutation, where the machine is operated at leading power factor. In this mode, the cycloconverter can be visualized as a step-down frequency changer from output to input, with power flowing from the low-frequency to the high-frequency side.

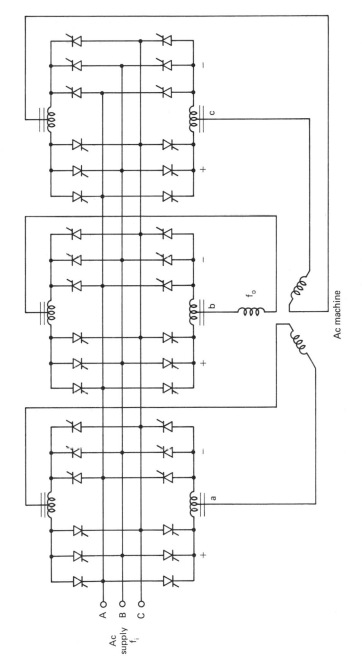

Figure 3.51 36-thyristor bridge cycloconverter.

Load and Source Harmonics

The fabricated output voltage wave of a cycloconverter contains harmonics which if impressed across the machine cause additional heating and torque pulsation problems. The harmonics are generally influenced by the following:

- Circulating or noncirculating mode of operation
- Pulse number
- Modulation factor of output voltage
- Output-to-input frequency ratio
- Load power factor
- Continuous or discontinuous conduction
- Commutation overlap effect
- Feedback control method and its bandwidth

It was mentioned before that a three-phase phase-controlled converter contains triplen harmonics at the output. Since in a cycloconverter the firing angles are sinusoidally modulated, it is expected that the output will contain sidebands or beat-frequency components in the form $Mf_i \pm Nf_o$. Table 3.1 shows a summary of output-voltage harmonics of 18-thyristor and 36-thyristor circuits in the noncirculating current mode.

It is evident that some of the lower sideband frequencies, although having reduced voltage magnitude, may fall below the fundamental frequency and cause reasonably large subharmonic currents. The problem may be especially severe if

TABLE 3.1 OUTPUT-VOLTAGE HARMONICS IN NONCIRCULATING CURRENT MODE

18-Thyristor			
$3f_i \pm 2f_o$	$6f_i \pm f_o$	$9f_i \pm 2f_o$	$\cdots$
$3f_i \pm 4f_o$	$6f_i \pm 3f_o$	$9f_i \pm 4f_o$	
$3f_i \pm 6f_o$	$6f_i \pm 5f_o$	$9f_i \pm 6f_o$	
$3f_i \pm 8f_o$	$6f_i \pm 7f_o$	$9f_i \pm 8f_o$	
$\cdot$	$\cdot$	$\cdot$	
$\cdot$	$\cdot$	$\cdot$	

36-Thryistor			
$6f_i \pm f_o$	$12f_i \pm f_o$	$18f_i \pm f_o$	$\cdots$
$6f_i \pm 3f_o$	$12f_i \pm 3f_o$	$18f_i \pm 3f_o$	
$6f_i \pm 5f_o$	$12f_i \pm 5f_o$	$18f_i \pm 5f_o$	
$6f_i \pm 7f_o$	$12f_i \pm 7f_o$	$18f_i \pm 7f_o$	
$\cdot$	$\cdot$	$\cdot$	
$\cdot$	$\cdot$	$\cdot$	

the frequency ratio is large.   These subharmonics can be attenuated by feedback control.

In the circulating current mode of operation, the cycloconverter output voltage wave is somewhat smoother, with the harmonic amplitudes considerably attenuated.   The voltage wave shape (i.e., the harmonics) is not affected by the load power factor angle.   It can be shown that outer sidebands in the spectrum are terminated at a definite term, thus eliminating the possibility of subharmonic currents up to a reasonably high frequency ratio.   For example, in an 18-thyristor cycloconverter, the limiting sidebands are $3f_i \pm 4f_o$, $6f_i \pm 7f_o$, $9f_i \pm 10f_o$, and so on, and the corresponding terms in the bridge circuit are $6f_i \pm 7f_o$, $12f_i \pm 13f_o$, $18f_i \pm 19f_o$, and so on (Ref. 1).

The input line current wave of the cycloconverter contains harmonics since it is fabricated piece by piece from the output current waves.   It was mentioned before that for a three-phase p-pulse converter, the input line current contains $(np \pm 1)f_i$ harmonics, where n is an integer.   Therefore, with firing angle modulation it is expected to have harmonics with sidebands in the form $(np \pm 1)f_i \pm mf_o$.   For example, in a 18-thyristor circuit some of the harmonics are $f_i \pm 6f_o$, $f_i \pm 12f_o$, $2f_i \pm 3f_o$, $2f_i \pm 6f_o$, $4f_i \pm 3f_o$, $4f_i \pm 6f_o$, $5f_i$, and so on, and for the bridge circuit these are $f_i \pm 6f_o$, $f_i \pm 12f_o$, $5f_i$, $5f_i \pm 6f_o$, $7f_i$, $7f_i \pm 6f_o$, and so on.

Input Displacement Factor

A cycloconverter demands a lagging reactive current at the input because of the inherent phase-control mechanism for output voltage fabrication.   Even with a load power factor of unity and a unity modulation factor ($m_f = 1$), the input consumes reactive current, which is determined by the mean firing angle.   The reactive current requirement increases as the load power factor decreases or the modulation factor decreases.   Again, the input reactive current is always lagging irrespective of lagging or leading power factor of the load.   The large lagging reactive current requirement at the input together with the harmonic distortion makes the input power factor poor, which is indeed a great disadvantage of cycloconverter application.

It is not difficult to derive a quantitative expression of input displacement factor of a cycloconverter.   Consider that the positive converter of phase a of an 18-thyristor cycloconverter is conducting only.   Figure 3.52 shows the input phase current at retard angle α_P with the corresponding phase voltage.   Assume that the output frequency ratio is low so that the current pulse width always remains 120° wide.   The Fourier series of the current wave can be given as

$$
\begin{aligned}
i = \frac{i_o}{3} + \frac{\sqrt{3}}{\pi} i_o &\left[\sin(\omega_i t - \alpha_P) - \frac{1}{2}\cos(\omega_i t - \alpha_P) \right. \\
&\left. - \frac{1}{4}\cos 4(\omega_i t - \alpha_P) - \frac{1}{5}\sin 5(\omega_i t - \alpha_P) - \cdots \right]
\end{aligned} \tag{3.67}
$$

where i is the instantaneous value of input current, i_o the instantaneous value of output current, and ω_i the supply frequency.   Note that the fundamental of input

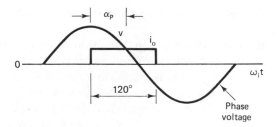

Figure 3.52   Input voltage and current relations of 18-thyristor cycloconverter (one phase of positive converter only).

current lags the phase voltage by the angle of phase retard. Since the harmonics do not contribute to the real and reactive power when the supply voltage is a sine wave, the instantaneous real power and reactive power of the supply are

$$P'_i = 3V \left(\frac{\sqrt{3} i_o}{\sqrt{2} \pi} \right) \cos \alpha_P = (1.17V \cos \alpha_p) i_o \tag{3.68}$$

$$Q'_i = 3V \left(\frac{\sqrt{3} i_o}{\sqrt{2} \pi} \right) \sin \alpha_P = (1.17V \sin \alpha_P) i_o \tag{3.69}$$

where V is the rms supply phase voltage. Since the angle α_P is continuously modulated, P'_i and Q'_i in the equations above are also being modulated accordingly. It is therefore necessary to average P'_i and Q'_i to determine loading on the source. The mean output voltage given by $1.17V \cos \alpha_P$ is plotted in relation to output current i_o in Fig. 3.53(a), where ϕ is the load power factor angle. The voltage wave $1.17V \sin \alpha_P$ can be determined from the $1.17V \cos \alpha_P$ wave by a phase shift of 90° and noting that the α_P range is $0 < \alpha_P < 180°$. The $1.17V \sin \alpha_P$ wave is multiplied by the i_o wave in Fig. 3.53(b) to determine the instantaneous input reactive power, which can be averaged to get the average input reactive power as follows:

$$Q_i = \frac{1}{\pi} \int_0^\pi Q'_i \, d\omega_i t = \frac{2P_o}{\pi} \cos^2 \phi + \frac{2Q_o}{\pi} \left(\phi + \frac{1}{2} \sin 2\phi \right) \tag{3.70}$$

where P_o and Q_o are the cycloconverter active and reactive power, respectively, at the output. Since the active power at the output is the same as that of the input, we can write

$$P_i + jQ_i = P_i + j\frac{2}{\pi} \left[P_i \cos^2 \phi + Q_o(\phi + \frac{1}{2} \sin 2\phi) \right] \tag{3.71}$$

and therefore the input displacement factor is

$$\text{Dis.F.} = \frac{P_i}{P_i + jQ_i}$$

$$= \frac{1}{1 + j(2/\pi) \left[\cos^2 \phi + (Q_o/P_i) \, (\phi + \frac{1}{2} \sin 2\phi) \right]} \tag{3.72}$$

$$= \frac{1}{1 + j \, (2/\pi) \, (1 + \phi \tan \phi)}$$

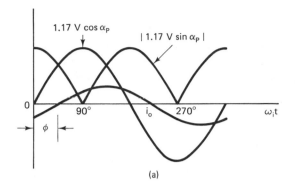

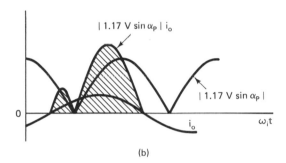

Figure 3.53   Reactive loading on source during conduction of positive converter.

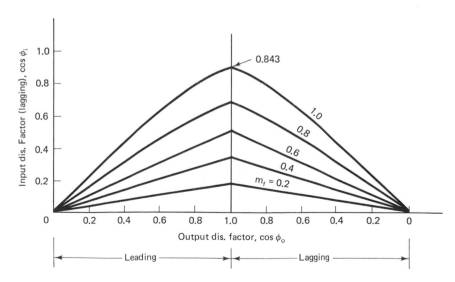

Figure 3.54   Input and output displacement factor relations. From *Thyristor Phase-Controlled Converters and Cycloconverters* by B. R. Pelly © 1971, John Wiley and Sons, Inc.

where Q_o/P_i is substituted for by $\tan \phi$.   Equation (3.72) is derived by considering the positive converter only.   If more converters are placed on the source, the active and reactive powers reflect to the input equally so that the displacement factor expression remains unchanged.   Equation (3.72) has been derived for a modulation factor of unity.   Since the modulation factor does not affect the current demanded by the load but affects the displacement angle directly, we can write

$$\text{Dis.F.} = \frac{m_f}{1 + j\,(2/\pi)\,(1 + \phi \tan \phi)} \tag{3.73}$$

Equation (3.73) has been plotted in Fig. 3.54 for different m_f and for both lagging and leading load power factor.   It is obvious from the figure that the theoretically maximum input displacement factor is 0.843 for unity power factor load and with unity modulation factor.

REFERENCES

1. B. R. Pelly, *Thyristor Phase-Controlled Converters and Cycloconverters*, Wiley, New York, 1971.

2. V. R. Stefanovic, "Power Factor Improvement with a Modified Phase-Controlled Converter," *IEEE Trans. Ind. Appl.*, Vol. IA-15, pp. 193–201, Mar.–Apr. 1979.

3. B. K. Bose and K. J. Jentzen, "Digital Speed Control of a DC Motor with Phase-Locked Loop Regulation," *IEEE Trans. Ind. Electron. Control Instrum.*, Vol. IECI-25, pp. 10–13, Feb. 1978.

4. E. A. Parrish and E. S. McVey "A Theoretical Model for Single-Phase Silicon-Controlled Rectifier Systems," *IEEE Trans. Autom. Control*, Vol. AC-12, pp. 577–579, Oct. 1967.

5. T. H. Barton and R. S. Birtch, "A 5-kW Low-Frequency Power Amplifier of Improved Frequency Response," *IEEE Trans. Ind. Electron. Control Instrum.*, Vol. IECI-14, pp. 33–39, Apr. 1967.

6. P. M. Espelage and B. K. Bose, "High Frequency Link Power Conversion," *IEEE Trans. Ind. Appl.*, Vol. IA-13, pp. 387–394, Sept.–Oct. 1977.

7. E. F. W. Alexanderson and A. H. Mittag, "The Thyration Motor," *Electr. Eng.*, New York, Vol. 53, pp. 1517–1520, 1934.

8. W. Slabiak and L. J. Lawson, "Precise Control of a Three-Phase Squirrel-Cage Induction Motor Using Practical Cycloconverter," *IEEE Trans. Ind. Appl.*, Vol. IA-2, pp. 274–280, 1966.

9. D. L. Plette and H. G. Carlson, "Performance of a Variable Speed Constant Frequency Electrical System," *IEEE Trans. Aerospace*, Vol. 2, pp. 957–970, 1964.

10. Y. Tamura, S. Tanaka, and S. Tadakuma, "Control Method and Upper Limit of Output Frequency in Circulating-Current Type Cycloconverter," *Conf. Rec. IEEE/IAS Int. Sem. Power Conv. Conf.*, pp. 313–323, 1982.

4

VOLTAGE-FED INVERTERS

4.0 INTRODUCTION

The class of converters that converts dc power to ac power is defined as inverters, and in general there are two types of inverters: voltage-fed inverters and current-fed inverters. The current-fed inverter using the phase-control line commutation principle was introduced in Chapter 3 and more of these types of inverter are discussed in Chapter 5. A voltage-fed inverter is characterized by a stiff dc voltage supply at the input (i.e., the voltage source ideally has zero Thévenin impedance). The dc supply voltage may be fixed or variable, and may be obtained from a utility power supply or by a rotating alternator through rectification or a battery, fuel cell, photovoltaic array, or magneto hydrodynamic (MHD) generator. The input ac may be single phase or polyphase and the output voltage and frequency may be constant or variable. The inverter applications may include adjustable-speed ac drives, regulated voltage and frequency power supplies, uninterruptible power supplies, induction heating, lagging or leading VAR generation, and so on.

In a voltage-fed inverter, the power semiconductor devices always remain forward biased due to the dc supply voltage, and therefore some type of forced commutation is mandatory when using thyristors. Alternatively, self-commutation with base or gate drive is possible when using GTOs, power transistors, IGTs, or power MOS devices.

In this chapter we discuss various types of voltage-fed inverters, analyze their performance, review commutation methods, develop an inverter–machine model, discuss the inverter–machine interface, and describe the control methods. Again, the nonideal characteristics of the switching devices will be ignored.

4.1 SQUARE-WAVE INVERTER

The power circuit of a three-phase bridge inverter using thyristors is shown in Fig. 4.1, where commutation and snubber circuits are omitted for simplicity. The synthesis of output voltage waves is shown in Fig. 4.2. The inverter consists of three half-bridge units where the upper and lower thyristors of each unit are switched on and off alternately for 180° intervals. The three half-bridges are phase-shifted by 120°, as is evident in the synthesis of three-phase voltage waves shown in Fig. 4.2. The dc supply voltage is assumed center-tapped for convenience of waveform synthesis but is not necessary. It is normally obtained from a utility power supply through a bridge rectifier and an *LC* filter to establish a stiff dc voltage source. The inverter output voltage wave shapes are determined by the circuit configuration and switching pattern but are not affected by the load condition. These waves are rich in harmonics, but the current waves are somewhat smoother, due to the filtering effect of the load. The bypass diodes of the inverter permit reverse current flow during reactive power flow and regeneration, and clamp the load voltage to that of the input level. These devices also participate in forced commutation, which is discussed later.

The phase voltages with respect to the dc center tap can be described by Fourier series as follows:

$$v_{a0} = \frac{4V}{\pi}\left(\cos \omega t + \frac{1}{3}\cos 3\omega t + \frac{1}{5}\cos 5\omega t + \cdots \right) \tag{4.1}$$

$$v_{b0} = \frac{4V}{\pi}\left[\cos(\omega t - 120°) + \frac{1}{3}\cos 3(\omega t - 120°) \right.$$
$$\left. + \frac{1}{5}\cos 5(\omega t - 120°) + \cdots \right] \tag{4.2}$$

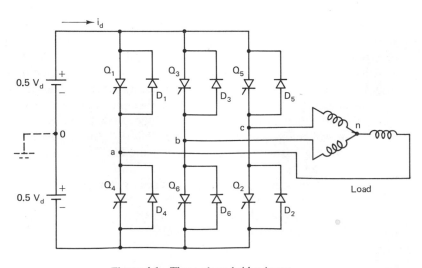

Figure 4.1 Three-phase bridge inverter.

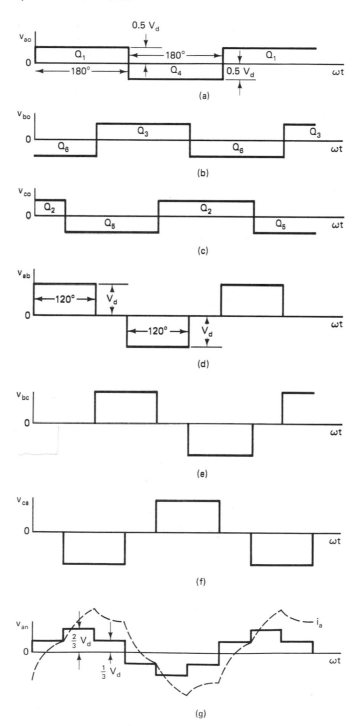

Figure 4.2 Synthesis of voltage waves in bridge inverter.

$$v_{c0} = \frac{4V}{\pi}\left[\cos(\omega t + 120°) + \frac{1}{3}\cos 3(\omega t + 120°)\right.$$
$$\left. + \frac{1}{5}\cos 5(\omega t + 120°) + \cdots\right] \tag{4.3}$$

where $V = 0.5V_d$.

The line voltages can therefore be constructed as

$$v_{ab} = v_{a0} - v_{b0} \tag{4.4}$$

$$v_{bc} = v_{b0} - v_{c0} \tag{4.5}$$

$$v_{ca} = v_{c0} - v_{a0} \tag{4.6}$$

The line voltage Fourier series can be determined by combining equations (4.1), (4.2), and (4.3) as

$$v_{ab} = \sqrt{3}\,\frac{4V}{\pi}\left[\cos(\omega t + 30°) + 0 - \frac{1}{5}\cos 5(\omega t + 30°)\right.$$
$$\left. - \frac{1}{7}\cos 7(\omega t + 30°) + \cdots\right] \tag{4.7}$$

$$v_{bc} = \sqrt{3}\,\frac{4V}{\pi}\left[\cos(\omega t - 90) + 0 - \frac{1}{5}\cos 5(\omega t - 90°)\right.$$
$$\left. - \frac{1}{7}\cos 7(\omega t - 90°) + \cdots\right] \tag{4.8}$$

$$v_{ca} = \sqrt{3}\,\frac{4V}{\pi}\left[\cos(\omega t + 150°) + 0 - \frac{1}{5}\cos 5(\omega t + 150°)\right.$$
$$\left. - \frac{1}{7}\cos 7(\omega t + 150°) + \cdots\right] \tag{4.9}$$

The line voltage waves have a characteristic six-stepped wave shape with the presence of odd harmonics $6n \pm 1$, where n = integer. The fundamental and harmonic voltages are balanced and mutually phase shifted by 120°. The voltage waves are dual to the input current waves of the bridge rectifier shown in Fig. 3.16. This type of inverter is defined as a square or stepped-wave inverter because of the characteristic wave shapes. If the load is delta-connected, the phase and line voltages are identical and the problem of circulating current does not exist because of the absence of triplen harmonics.

For the wye-connected load with isolated neutral as shown in Fig. 4.1, we can write

$$v_{a0} = v_{an} + v_{n0} \tag{4.10}$$

$$v_{b0} = v_{bn} + v_{n0} \tag{4.11}$$

$$v_{c0} = v_{cn} + v_{n0} \tag{4.12}$$

Since for balanced three-phase supply $v_{an} + v_{bn} + v_{cn} = 0$, adding the equations above, we get

$$3v_{n0} + 0 = v_{a0} + v_{b0} + v_{c0}$$

or

$$v_{n0} = \tfrac{1}{3}(v_{a0} + v_{b0} + v_{c0}) \tag{4.13}$$

Therefore, substituting equation (4.13) in equations (4.10), (4.11), and (4.12) yields

$$v_{an} = v_{a0} - v_{n0} = \tfrac{2}{3}v_{a0} - \tfrac{1}{3}(v_{b0} + v_{c0}) \tag{4.14}$$

$$v_{bn} = v_{b0} - v_{n0} = \tfrac{2}{3}v_{b0} - \tfrac{1}{3}(v_{a0} + v_{c0}) \tag{4.15}$$

$$v_{cn} = v_{c0} - v_{n0} = \tfrac{2}{3}v_{c0} - \tfrac{1}{3}(v_{a0} + v_{b0}) \tag{4.16}$$

The phase voltage can be constructed graphically as shown in Fig. 4.2(g) or described by the Fourier series. It also has a six-stepped wave shape but is phase shifted from line voltage by 30°.

For a linear and balanced three-phase load, individual line current components can be solved for each component of the voltage Fourier series and then the resultant can be derived by superposition principle. A typical line current wave with inductive load is shown on the v_{an} wave of Fig. 4.2.

Figure 4.3(a) shows the phase voltage and line current waves assuming perfect load filtering (i.e., the harmonic currents have been neglected). During the interval ωt_1, the phase voltage is positive but the line current is negative (i.e., the reactive current is flowing to the source through the feedback diode D_1). But in interval ωt_2, the thyristor Q_1 is carrying the active load current. The next half-cycle is symmetrical and the respective conduction intervals of D_4 and Q_4 are shown. It may be inferred that if the load is purely resistive or has unity displacement factor ($\phi = 0$), each thyristor conducts for 180°. It can be shown that for the range $-90° < \phi < +90°$, the inverter is active (i.e., the dc power is flowing to the ac); on the other hand, for $+90° < \phi < -90°$, the inverter is regenerative (i.e., the power is flowing from ac to dc side). Such a controlled rectification mode is entirely possible in an inverter and this permits regenerative or dynamic braking of an ac machine, which is discussed later. In an extreme condition, if $\phi = 180°$, the inverter operates as a diode rectifier with the diodes only conducting.

Input Ripple

The inverter input voltage and current will deviate from ideal dc values due to the presence of ripple components. The ripple may be introduced by a utility line-side rectifier due to the finite pulse number and economical size of the filter, or it may be introduced by the inverter itself. The distortion of dc voltage will affect the output-voltage waves and will correspondingly deteriorate the line current waves. The distortion introduced by the inverter itself can be approximately calculated by considering the instantaneous power balance between input and output, since there is no energy storage in the switching elements. It can be shown that for the ideal sine-wave voltage and current waves at the output, the instan-

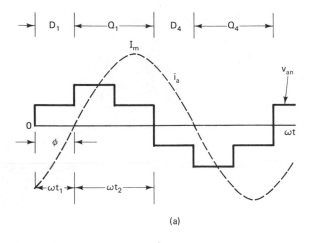

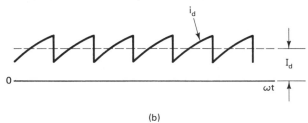

(a)

(b)

Figure 4.3 (a) Output phase voltage and current waves assuming perfect filtering; (b) input dc current wave.

taneous power given by

$$P = v_a i_a + v_b i_b + v_c i_c \tag{4.17}$$

is always constant irrespective of the load power factor condition. The reactive currents will circulate through the inverter and the input current will not contain any harmonics. For the stepped-voltage and sine current waves, the input current wave as shown in Fig. 4.3(b) can be derived from an instantaneous power balance equation. Considering the duality of voltage and current waves between a phase-controlled converter and a voltage-fed inverter, the input current wave i_d is identical to the voltage wave shape at rectifier output for $\alpha = \pi - \phi$. Once the worst-case ripple current is determined at the input, the approximate filter capacitor size can be calculated. The accurate determination of harmonics both at the input and output is quite involved and requires detailed computer simulation.

Voltage and Current Ratings

For the specified output kVA requirement, a suitable power switching device is to be selected and its voltage and current ratings are to be designed. The devices have to withstand the maximum voltage V_d in the forward direction on which the margin due to commutation overshoot, typically 50% overvoltage, is added. The diodes are of the fast-recovery type and have the same voltage rating as thyristors.

From Fig. 4.3, the peak thyristor current is I_m and the rms and average currents can be derived as

$$I_{rms}(Q_1) = I_{rms}(Q_4) = \sqrt{\frac{1}{\pi} \int_\phi^\pi i_a^2 \, d\omega t} \qquad (4.18)$$

$$I_{rms}(D_1) = I_{rms}(D_2) = \sqrt{\frac{1}{\pi} \int_0^\phi i_a^2 \, d\omega t} \qquad (4.19)$$

$$I_{av}(Q_1) = I_{av}(Q_4) = \frac{1}{\pi} \int_\phi^\pi i_a \, d\omega t \qquad (4.20)$$

$$I_{av}(D_1) = I_{av}(D_4) = \frac{1}{\pi} \int_0^\phi i_a \, d\omega t \qquad (4.21)$$

Twelve-Step Inverter

For the large power rating of a six-stepped bridge inverter, the devices can be connected in series–parallel combination. With series–parallel operation, the devices require matching, and some amount of voltage and current derating of the inverter is essential. In addition to design complexity, the matched devices are not easily available. Alternatively, multiple three-phase bridge inverter modules using single devices can be connected in parallel through center-tapped reactors to increase the current (i.e., the power rating).

In large power application, it is desirable that the inverter output approach a sine voltage wave because it will demand reduced filter size on both the dc and ac sides. With an ac machine load on the inverter, where an additional filter is not desirable, a smooth waveform results in reduction of harmonic current heating and torque pulsation effects. The lower-order harmonics of a six-stepped wave (i.e., the 5th and 7th) can be neutralized by synthesizing a 12-stepped waveform as discussed in Chapter 3. The significant harmonics of multistepped waveforms are $Kn \pm 1$, where K = number of steps (6, 12, 24, etc.) and n = integer. In fact, the circuit of Fig. 3.33 can be used directly to design a 12-stepped voltage-fed inverter. A more practical 12-stepped inverter using two parallel bridges is shown in Fig. 4.4(a) and part (b) explains the phasor diagram for 12-stepped wave synthesis. The component bridge inverters are operated in six-stepped mode but are mutually phase shifted by 30° and each inverter is connected to the primary delta winding of the respective transformers as shown. The upper transformer has one secondary winding for each phase, whereas the lower transformer has two secondary windings per phase and the winding ratios are as indicated in the figure. The secondary windings are interconnected to satisfy the phasor diagram shown in Fig. 4.4(b).

The output phase voltage is obtained by series connection of three secondary windings and the phase A voltage can be given as

$$v_{NA} = v_{ab} + v_{de} - v_{ef} \qquad (4.22)$$

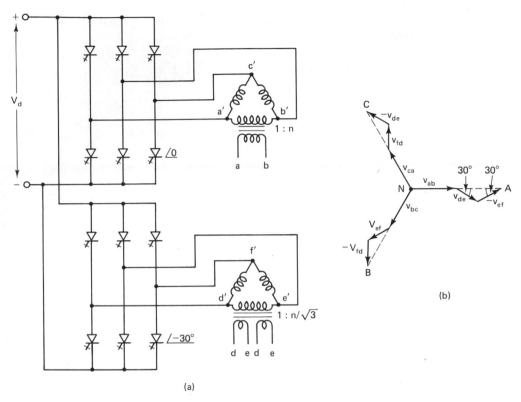

(a)

Figure 4.4   Principle of dual bridge 12-step inverter (bypass diodes omitted).

The lower bridge is considered to operate at 30° phase lag with respect to the upper bridge and the line voltage $v_{a'b'}$ is assumed to be the reference phase. Therefore, the component voltage Fourier series expressions can be given as

$$v_{ab} = \frac{2\sqrt{3}\, nV_d}{\pi} \left(\cos \omega t + \frac{1}{5} \cos 5\omega t + \frac{1}{7} \cos 7\omega t + \cdots \right) \tag{4.23}$$

$$v_{de} = \frac{2nV_d}{\pi} \left[\cos(\omega t - 30°) + \frac{1}{5} \cos 5(\omega t - 30°) \right.$$
$$\left. + \frac{1}{7} \cos 7(\omega t - 30°) + \cdots \right] \tag{4.24}$$

$$v_{ef} = \frac{2nV_d}{\pi} \left[\cos(\omega t - 150°) + \frac{1}{5} \cos 5(\omega t - 150°) \right.$$
$$\left. + \frac{1}{7} \cos 7(\omega t - 150°) + \cdots \right] \tag{4.25}$$

From the equations above, it can be shown that

$$v_{ab(5th)} + v_{de(5th)} - v_{ef(5th)} = 0 \tag{4.26}$$

$$v_{ab(7th)} + v_{de(7th)} - v_{ef(7th)} = 0 \tag{4.27}$$

and the fundamental phase voltage can be given as

$$v_{\text{NA}(f)} = v_{ab(f)} + v_{de(f)} - v_{ef(f)} = \frac{4\sqrt{3}\,nV_d}{\pi} \cos \omega t \tag{4.28}$$

which is also evident from the phasor diagram. The lowest harmonics in the phase voltage are 11, 13, and so on, and the phase voltage Fourier series can be given as

$$v_{\text{NA}} = \frac{4\sqrt{3}\,nV_d}{\pi} \left(\cos \omega t + \frac{1}{11} \cos 11\, \omega t + \frac{1}{13} \cos 13\, \omega t + \cdots \right) \tag{4.29}$$

The principle above can be extended to construct a 24-stepped inverter which will require four bridges and four transformers.

The multistepped inverter with harmonic neutralization as discussed above requires a bulky transformer at the output and the inverter power is derated as a result of the phase shift. The transformer will not saturate if the volt/hertz ratio is maintained constant, but saturation is difficult to avoid near zero-frequency operation. However, the advantages are that the transformer permits isolation and voltage-level shift. For an ac machine load, the stator can be designed with 12-phase winding, thus eliminating the requirement for separate transformers.

4.2 VOLTAGE AND FREQUENCY CONTROL

The inverter output voltage and frequency are to be controlled continuously for machine drive applications as shown in Fig. 2.12. For a regulated ac power supply, the frequency is fixed but the voltage requires control due to supply and load variations. The inverter frequency control is very straightforward. If a fixed output frequency is desired, the reference frequency may be generated by a crystal oscillator. The frequency can then be subdivided to the desired frequency and the gate drive waves can be generated through logic and counter circuits, which are described later. For variable-frequency operation, an analog voltage may represent the frequency command, which is then converted to proportional frequency through a voltage-controlled oscillator. Alternatively, the reference frequency generation and its processing to generate the gate drive waves can be done entirely by microcomputer software. The stability of the inverter output frequency is determined by the stability of the reference signal frequency and is not affected by load and source variations.

The inverter output voltage can in general be controlled by the following two methods:

- Input voltage control or pulse amplitude modulation (PAM)
- Voltage control within the inverter by pulse width modulation (PWM)

The PWM method will be discussed in a separate section. For the voltage control of a square-wave inverter, only the former method is applicable. If the ac supply is rectified to dc, there may be two possible schemes: (1) a phase-controlled rectifier with filter and (2) a diode rectifier-chopper with filters. With a phase-controlled rectifier, the ac line-side harmonics are high and the power factor deteriorates at reduced voltage. In the latter case, the power is converted twice but has the advantage of high line-side power factor with near-unity displacement factor. If the primary power is dc, say, from a battery or photo-voltaic array, a chopper can also be used to control the inverter dc voltage. The schemes with chopper control are not favored; instead, PWM voltage control scheme is normally used. If ac power is obtained from an engine-driven alternator, it can be converted to dc by a diode rectifier and voltage can be controlled by regulation of alternator field current.

The variation of dc link voltage will not cause any problem in self-commutated inverter operation using GTO, transistor, or MOS devices. But for force-commutated inverters using thyristor devices, the inverter commutation becomes difficult at reduced dc link voltage, which is discussed later. It may be noted here that with economical filter size, the dc link voltage may be distorted and therefore the output voltage waves may deviate from the ideal shapes discussed so far.

4.3 COMMUTATION METHODS

Commutation can be defined as the turning off of a power semiconductor device from the state of conduction. The commutation methods can be classified as follows:

- Line commutation
- Load commutation
- Self-commutation
- Forced commutation*

Line commutation has been discussed in Chapter 3. The load commutation is similar to line commutation except that the commutating voltage is induced from the load circuit and it is discussed in Chapter 5. The self-commutation relates to GTO, transistor, and MOS devices where commutation is performed by gate or base drive signals. Forced commutation is discussed in detail in this section.

The thyristors in a voltage-fed inverter remain forward biased by the dc supply voltage and therefore are to be commutated by what is known as forced commutation. In forced commutation, a precharged capacitor in correct polarity creates a voltage or current transient around the conducting thyristor. This causes diversion of the thyristor current and an inverse voltage is impressed across the device, helping to turn it off. Many different forced commutation circuits have been described in the literature. The popularly used McMurray commutation technique will be described here in some detail, and later a few more common forced commutation methods are reviewed briefly.

---

*Forced commutation is often grouped under self-commutation in literature.

McMurray Inverter

The McMurray commutation technique using a half-bridge voltage-fed inverter is illustrated in Fig. 4.5. The half-bridge shown on the right is aided by the commutation circuit, which consists of auxiliary thyristors and diodes Q_{1A}, Q_{4A} and D_{1A}, D_{4A}, respectively, and circuit components L_c, C_c, and R_d. The inductance L_1 is the di/dt limiting inductance for the main and auxiliary thyristors. A full three-phase inverter will consist of three such identical units. A large filter capacitor shown at the input can sink the harmonics fed back to the source.

Figure 4.5 shows the salient modes during commutation and Fig. 4.6 shows the corresponding voltage and current waveforms. Assume that initially the thy-

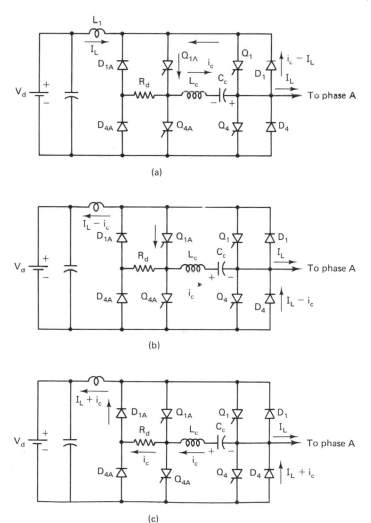

Figure 4.5   Modes of operation of McMurray inverter: (a) mode 1: $0-t_2$ interval; (b) mode 3: t_3-t_4 interval; (c) mode 4: t_4-t_5 interval.

ristor Q_1 is conducting the full line current I_L and it is to be force commutated so that after completion of commutation the diode D_4 takes the current I_L as shown in Fig. 4.6(a).   The load is assumed to be inductive and the magnitude of I_L is assumed to remain constant during the short commutation interval, typically 100 to 200 µs.   The circuit of Fig. 4.5 is commonly known as a modified McMurray inverter because in the original version the elements D_{1A}, D_{4A}, and R_d were not present.

To commutate Q_1 successfully, the capacitor C_c is assumed to be initially charged to voltage V_d with the polarity shown.   The initial charge can be built at startup condition by firing Q_1 and Q_{4A} together. The commutation is initiated by firing the thyristor Q_{1A}.   The modes of commutation can be summarized as follows:

Mode 1.   In the beginning, the resonant current i_c builds up in the circuit Q_{1A}, L_c, C_c, Q_1 and subtracts from the current I_L flowing in Q_1.   At time t_1, the current in Q_1 is zero and the excess resonance current circulates through the diode

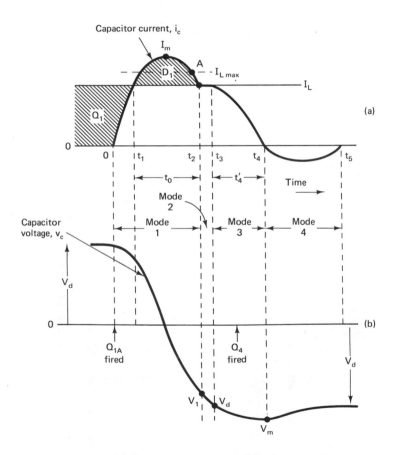

Figure 4.6   (a) Commutation currents; (b) voltage waveforms.

D_1 shown in Fig. 4.6(a). The capacitor voltage discharges to zero near the peak current I_m and then charges in the reverse direction. At time t_2, mode 1 of commutation ends when $i_c = I_L$ and capacitor charges to voltage V_1. The inverse diode drop during $t_0 = t_2 - t_1$ helps to turn off the thyristor Q_1.

Mode 2.   In mode 2, $i_c = I_L$ and the constant current flowing in the path Q_{1A}, L_c, and C_c linearly charges the capacitor until $v_c = V_d$.

Mode 3.   As the capacitor charges to voltage V_d, D_4 comes into conduction and the load current is shared between the resonance circuit and D_4. The supply voltage V_d is impressed across the resonance circuit and the trapped energy in L_c overcharges the capacitor to peak voltage V_m. The thyristor Q_4 is fired typically at half-time period $0.5\,T = \pi\sqrt{L_cC_c}$ after initiation of commutation, but it appears redundant due to conduction of bypass diode D_4. Mode 3 ends at time t_4 when $i_c = 0$ and the full-line current is taken by D_4.

Mode 4.   In mode 4, the overcharge on C_c bleeds off to the source through R_d in the path D_{1A}, R_d, L_c, C_c, and D_4. The circuit is normally critically damped so that v_c gradually falls off to V_d at time t_5 when the commutation process is complete.

The reactive current in D_4 eventually falls to zero and then Q_4 resumes conduction automatically. The commutation process from Q_4 to D_1 is identical to that described above except that the elements Q_{4A} and D_{4A} become active. The commutation will be satisfactory if I_L is varied from zero up to a maximum value so that the t_0 interval is sufficient. At no-load conditions, mode 2 will vanish and mode 1 conduction will extend for the period $0.5\,T = \pi\sqrt{L_cC_c}$. In mode 3, Q_4 will conduct instead of D_4 and again the duration is $t_4 - t_3 = 0.5\,T$.

The commutation process as described above is known as necessary commutation. If in the beginning, the diode D_1 is carrying negative I_L and the conduction is to be transferred to Q_4, the commutation process above is repeated and it is known as redundant commutation. The reason for redundant commutation is to reverse the v_c polarity so that in the subsequent interval, Q_4 can be commutated successfully. The redundant commutation occurs during the regenerative mode or PWM voltage control, which is described later.

McMurray inverter analysis.   Consider a general series resonance circuit with the impressed dc voltage V_d and assume that the circuit has initial capacitor voltage V_0 and initial inductor current I_0. The following circuit equations can be written:

$$L_c\frac{di_c}{dt} + \frac{1}{C_c}\int_0^t i_c\,dt + V_0 + R_c i_c = V_d \tag{4.30}$$

$$v_c = V_0 + \frac{1}{C_c}\int_0^t i_c\,dt \tag{4.31}$$

Laplace-transforming equations (4.30) and (4.31) yield

$$I_c(S) = \frac{(V_d - V_0)/L_c + L_c I_0}{S^2 + (R_c/L_c)S + 1/L_c C_c} \tag{4.32}$$

$$v_c(S) = \frac{V_0}{S} + \frac{(V_d - V_0)/L_c C_c S + I_0/C_c}{S^2 + (R_c/L_c)S + 1/L_c C_c} \tag{4.33}$$

Equations (4.32) and (4.33) can be solved in the time domain as

$$i_c = (V_d - V_0) \sqrt{\frac{C_c}{L_c}} \frac{1}{\sqrt{1 - \rho^2}} e^{-\rho\omega_0 t} \sin \omega t$$

$$- I_0 \frac{1}{\sqrt{1 - \rho^2}} e^{-\rho\omega_0 t} \sin(\omega t - \phi) \tag{4.34}$$

$$v_c = V_d - \frac{V_d - V_0}{\sqrt{1 - \rho^2}} e^{-\rho\omega_0 t} \sin(\omega t + \phi) + \frac{L_c I_0 \omega_0}{\sqrt{1 - \rho^2}} e^{-\rho\omega_0 t} \sin \omega t \tag{4.35}$$

where

$$\omega_0 = \frac{1}{\sqrt{L_c C_c}}, \qquad \rho = \frac{R_c}{2} \sqrt{\frac{C_c}{L_c}}$$

$$\omega = \omega_0 \sqrt{1 - \rho^2}, \qquad \phi = \tan^{-1} \frac{\sqrt{1 - \rho^2}}{\rho}$$

If the circuit has low loss so that it is highly underdamped, then $\rho \approx 0$, $\omega_0 \approx \omega$, and $\phi \approx 90°$. Therefore,

$$i_c = \left[(V_d - V_0) \sqrt{\frac{C_c}{L_c}} \sin \omega_0 t + I_0 \cos \omega_0 t \right] e^{-\omega_0 t/2Q} \tag{4.36}$$

$$v_c = V_d - \left[(V_d - V_0) \cos \omega_0 t - L_c I_0 \omega_0 \sin \omega_0 t \right] e^{-\omega_0 t/2Q} \tag{4.37}$$

where $Q = \omega_0 L_c/R_c = 1/2\rho$. In the beginning of mode 1 of the McMurray inverter, $V_d = 0$, $I_0 = 0$, and $V_0 = -V_d$. Therefore, the mode 1 equations are

$$i_c = V_d \sqrt{\frac{C_c}{L_c}} e^{-\omega_0 t/2Q} \sin \omega_0 t = I_m \sin \omega_0 t \tag{4.38}$$

$$v_c = -V_d e^{-\omega_0 t/2Q} \cos \omega_0 t \tag{4.39}$$

At $t = t_1$, $i_c = I_L$ and equation (4.38) can be solved to determine t_1. At the end of mode 1, $i_c = I_L$ and $v_c = V_1$. Therefore,

$$I_L = V_d \sqrt{\frac{C_c}{L_c}} \, e^{-\omega_0 t_2/2Q} \sin \omega_0 t_2 \tag{4.40}$$

$$V_1 = -V_d e^{-\omega_0 t_2/2Q} \cos \omega_0 t_2 \tag{4.41}$$

The time t_2 can be solved from equation (4.40), and correspondingly V_1 can be determined from equation (4.41).

In mode 2,

$$i_c = I_L \tag{4.42}$$

$$v_c = V_1 + \int_{t_2}^{t_2+t} I_L \, dt \tag{4.43}$$

At time t_3, $v_c = V_d$, so that mode 2 interval is

$$t_3 - t_2 = \frac{C_c(V_d - V_1)}{I_L} \tag{4.44}$$

The initial conditions in the mode 3 interval are $v_c = V_d$ and $i_c = I_L$. The final conditions are $v_c = V_m$ and $i_c = 0$. Therefore, the equations at the end of mode 3 are

$$0 = (I_L \cos \omega_0 t_4')e^{-\omega_0 t_4'/2Q} \tag{4.45}$$

$$V_m = V_d + (LI_L\omega_0 \sin \omega_0 t_4')e^{-\omega_0 t_4'/2Q} \tag{4.46}$$

where $L = L_1 + L_c$ and time t_4' is counted from the beginning of mode 3. The values of t_4 and V_m can be solved from these equations.

Mode 4 is critically damped and the interval can be solved from original equations (4.34) and (4.35).

Commutation loss. The power loss is an important criterion for the selection of a particular commutation circuit. In the McMurray inverter, we will assume that the losses are principally confined in mode 1 and mode 4. The energy loss in mode 1 can be given as

$$W_1 = \tfrac{1}{2} C_c(V_d^2 - V_1^2) \tag{4.47}$$

where V_1 is given from equation (4.41). If $\omega t_2 \approx \pi$ as in the no-load condition,

$$V_1 = V_d e^{-\pi/2Q} \tag{4.48}$$

Substituting equation (4.48) in (4.47) yields

$$W_1 = \tfrac{1}{2} C_c V_d^2(1 - e^{-\pi/Q}) \tag{4.49}$$

In mode 4, part of the excess capacitor energy is dissipated in R_d and part of it is returned to the source. The energy returned to the source is

$$W_s = V_d \int_{t_4}^{t_5} i_c \, dt$$

$$= V_d C_c (V_m - V_d)$$

(4.50)

Therefore, the energy dissipation in R_d is

$$W_4 = \tfrac{1}{2} C_c (V_m^2 - V_d^2) - V_d C_c (V_m - V_d)$$

$$= C_c (V_m - V_d) \left(V_d - \frac{V_m + V_d}{2} \right)$$

(4.51)

The total commutation loss in watts can be given as

$$P_c = m(W_1 + W_4)$$

(4.52)

where m is the number of commutations per second. For a six-stepped bridge inverter $m = 6 \times$ fundamental frequency.

Commutation circuit design.   A simple illustrative example will be given as the design procedure is described.   Let $I_{L\max} = 500$ A. Select $I_m/I_{L\max} = 1.5$ so that $I_m = 750$ A. Select $V_d = 500$ V and $di/dt = 120$ A/µs for the main thyristors.

Solving $V_d = L_1(di/dt)$ gives $L_1 = 4.17$ µH.   The commutation circuit is to be designed on the basis of minimum V_d, because I_m varies directly with V_d.   Again, selection of L_c and C_c values are based on least stored energy for commutation. For $I_m/I_{L\max} = 1.5$, it can be shown that $t_0 \simeq 1.68\sqrt{L_c C_c}$. (Ref. 12)

For thyristor $t_q = 20$ µs, select $t_0 = 30$ µs. Solve L_c and C_c from the following relations:

$$I_m = V_d \frac{C_c}{L_c}$$

$$t_0 = 1.68 \sqrt{L_c C_c}$$

which gives $C_c = 26.79$ µF and $L_c = 11.9$ µH.   Select R_d for critical damping of resonance circuit;

$$\sqrt{\frac{1}{LC_c} - \frac{R_d^2}{4L_c^2}} = 0$$

which gives

$$R_d = 2\sqrt{\frac{L}{C_c}} = 2\sqrt{\frac{11.9 + 4.17}{26.79}} = 1.76 \ \Omega$$

Adaptive triggering.   In Fig. 4.6, the main thyristor Q_4 was fired with a fixed time delay (i.e., half the resonance time period after firing the auxiliary thyristor Q_{1A}). Instead, firing of Q_4 could have been advanced to the fixed point

A, which corresponds to the maximum load current $I_{L\max}$ for the safe commutation. The advantages of advance triggering are that the commutation clearing time ($0 \to t_5$) is reduced and inverter reliability is improved by locking the circuit at maximum commutating capability irrespective of load transients. However, the disadvantage is that the commutation energy loss is high and always corresponds to the worst-case load current. Low energy loss that is comparable to normal triggering but with faster clearing time can be obtained by providing adaptive triggering. In this method, Q_4 is always fired at the end of mode 1 (i.e., the crossover point between capacitor current and load current). Figure 4.7 shows commutation current waves in adaptive triggering.

Several other commutation circuits to be discussed here are shown in Fig. 4.8.

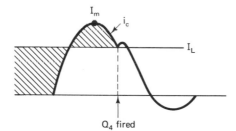

Figure 4.7 Commutation current waves with adaptive triggering.

McMurray–Bedford Inverter

The McMurray–Bedford circuit [Fig. 4.8(a)] works on the principle of complementary voltage commutation compared to the current commutation principle as described before. The devices Q_1, Q_4 and D_1, D_4 are the main thyristors and feedback diodes, respectively of the half-bridge inverter. The tapped inductor in series with the thyristors, which has self-inductance L_c for each section, is the commutating inductance, and C_c is the commutating capacitance as shown. The autotransformer and the diodes D_A, D_B are used for the recovery of trapped energy in L_c after commutation. Assume, for example, that the thyristor Q_1 is initially conducting the load current that is to be commutated, and the conduction is to be transferred to diode D_4. The commutation is initiated by firing the complementary thyristor Q_4. The initial line voltage in the lower capacitor is induced in the upper section of the tapped inductor with the cathode of Q_1 as positive. This turns off Q_1 and its line current is transferred to Q_4 by the conservation of flux linkage principle. The resonance component of Q_4 current initially builds up and then decays, but during decay diodes D_4 and D_A conduct, feeding back the trapped energy in L_c to the source. Eventually, the current in Q_4 falls to zero and the full-load current is taken by D_4.

The circuit is somewhat simple in the sense that no auxiliary thyristors are used; but the weight of autotransformer required for energy recovery is added. The commutating L_c and C_c components are somewhat larger, and close coupling of the inductor in a large power inverter becomes a difficult problem.

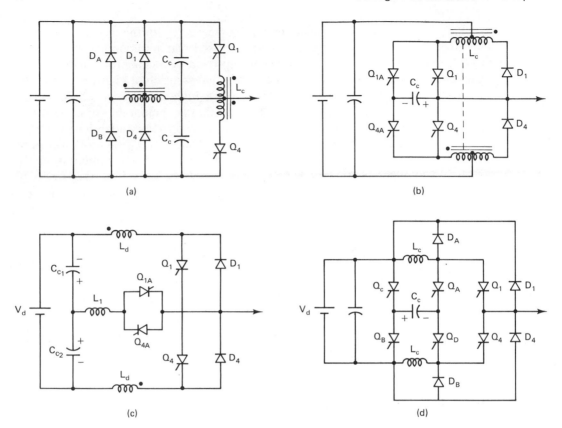

Figure 4.8   Several commutation circuits: (a) McMurray–Bedford; (b) Verhoef; (c) ac switched; (d) dc side.

Verhoef Inverter

The Verhoef inverter shown in Fig. 4.8(b) is a voltage-commutated inverter with auxiliary thyristors, and the supply voltage is connected to the center tap of the commutating inductance L_c, which is in series with the main thyristor and feedback diode. All four sections of the tapped inductors are closely coupled so that the current initially flowing in one section can easily be transferred to the coupling section when the primary current is turned off. Consider that initially the main thyristor Q_1 is carrying the load current that is to be transferred to feedback diode D_4. The capacitor C_c is initially charged to the supply voltage with the polarity shown. The commutation is initiated by firing the auxiliary thyristor Q_{1A}, which turns off Q_1 immediately and takes its load current. The resonant current flows in the loop Q_{1A}, C_c, D_1, L_c, and the capacitor discharges and then charges in the reverse direction. The commutation ends when C_c charges to the full voltage in the reverse direction and the load current automatically transfers from Q_{1A} to D_4 due to the coupling effect.

In this inverter, ideally there is no trapped energy due to close coupling of opposite sections of the coils, and therefore the related loss is eliminated. However, steady-state current flowing through inductor sections causes a higher loss. The large inductors make the inverter bulky, but the beneficial effect is that the current due to short circuit (shoot-through) fault tends to be limited.

AC Switched Inverter

The ac switched inverter shown in Figure 4.8(c) is a current commutated inverter with auxiliary thyristor triggering, and the performance is somewhat analogous to McMurray inverter. The commutation circuit consists of an inverse-parallel thyristor switch, series inductor L_1, split commutating capacitors, and line inductors L_d. The effective commutating parameters are given by $C_c = C_{c1} \| C_{c2}$ and $L_c = L_1 + L_d$. The L_d of each side may be coupled with the polarity shown. The equivalent McMurray configuration can be obtained by exchanging the location of C_{c1} and C_{c2} with the respective auxiliary thyristor. Assume initially that Q_1 is conducting and C_{c1} is charged in the polarity shown by the previous commutation. The commutation is initiated by firing Q_{1A}, which causes the resonant current to flow in the loop L_1, Q_{1A}, D_1, L_d and thus turns off Q_1. When the resonant current tends to fall below the load current, the constant-current charging interval starts, as in the McMurray inverter. Then Q_4 is fired to complete charging of C_{c1} and the load current is eventually transferred to diode D_4. At the end of commutation, C_{c1} overcharges ($>V_d$), leaving C_{c2} with negative polarity so that the subsequent commutation can occur satisfactorily.

The advantage of the circuit is that the trapped energy is eliminated and therefore the efficiency tends to improve. However, this advantage is offset by higher losses during Q_4 conduction when the peak current reaches a high value. The main and auxiliary thyristors are to be designed for higher pulse current duty and the circuit commutation time is somewhat longer. The advantage of splitting L_c is that the line inductor L_d can limit the short-through fault current.

DC Side-Commutated Inverter

The principle of this type of inverter [Fig. 4.8(d)] is different in the sense that all the thyristors in the upper group or lower group of the bridge inverter are commutated simultaneously by a common dc side-commutation circuit. In a square-wave inverter, it was shown that three main thyristors may conduct at any instant: two from the upper group and one from the lower group, or one from the upper group and two from the lower group. In an upper group commutation, for example, all upper thyristors are turned off simultaneously, diverting the line currents temporarily through the feedback diodes until the incoming thyristors are fired again after commutation. In Figure 4.8(d), only a half-bridge circuit is shown for illustration. Assume that initially the upper thyristor Q_1 is conducting, which is to be turned off, and the commutating capacitor C_c is initially charged to the supply voltage with the polarity shown. To initiate commutation, the auxiliary thyristors Q_A and Q_B are fired. The capacitor voltage is impressed across Q_1 with its cathode positive and it is turned off, diverting the line current through D_4. The capacitor voltage in series with the supply voltage builds up a resonance current in the upper commutating inductance through the path Q_A, C_c, Q_B, V_d, and L_c. When the capacitor charges to negative supply voltage, Q_A and Q_B turn off and the trapped current in L_c free-wheels through D_A. The commutation is complete when the free-wheeling current falls to zero.

The dc side-commutation principle is simple and economical, but dissipation of the full trapped energy by free-wheeling lowers the circuit efficiency and lengthens the commutation interval. Modification of the circuit is possible where the trapped energy is fed back to the source by transformer coupling the L_c inductor.

4.4 PWM INVERTERS

The stepped-wave inverters discussed in Section 4.2 have several advantages and limitations. The inverter control logic is somewhat simple and the switching losses are low, due to the limited number of switchings per cycle of fundamental frequency. One difficult problem is that the commutation capability of the inverter is reduced as dc link voltage is reduced to control the output voltage. Of course, the problem can be solved by coupling fixed auxiliary voltage for commutation purposes. The commutation circuit and the associated problems do not exist for self-commutated inverters which use transistors and GTOs and therefore cost, weight, efficiency, and voltage control range are improved. The speed control typically beyond a 10:1 range becomes a problem with a six-stepped inverter, because at low voltage harmonic currents become excessive, causing machine heating and torque pulsation problems. In addition, the utility line power factor deteriorates due to phase shift control, and a system stability problem may arise at low speed due to the low-pass filter in the dc link.

The problems noted above can be solved using a pulse-width-modulated (PWM) inverter, discussed in this section. The PWM inverter is supplied in the front end by a diode bridge rectifier and LC filter for general industrial applications. The fundamental frequency output voltage is controlled electronically within the inverter by a multiple pulse-width-modulation technique. The inverter topology remains the same as in Fig. 4.1, but the devices are switched on and off many times within a cycle to control the output voltage, which is normally low in harmonic content. The following general PWM types will be discussed:

- Sinusoidal PWM
- Harmonic elimination principle
- Adaptive current control PWM
- Phase-shift PWM

Sinusoidal PWM

The sinusoidal PWM technique is popular in industrial applications and is reviewed extensively in the literature. Figure 4.9 shows the general principle of PWM where an isoceles triangle carrier wave is compared with a fundamental-frequency sine

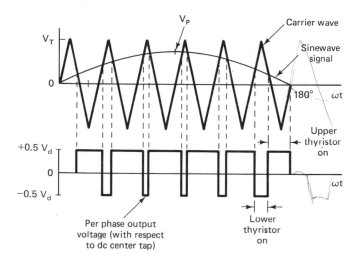

Figure 4.9 Principle of sinusoidal pulse width modulation with natural sampling (only half-cycle is shown).

modulating wave, and the natural points of intersection determine the switching points of power devices of a half-bridge inverter as shown in the figure. The technique is also known as the triangulation, subharmonic, or suboscillation method. A common carrier can be used for all the three phases. The typical wave shapes of line voltage and phase voltage with respect to the neutral are shown in Fig. 4.10.

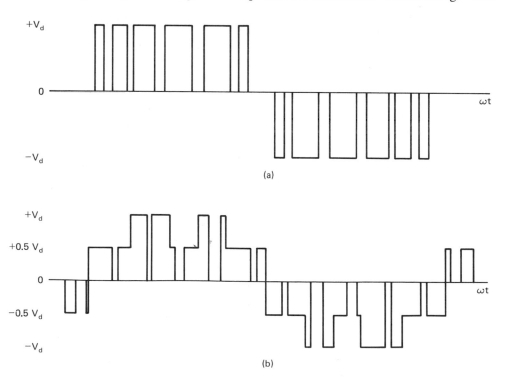

Figure 4.10 Line and phase voltage waves of PWM inverter: (a) line voltage, (b) phase voltage.

The alternate pulse and notch widths of half-bridge inverter output are sinusoidally modulated, and the waveform contains a fundamental component of which the frequency and amplitude can be varied by varying the frequency and voltage, respectively, of the modulating wave. A Fourier analysis of the output wave is quite involved but can be shown to be of the following form:

$$v(t) = m \frac{V_d}{2} \sin (\omega_s t + \phi) + \text{Bessel function harmonic terms} \qquad (4.53)$$

where m is the modulation index, ω_s the fundamental frequency (same as the modulating frequency), ϕ the phase shift of output, depending on the position of the modulating wave.

The modulation index is again defined as $m = V_p/V_T$, where V_p is the peak value of the modulating wave and V_T the peak value of the carrier wave. Ideally, m can vary between zero and 1 to give a linear relationship between the modulating and the output voltage. For $m = 1$, the maximum value of fundamental peak voltage is $0.5V_d$, which is 78.5% of the peak voltage $(4V_d/2\pi)$ of the square wave. It can be shown that the maximum voltage in the linear range can be increased to some extent by mixing triplen harmonics with the modulating wave. At $m = 0$, the output is a square wave with symmetrical pulse and notch widths. As m approaches 1, the notch width near the center of the half-cycle tends to vanish. For successful operation of the inverter, minimum pulse and notch widths are to be maintained for commutation and snubber relaxation. Similarly, a minimum lock-out or dwell time is required between switching of the upper and lower devices when both devices are turned off to prevent bus shoot-through fault.

The PWM output wave contains carrier-frequency-related harmonics with modulating frequency-related sidebands in the form of $M\omega_c \pm N\omega_s$, where ω_c is the carrier frequency, ω_s the modulating frequency, M and N are integers, and $M + N$ is an odd integer. For a carrier-to-modulating frequency ratio $P = 15$, a summary of the output harmonics is given in Table 4.1.

It can be shown that the amplitudes of the harmonics are independent of P and diminish with higher values of M and N. With a high carrier-frequency ratio P, the inverter line current harmonics will be well filtered by nominal leakage inductance of the machine and the current will approach a sine wave. By selecting P as a multiple of 3, triplen frequency-related currents can be avoided in the line. The selection of P depends on the balance between inverter loss and machine harmonic losses. Higher value of P (i.e., a higher number of commutations per second) increases the inverter switching loss but reduces the machine harmonic losses.

A popular PWM principle based on the uniform sampling technique is shown in Fig. 4.11. In natural sampling as described earlier, the switching instants are determined by intrinsic natural selection of the sampling points. In a uniform sampling method which is based on the sample-and-hold principle, the sampling frequency is equal to the carrier frequency. In the former method, the pulse is asymmetrical about the trough of the carrier, whereas in the latter the pulse is always symmetrical, as indicated in Fig. 4.11. The uniform sampling method, which is easily adaptable to microcomputer implementation, gives a significant

TABLE 4.1 SUMMARY OF
OUTPUT HARMONICS IN PWM
INVERTER

M	Harmonics
1	$15\omega_c$
	$15\omega_c \pm 2\omega_s$
	$15\omega_c \pm 4\omega_s$
	$15\omega_c \pm 6\omega_s$
	etc.
2	$30\omega_c \pm \omega_s$
	$30\omega_c \pm 3\omega_s$
	$30\omega_c \pm 5\omega_s$
	etc.
3	$45\omega_c$
	$45\omega_c \pm 2\omega_s$
	$45\omega_c \pm 4\omega_s$
	$45\omega_c \pm 6\omega_s$
	etc.

improvement in low-frequency harmonics and eliminates subharmonics in the free-running mode, as discussed later.

The fundamental output voltage can be increased beyond the linear range by increasing the modulation factor m beyond unity until the maximum value is obtained at the square-wave output. As the linear-to-nonlinear transition region begins, the notches near the middle of the wave drop abruptly, causing a current surge problem. The current surge causes a pulsed torque problem and may cause commutation failure in force-commutated inverters. This is especially objectionable to a transistor inverter, which is peak current sensitive. Besides nonlinear voltage transfer characteristics, the fundamental-frequency-related harmonics (e.g., 3rd, 5th, 7th, etc.) begin to appear in the transition region, causing a considerable increase in machine losses.

Instead of the sine modulating wave, a square or trapezoidal modulating wave can be considered, which gives symmetrical pulse widths at the inverter output. The inverter output voltage can be controlled linearly in the range zero to square-

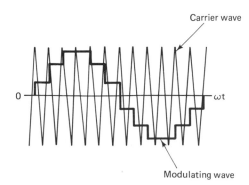

Carrier wave

Modulating wave

Figure 4.11 Uniform sampling PWM principle.

wave by varying the amplitude of the modulating wave. No notch dropping problem is encountered except at the transition to square wave. The harmonic output of square-wave PWM is considerably worse than that of sine-wave PWM, but generation of modulating waves is simple.

Frequency relation.   For variable-speed drive applications, the inverter output voltage and frequency are to be varied in the relation shown in Fig. 2.12. In the constant-power region, the maximum voltage can be obtained by operating the inverter in the square-wave mode, but in the constant-torque region, the voltage can be controlled using the PWM principle. It is desirable to operate the inverter with an integral ratio of carrier-to-modulating frequency, with the modulating wave remaining synchronized with the carrier wave in the entire region. A fixed ratio P causes a low carrier frequency as the fundamental frequency goes down, which is not desirable from the machine-harmonic-loss point of view. A practical carrier-to-fundamental frequency relation of a transistor inverter is shown in Fig. 4.12. At a low fundamental frequency, the carrier frequency is maintained constant and the inverter operates in the free-running or asynchronous mode. The ratio P may be nonintegral and the phase may continually drift. This gives rise to a subharmonic problem with drifting dc offset, but these harmful effects can be ignored because of a large frequency ratio. The free-running region is followed by the synchronized region, where P is varied in steps as shown so that the maximum and minimum carrier frequencies remain bounded within a definite zone. Near the

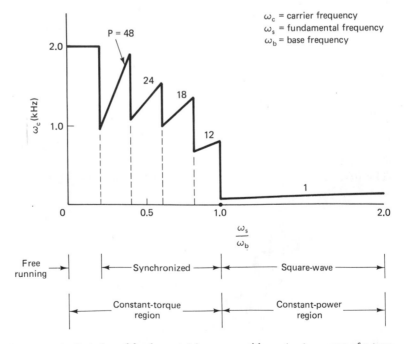

Figure 4.12   Relation of fundamental frequency with carrier frequency of a transistor inverter.

base frequency, a transition occurs to the square-wave mode, where the carrier frequency is assumed to be the same as the fundamental frequency. The control should be designed carefully so that at the jump of carrier frequency there is no voltage jump problem, and chattering between adjacent *P*'s should be avoided by providing a narrow hysterisis band at the critical points.

Harmonic Elimination Method

The undesirable harmonics of a square wave can be eliminated and the fundamental voltage component can be controlled as well by what is known as the harmonic elimination method. In this method, notches are created on the square wave at predetermined angles, as shown in Fig. 4.13. In the figure, half-cycle output is shown with quarter-wave symmetry. It can be shown that the four notch angles α_1, α_2, α_3, and α_4 can be controlled to eliminate three harmonic components and control the fundamental voltage. A larger number of harmonic components can be eliminated if the waveform can accommodate additional notch angles.

Theory.   The general Fourier series of the wave can be given as

$$v(t) = \sum_{n=1}^{\infty} (a_n \cos n\omega t + b_n \sin n\omega t) \tag{4.54}$$

where

$$a_n = \frac{1}{\pi} \int_0^{2\pi} v(t) \cos n\omega t \, d\omega t$$

$$b_n = \frac{1}{\pi} \int_0^{2\pi} v(t) \sin n\omega t \, d\omega t$$

For a waveform with quarter-cycle symmetry, only the odd harmonics with sine components will exist. Therefore, the coefficients are given as

$$a_n = 0 \tag{4.55}$$

$$b_n = \frac{4}{\pi} \int_0^{\pi/2} v(t) \sin n\omega t \, d\omega t \tag{4.56}$$

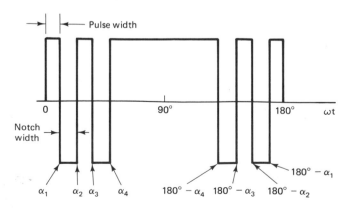

Figure 4.13 Voltage wave in harmonic elimination method.

Assuming that the wave has unit amplitude i.e., $v(t) = \pm1$, b_n can be expanded as

$$b_n = \frac{4}{\pi} \left[\int_0^{\alpha_1} (+1) \sin n\omega t \, d\omega t + \int_{\alpha_1}^{\alpha_2} (-1) \sin n\omega t \, d\omega t \right.$$

$$+ \int_{\alpha_2}^{\alpha_3} (+1) \sin n\omega t \, d\omega t + \cdots + \int_{\alpha_{K-1}}^{\alpha_K} (-1)^{K-1} \sin n\omega t \, d\omega t \qquad (4.57)$$

$$\left. + \int_{\alpha_K}^{\pi/2} (+1) \sin n\omega t \, d\omega t \right]$$

Using the relation

$$\int_{\theta_1}^{\theta_2} \sin n\omega t \, d\omega t = \frac{1}{n} (\cos n\theta_1 - \cos n\theta_2)$$

the first and last terms are

$$\int_0^{\alpha_1} (+1) \sin n\omega t \, d\omega t = \frac{1}{n} (1 - \cos n\alpha_1) \qquad (4.58)$$

$$\int_{\alpha_K}^{\pi/2} (+1) \sin n\omega t \, d\omega t = \frac{1}{n} \cos n\alpha_K \qquad (4.59)$$

Integrating the other components of equation (4.57) and substituting equations (4.58) and (4.59) in it yields

$$b_n = \frac{4}{n\pi} [1 + 2 (-\cos n\alpha_1 + \cos n\alpha_2 - \cdots + \cos n\alpha_K)]$$

$$= \frac{4}{n\pi} \left(1 + 2 \sum_{K=1}^{K} (-1)^K \cos n\alpha_K \right) \qquad (4.60)$$

It may be noted that equation (4.60) contains K variables (i.e., $\alpha_1, \alpha_2, \alpha_3, \ldots, \alpha_K$) and K number of simultaneous equations are required to solve their values. With K number of α angles, the fundamental voltage can be controlled and $K - 1$ harmonics can be eliminated.

Consider, for example, that the 5th and 7th harmonics are to be eliminated and the fundamental voltage is to be controlled. The 3rd and other triplen harmonics can be ignored if the machine has a star winding with an isolated neutral. Here $K = 3$ and the simultaneous equations can be written from equation (4.60) as

$$\text{Fundamental: } b_1 = \frac{4}{\pi} (1 - 2 \cos \alpha_1 + 2 \cos \alpha_2 - 2 \cos \alpha_3) \qquad (4.61)$$

$$\text{5th harmonic: } b_5 = \frac{4}{5\pi} (1 - 2 \cos 5\alpha_1 + 2 \cos 5\alpha_2 - 2 \cos 5\alpha_3) = 0 \qquad (4.62)$$

$$\text{7th harmonic: } b_7 = \frac{4}{7\pi} (1 - 2 \cos 7\alpha_1 + 2 \cos 7\alpha_2 - 2 \cos 7\alpha_3) = 0 \qquad (4.63)$$

The nonlinear transcendental equations above can be solved numerically for the specified fundamental amplitude and α_1, α_2, and α_3 can be determined. The α angles are solved at different output voltages and plotted in Fig. 4.14. Also shown in the figure are the lower-order significant harmonics (i.e., 11th and 13th), which have been considerably boosted as a result of lower-order harmonics elimination. The effect of these harmonics will be small, however, because of the large separation from the fundamental. Also note in Fig. 4.14 that the 5th and 7th harmonics can be eliminated up to a voltage level of 93.34% (100% corresponds to the square wave) where $\alpha_1 = 0$. The single notch remaining on either side can be symmetrically shifted toward the edge and then dropped so that the voltage jump remains within the specified limit. This segment of α angle table for the voltage jump within 1% is illustrated in Table 4.2. A small amount of the 5th and 7th harmonic voltages will reappear in this region but can be ignored in favor of limiting voltage jump.

The harmonic elimination method can be conveniently implemented with a microcomputer using a look-up table of notch angles. At a certain command voltage V_s, the angles are retrieved from a look-up table and the corresponding pulse widths are generated in the time domain with the help of down-counters. Figure 4.15 shows the spectrum analyzer output of a microcomputer-based modulator at $V_s = 50\%$ and $f_s = 100$ Hz, where α angles are stored with two-decimal accuracy.

As the fundamental frequency decreases, the number of notch angles can be increased so that the higher number of harmonics may be eliminated. Again the number of notch angles per half-cycle or the number of commutations per second

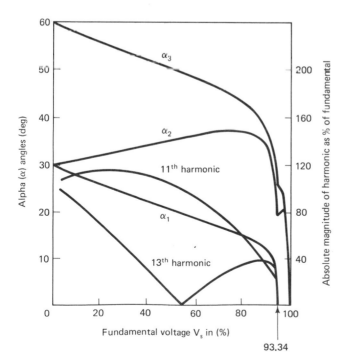

Figure 4.14 Notch angle relation with fundamental output voltage for 5th and 7th harmonic elimination. From *IEEE Press* B.K. Bose, © 1974 IEEE.

TABLE 4.2   α ANGLE TABLE FOR V_s FROM 93 TO 100%

V_s	α_1	α_2	α_3
93	0	15.94	22.03
94	0	16.17	21.56
95	0	16.41	20.86
96	0	16.88	20.39
97	0	17.34	19.92
98	0	11.02	13.59
99	0	4.69	7.27
100	0	0	0
(square wave)			

is to be determined by the switching losses of the inverter. One difficulty of a large number of notch angles at low frequency is that the look-up table of α angles for any voltage wave pattern tends to be unusually large. For this reason, a hybrid PWM scheme where the low-frequency, low-voltage region uses a sinusoidal PWM method, whereas the high-frequency, high-voltage region uses the harmonic elimination method, appears to be very attractive. In this scheme, the voltage jump is controlled accurately within the whole region and harmonic loss due to a sinusoidal PWM transition mode is substantially reduced. The harmonic elimination method can be extended to the constant-power region, where voltage control may be desirable in high-performance drive systems.

Minimum Ripple Current Method

One disadvantage of the harmonic elimination method is that the elimination of lower-order harmonics considerably boosts the lower-order significant harmonics, as shown in Fig. 4.15. Since the harmonics loss in a machine is dictated by the

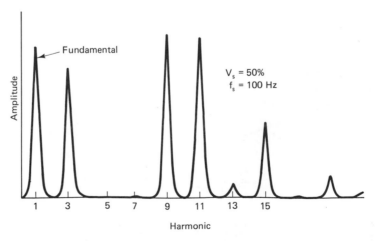

Figure 4.15 Spectrum analyzer output of voltage wave with 5th and 7th harmonic elimination.

rms ripple current, it is this parameter that should be minimized instead of paying attention to an individual harmonic. This, of course, assumes that the skin effect of machine parameters is negligible, which may not be true, especially for the rotor winding. It was shown in Chapter 2 that the effective leakage inductance of the passive equivalent circuit (Fig. 2.14) determines the harmonic current corresponding to any harmonic voltage. Therefore, the expression of rms ripple current can be given as

$$I_{ripple} = \sqrt{I_3^2 + I_5^2 + I_7^2 + I_9^2 + \cdots}$$

$$= \sqrt{\frac{I_{3m}^2}{2} + \frac{I_{5m}^2}{2} + \frac{I_{7m}^2}{2} + \frac{I_{9m}^2}{2} + \cdots} \tag{4.64}$$

$$= \sqrt{\frac{1}{2} \sum_{n=3}^{\infty} \left(\frac{V_n}{n\omega_s L}\right)^2}$$

where

I_3, I_5, etc. = rms harmonic currents

I_{3m}, I_{5m}, etc. = peak value of harmonic currents

n = order of harmonic

V_n = peak value of nth-order harmonic

L = effective leakage inductance of the machine per phase

ω_s = fundamental frequency

The corresponding harmonic copper loss is

$$P_L = 3I_{ripple}^2 R \tag{4.65}$$

where R is the effective resistance of the machine per phase.

For a given number of notch angles, the expression of V_n is given by equation (4.60). Substituting it in equation (4.64), I_{ripple} is found as a function of notch angles. The notch angles can then be iterated in a computer program so as to minimize I_{ripple}. The modified look-up table of α angles based on harmonic loss minimization is more desirable than that of the harmonic elimination method. Figure 4.16 shows the voltage spectrum in the minimum ripple current method under the same conditions as in Fig. 4.15. Note that the 7th harmonic has reappeared, but the 11th harmonic is considerably lower. Again, the triplen harmonic voltages can be ignored.

Adaptive Current Control PWM

The discussion so far on the harmonics of the PWM output voltage wave has been based on the assumption that the dc link voltage V_d is ideally filtered. The condition is far from true in rectifier-supplied dc, where a considerable amount of ripple may be present due to finite LC or C filter. The adaptive or hysteresis

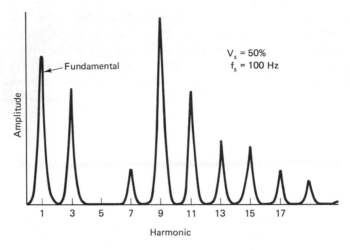

Figure 4.16 Spectrum analyzer output of voltage wave with minimum ripple current method.

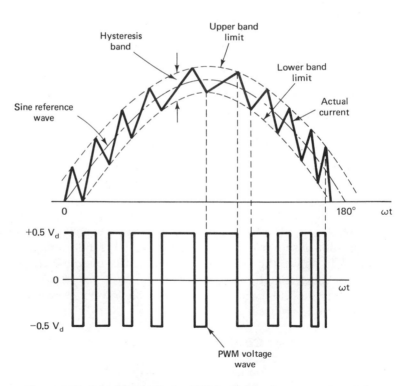

Figure 4.17 Principle of adaptive PWM with bang-bang current control.

band current control PWM technique explained here may be adopted to overcome this problem.   The technique is based on current control as explained in Fig. 4.17, in contrast to voltage control, which was discussed before.   The control circuit generates the sine reference current wave of desired magnitude and frequency, which is compared with the actual phase current as shown in Fig. 4.18.   As the current exceeds a prescribed hysteresis band, the upper transistor in the half-bridge is turned off and the lower transistor is turned on.   As a result, the output voltage is transitioned from $+0.5V_d$ to $-0.5V_d$ and the current starts to decay.   As the current crosses the lower band limit, the upper transistor is turned on, switching off the lower transistor.   A prescribed lock-out time t_L is provided at each transition to prevent a shoot-through fault.   The actual current wave is thus forced to track the sine reference wave within the desired hysteresis band by back-and-forth switching of upper and lower transistors.   The inverter then essentially becomes a current source type instead of a voltage source type, and peak-to-peak current ripple is controlled adaptively within the hysteresis band irrespective of V_d fluctuation.   The rms ripple current, which is indirectly related to peak-to-peak ripple current, is thus closely controlled, minimizing machine heating.   The control of instantaneous peak current is a great advantage for transistor-type devices which are peak current sensitive.   The current control PWM mode can be smoothly transitioned to a square-wave voltage mode in the constant-power region.   In a low-speed region where machine counter emf is low, there is no difficulty in current controller tracking.   But in high speed, the current controller will saturate in part of the cycle because of higher counter emf.   In this condition, the fundamental current magnitude will be less and its phase will deviate from that of the commanded current. The slope of the current wave can be given as

$$\frac{di}{dt} = \frac{0.5V_d - V_{cm}\sin\omega_s t}{L} \tag{4.66}$$

where $V_{cm}\sin\omega_s t$ is the sinusoidally varying counter emf and L is the effective

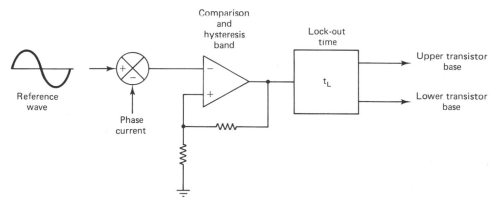

Figure 4.18   Adaptive PWM control block diagram.

leakage inductance. The hysteresis band can be adapted to control the switching frequency of the inverter, which will correspondingly vary the peak-to-peak ripple.

Phase-Shift PWM

If a number of bridge inverters are coupled at the output by transformers, the output voltage can be controlled by adjusting the phase-shift angle between the bridges. For example, in the dual-bridge inverter shown in Fig. 4.4(a), the output fundamental voltage can be controlled if the lower bridge phase shift angle is deviated from 30°. Such phase-shift voltage control is simple, but the penalty is that the 12-step waveform at the output is lost. However, since six-stepped waveforms are being mixed at different phase angles, the output retains the characteristics of the six-step inverter (i.e., the output contains 5th, 7th, 11th, 13th, etc., significant harmonics).

A general expression of fundamental output voltage can then be derived as a function of phase shift angle θ, where θ is the lagging angle of the lower bridge. From equations (4.23), (4.24), and (4.25),

$$v_{ab(f)} = \frac{2\sqrt{3}\,nV_d}{\pi} \cos \omega t \tag{4.67}$$

$$v_{de(f)} = \frac{2nV_d}{\pi} \cos (\omega t - \theta) \tag{4.68}$$

$$v_{ef(f)} = \frac{2nV_d}{\pi} \cos [\omega t - (120 + \theta)] \tag{4.69}$$

Therefore, the fundamental voltage phasor as shown in Fig. 4.4(b) can be derived as

$$v_{NA(f)} = v_{ab(f)} + v_{de(f)} - v_{ef(f)}$$

$$= \frac{2nV_d}{\pi} \left\{ \sqrt{3} \cos \omega t + \cos (\omega t - \theta) + \cos [\omega t - (120 + \theta)] \right\} \tag{4.70}$$

$$= A \sin \omega t + B \cos \omega t$$

$$= R \sin (\omega t + \theta)$$

where

$$R = A^2 + B^2$$

$$\theta = \tan^{-1} \frac{B}{A}$$

$$A = \sqrt{3} + 1.5 \cos \theta + \sqrt{3}/2 \sin \theta$$

$$B = 1.5 \sin \theta - \sqrt{3}/2 \cos \theta$$

The fundamental voltage can be varied between zero and a maximum value by controlling the θ angle. It can be shown that the output voltage vanishes for $\theta = 210°$ and becomes maximum for $\theta = 30°$, which is the case for the 12-stepped wave.

4.5 DYNAMIC AND REGENERATIVE BRAKING

In adjustable-speed ac drives, the machines may be subjected to electrical braking for reduction of speed. In electrical braking, the motor is operated in the generating mode and the kinetic energy stored in the system inertia is converted to electrical energy. The energy is then either dissipated in a resistor, used in parallel drive systems, or recovered in the power supply. The former is known as dynamic braking and the latter two are known as regenerative braking. An induction motor can run as a generator in supersynchronous speed, which can be made possible by lowering the inverter frequency below the machine speed ($\omega_e < \omega_r$). The same condition is attained when the machine drives an overhauling type of load. A synchronous motor can run as a generator if the power angle δ is transitioned from a negative to a positive value. In four-quadrant speed control, the machine speed is brought down to zero by electrical braking and then the phase sequence of the inverter is reversed to reverse the direction of machine rotation.

For a square-wave inverter drive, if the load is assumed to have a unity displacement factor, then during generator operation, the voltage and current will be at 180° phase difference and the feedback diodes of each half-bridge will conduct alternately for 180° duration. For a typical lagging displacement factor load, 180° duration will be shared between the diode and thyristor as explained in Fig. 4.3(a). Whatever may be the case, the inverter will operate as a rectifier and active power from the generator will flow back to the dc link. If the inverter is supplied from a battery, the reverse dc current will charge the battery. With rectifier-supplied dc, because the current cannot flow backward through the rectifier, it charges the filter capacitor, raising the dc link voltage. A dynamic braking resistor can be switched on and off within a hysteresis band of dc link voltage to dissipate this braking power. Otherwise, an inverse-parallel converter can be connected to transfer dc power to the ac supply line. It is assumed, of course, that the ac line can absorb this reverse power.

The dynamic braking capability of the PWM inverter and the square-wave inverter are compared in Fig. 4.19. The dynamic braking power is related to the dc link voltage and is given by V_d^2/R, where R is the braking resistor. In a PWM inverter with constant dc link voltage, the braking power is given by the dashed horizontal line shown in part (a). In a square-wave inverter, the dc voltage varies linearly with frequency (i.e., speed) in the constant-torque region and therefore the braking capability is given by a parabola, as shown in part (b). The full dc voltage is reached in the constant-power region and the braking capability becomes identical to that of the PWM inverter. The inverter–machine power capability is also plotted in Fig. 4.19. Ideally, the curve is identical to that in motoring and is given by a fixed-slope straight line in the constant-torque region and a horizontal line in the constant-power region. It is evident from the curves that in a PWM inverter, the dynamic braking resistor can absorb the full braking power, but in a square-wave inverter it falls short by magnitude ΔP in the constant-torque region. This excess power will boost the dc voltage and cause overexcitation in the machine unless the control backs down the braking power to match the dissipation capability.

The additional inverse-parallel converter for regenerative braking adds to the

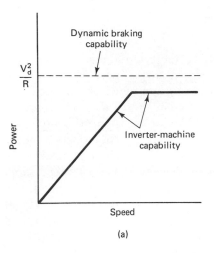

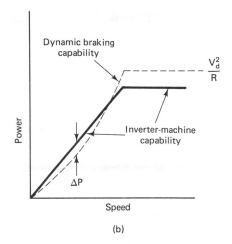

Figure 4.19 Dynamic braking capacity in (a) PWM and (b) square-wave inverters. From *IEEE Press*, B.K. Bose, © 1972 IEEE.

cost and weight and therefore should be carefully considered against the energy saving.   A free power flow in either direction is possible by using dual PWM inverters, as shown in Fig. 4.20.   During motoring operation, the power flows from line to motor, with the front-end inverter acting as a rectifier, but during regeneration the inverters reverse their roles.   One advantage of this configuration is that the line current is nearly sinusoidal and the displacement factor can be programmed to be unity.

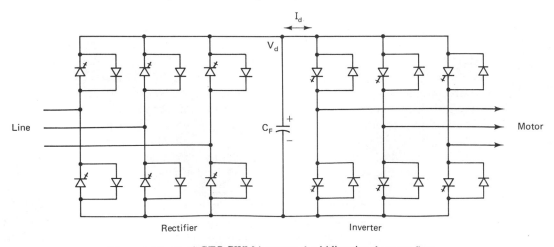

Figure 4.20   Dual GTO PWM inverters for bidirectional power flow.

4.6 MODELING AND SIMULATION

A variable-speed ac machine coupled to the inverter or rectifier–inverter system may give rise to complex stability problems. The machine is a nonlinear multivariable system and may interact dynamically with the source impedances of converters. The discrete-time effect of the converters may also compound the problem. Besides, the system configuration may be more complex; for example, a single inverter may supply a number of machines in parallel or a single rectifier can be connected to a number of parallel inverters. The analytical study of such systems is difficult and therefore computer simulation study becomes essential. The simulation study also becomes useful to study the feasibility of a new control strategy, the effect of harmonics, and the variation of system parameters.

For simulation study, a model of the system is required and the complexity of the model will depend on the goal of simulation study. In general, an electrical network containing active, passive, linear, and nonlinear elements can be described as a graph by nodes and branches, and a program such as SPICE can be applied in digital simulation. Although such a simulation study for a drive system is quite involved, it gives not only the performance on dynamics, but phase unbalance, fault condition, and so on, can also be studied in detail. More conveniently, the dynamics of the system can be simulated by state-space equations which may be excited by the converters simulated as logical networks. The model of an induction motor in a d^e-q^e synchronously rotating reference frame is shown in Fig. 2.17. The stator and rotor phase currents can be reconstructed from the machine model currents by d^e-q^e and phase transformation. The stator phase currents can be converted to inverter dc current by the inverter switching network and applied to the state equation of a dc link filter. Then the rectifier switching network can be applied to the link with the line voltage and current waves.

A complete dynamic model of a drive system including the controller can be simulated in a digital computer, using, for example, a VAX-based SIMNON* program. This simulation program is so attractive that it is relevant to review its features here. SIMNON is a very user-friendly language for simulation of dynamic systems using an interactive graphics terminal. The system may be linear or nonlinear, and both continuous and discrete time systems can be simulated. The simulation of a discrete time system with prescribed sampling time is of particular importance to a microcomputer-based controlled system. SIMNON accepts description of a dynamic system in state space form, i.e., the continuous system is described by differential equations and the discrete time-system by difference equations. For simulation, a large system is normally resolved into small interconnected subsystems. Each subsystem, described by a set of equations and input/output signals, constitutes a routine. These routines are then interconnected by a connecting system routine. The results of SIMNON simulation are displayed as curves on the terminal, or in the form of a table. The files, which are small in size, can

*The SIMNON language was developed by the Lund Institute of Technology, Sweden.

be edited by a simple line-oriented editor built into SIMNON, or can be edited like a FORTRAN file. SIMNON can interface with FORTRAN files to expand its capabilities.

We discuss here a simplified model of a rectifier–inverter–induction motor system mainly for stability study, assuming that the inverter generates symmetrical three-phase square-wave voltages and that the rectifier is represented as a controllable dc voltage source.

The stationary frame voltages v_{qs}^s and v_{ds}^s in a Fourier series can be derived by substituting equations (2.51), (2.52), (4.14), (4.15), and (4.16) in equations (4.1), (4.2), and (4.3) as

$$
\begin{aligned}
v_{qs}^s &= \frac{2}{3} v_{a0} - \frac{1}{3} v_{b0} - \frac{1}{3} v_{c0} \\
&= \frac{2V_d}{\pi} \left(\cos \omega_e t + \frac{1}{5} \cos 5\omega_e t - \frac{1}{7} \cos 7\omega_e t + \cdots \right)
\end{aligned}
\tag{4.71}
$$

$$
\begin{aligned}
v_{ds}^s &= \frac{1}{\sqrt{3}} \left(v_{c0} - v_{b0} \right) \\
&= \frac{2V_d}{\pi} \left(-\sin \omega_e t + \frac{1}{5} \sin 5\omega_e t + \frac{1}{7} \sin 7\omega_e t + \cdots \right)
\end{aligned}
\tag{4.72}
$$

The corresponding rotating frame relations can be derived as

$$
\begin{aligned}
v_{qs} &= v_{qs}^s \cos \omega_e t - v_{ds}^s \sin \omega_e t \\
&= \frac{2V_d}{\pi} \left(1 + \frac{2}{35} \cos 6\omega_e t - \frac{2}{143} \cos 12\omega_e t + \cdots \right)
\end{aligned}
\tag{4.73}
$$

$$
\begin{aligned}
v_{ds} &= v_{qs}^s \sin \omega_e t + v_{ds}^s \cos \omega_e t \\
&= \frac{2V_d}{\pi} \left(\frac{12}{35} \sin 6 \omega_e t - \frac{24}{143} \sin 12 \omega_e t + \cdots \right)
\end{aligned}
\tag{4.74}
$$

assuming that at $t = 0$ the q axis and a axis are aligned. Equations (4.73) and (4.74) relate machine voltages with the inverter dc voltage and the switching functions expressed in Fourier series.

The relation between inverter dc current and machine currents can be derived by considering the instantaneous power balance relation between the input and output of the inverter. This assumption is valid because the inverter is a switching network and does not contain any energy storage element. The instantaneous power relation is given as

$$
V_d I_d = v_{an} i_a + v_{bn} i_b + v_{cn} i_c
\tag{4.75}
$$

Substituting the phase variables in terms of d^s–q^s variables by equations (2.49) to (2.51) and simplifying yields

$$
V_d I_d = \frac{3}{2} \left(v_{ds}^s i_{ds}^s + v_{qs}^s i_{qs}^s \right)
\tag{4.76}
$$

Again substituting d^s-q^s variables in terms with $d-q$ variables by equations (2.55) and (2.56) and simplifying yields

$$V_d I_d = \frac{3}{2} (v_{ds} i_{ds} + v_{qs} i_{qs}) \tag{4.77}$$

Substituting equations (4.73) and (4.74) in (4.77), the inverter dc current as a function of machine currents can be derived as

$$
\begin{aligned}
I_d = \frac{3}{\pi} i_{qs} &\left(1 + \frac{2}{35} \cos 6\omega_e t - \frac{2}{143} \cos 12\omega_e t + \cdots \right) \\
&+ \frac{3}{\pi} i_{ds} \left(\frac{12}{35} \sin 6\omega_e t - \frac{24}{143} \sin 12\omega_e t + \cdots \right)
\end{aligned} \tag{4.78}
$$

Defining the switching functions as

$$G_{qs} = 1 + \frac{2}{35} \cos 6\omega_e t - \frac{2}{243} \cos 12\omega_e t + \cdots \tag{4.79}$$

$$G_{ds} = \frac{12}{35} \sin 6\omega_e t - \frac{24}{243} \sin 12\omega_e t + \cdots \tag{4.80}$$

so that

$$v_{qs} = \frac{2}{\pi} V_d G_{qs} \tag{4.81}$$

$$v_{ds} = \frac{2}{\pi} V_d G_{ds} \tag{4.82}$$

$$I_d = \frac{3}{\pi} (i_{qs} G_{qs} + i_{ds} G_{ds}) \tag{4.83}$$

So far we have assumed that the inverter is supplied by a stiff voltage source V_d. In general, there will be an LC filter in the dc link and the supply may be from a rectifier in the front end. Then the dc link equation can be written as

$$V_R = L_d \frac{dI_R}{dt} + I_R R_d + V_d \tag{4.84}$$

$$I_R - I_d = C_d \frac{dV_d}{dt} \tag{4.85}$$

where V_R is the rectifier voltage, I_R the rectifier current, and L_d, R_d, and C_d are filter parameters. The equations above can be modified by multiplying with constants $2/\pi$ and $\pi/3$, respectively, as

$$\frac{2}{\pi} V_R = \left(\frac{6L_d}{\pi^2} \right) \frac{d}{dt} \left(\frac{\pi}{3} I_R \right) + \left(\frac{\pi}{3} I_R \right) \left(\frac{6}{\pi^2} R_d \right) + \frac{2}{\pi} V_d \tag{4.86}$$

$$\frac{\pi}{3} I_R - \frac{\pi}{3} I_d = \left(\frac{\pi^2 C_d}{6} \right) \frac{d}{dt} \left(\frac{2}{\pi} V_d \right) \tag{4.87}$$

or

$$V'_R = L'_d \frac{d}{dt} I'_R + I'_R R'_d + V'_d \tag{4.88}$$

$$I'_R - I'_d = C'_d \frac{dV'_d}{dt} \tag{4.89}$$

where the primed quantities replace the quantities with coefficients as shown.   In terms of the primed variables, equations (4.81), (4.82), and (4.83) can be written as

$$v_{qs} = V'_d G_{qs} \tag{4.90}$$

$$v_{ds} = V'_d G_{ds} \tag{4.91}$$

$$I'_d = i_{qs} G_{qs} + i_{ds} G_{ds} \tag{4.92}$$

The equivalent circuit in a synchronously rotating reference frame in terms of these equations is shown in Fig. 4.21.   The derivation of rotor speed ω_r from equations (2.83) and (2.79) is also shown in the figure.   In the model, the parameters V'_R,

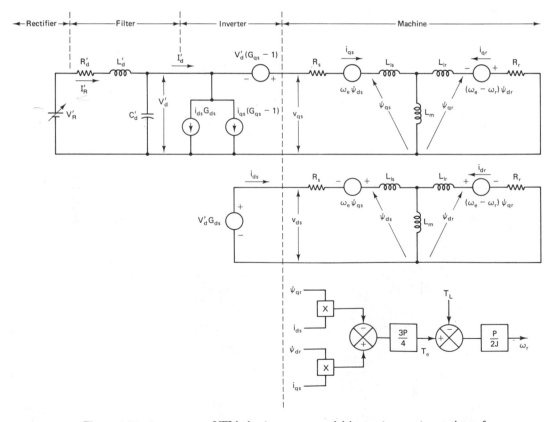

Figure 4.21   Square-wave VFI-induction motor model in synchronously rotating reference frame.

ω_e, and T_L are the impressed variables. The model can be further simplified by neglecting the harmonics; that is,

$$G_{qs} = 1$$

$$G_{ds} = 0$$

$$v_{qs} = V'_d$$

$$v_{ds} = 0$$

This simplification means that the machine is supplied by sinusoidal voltages, but the stability study due to machine filter interaction is adequate with this simple equivalent circuit.

An alternative approach to modeling is to linearize a rotating frame model of a converter–machine system at steady-state conditions using a small-signal perturbation principle, which is discussed in Chapter 7. The corresponding linear system model can be represented by transfer functions and the control system dynamics can be studied by using the Nyquist, Bode, or root-locus technique. Since the small-signal parameters differ at each operating point, the analysis should be carried over the whole region of operation to determine the worst-case performance.

The digital simulation of drive system has been mentioned so far, but several words about other types of computer simulation may be appropriate here. Although digital simulation is common because of the easy availability of digital computers, an analog computer can also be used for simulation. Since the machine–converter system is basically analog in nature, an analog simulation is easy and convenient and provides easy access to the analog variables. In a hybrid computer, which consists of analog and digital computers, the system can be appropriately partitioned for analog and digital simulation. Normally, the converter and the machine are simulated on an analog computer, but the control can be simulated on either an analog or a digital computer. A microcomputer-based control system, which is discrete time in nature and where complex computation and decision-making processes are involved, is simulated in the digital computer. The parity simulation technique has also been used. In this method, analog and digital simulation principles are combined to simulate individual components of the drive system to maintain their topological parity (i.e., a machine simulation looks like a machine at the terminal except that the voltages and currents are appropriately scaled). Such a simulation with modular building blocks can be done quickly by an inexperienced engineer. Real-time parity simulation of the power components can be driven by breadboard control hardware which subsequently can be used as a prototype.

4.7 INVERTER–MACHINE INTERFACE

Although the general practice is to use standard, off-the-shelf 60-Hz ac machines for variable-frequency inverter drives, it is desirable that the machine should be custom designed to match the inverter so that the integrated system has optimum performance and cost. For a voltage-fed inverter a high leakage inductance is preferred in the machine so that the resulting harmonic currents are low, but a compromise is necessary because at high leakage inductance the motor pull-out torque is reduced. With the inverter supply, the direct starting current of the induction motor is of no concern and therefore a high-resistance or double-cage rotor winding is not necessary. In fact, the rotor can use copper bars, so that the rotor copper loss is reduced substantially. Such a machine operates with very low slip and the induction motor behaves nearly as a synchronous machine in the stable operating region. The skin effect due to harmonics is an important factor in machine design. The skin effect can be ignored in stator winding but may be substantial in the rotor due to the bar structure. The bars may be designed as coffin-shaped with a wider area on the top so that the skin effect is minimized but low resistance at slip frequency is maintained. Besides reducing the copper loss, the hysteresis and eddy-current losses are also designed to be lower so that the machine operates at high efficiency. For a PWM inverter drive, the switching frequency can be adapted so that the composite inverter loss and machine loss are minimum in the whole range. Since the inverter cost is much higher than the machine cost and a high-efficiency machine saves the inverter power rating, the overall system cost is lowered. The other effects, such as harmonic torque pulsation, dc offset, shoot-through fault, and so on, should also receive proper consideration in the machine design.

An important consideration in the design of an inverter–machine system is that the thermal time constant of the power semiconductor elements is much shorter than that of the machine. Therefore, the inverter has to be designed for nearly peak power rating without leaving a margin for short-time increase in the power rating. This makes the inverter cost substantially more than the machine cost.

4.8 CONTROL CIRCUITS

Inverter control can be implemented either by dedicated analog/digital hardware or by microcomputer software. We will review here briefly the dedicated hardware control.

Square-Wave Inverter

Figure 4.22 shows the control block diagram of a six-stepped square-wave inverter and Fig. 4.23 explains its operation. The inverter voltage and frequency are commanded by the analog signals v_s and v_f, respectively. The signal v_s controls the gate firing angle of the phase-controlled rectifier in the front end to regulate the dc link voltage V_d. The system control is reviewed in detail in Chapter 7. The signal v_f is converted to a pulse train of frequency f_1 through a voltage-controlled oscillator (VCO) which drives a six-stage ring counter as shown in Fig. 4.22. The analog signal is scaled such that $f_1 = 6f_s$, where f_s is the inverter fundamental frequency. For a constant and precision frequency requirement of the inverter, such as in a UPS application, the frequency command can be generated by a crystal

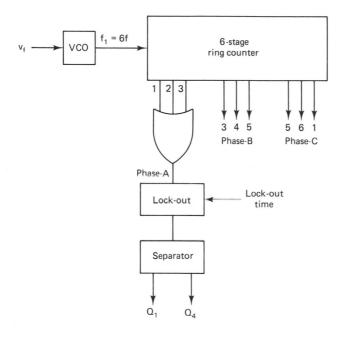

V_s ⟶ Rectifier α control

V_f ⟶ VCO ⟶ $f_1 = 6f$ ⟶ 6-stage ring counter

1 2 3

3 4 5
Phase-B

5 6 1
Phase-C

Phase-A

Lock-out ← Lock-out time

Separator

Q_1 Q_4

Figure 4.22   Control block diagram of square-wave inverter.

oscillator.   The adjacent three stages of the ring counter are coupled through OR gates with overlapping of 60° to generate the three-phase square waves at the fundamental frequency f_s.   These square waves are the basic logic drive signals of the respective phases of the inverter where the upper device should conduct in the positive half-cycle and the lower device should conduct in the negative half-cycle.

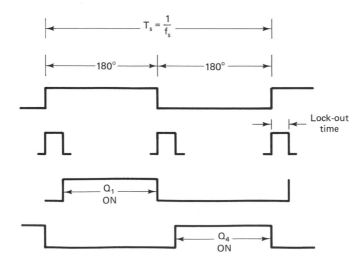

$T_s = \dfrac{1}{f_s}$

180°   180°

Lock-out time

Q_1 ON

Q_4 ON

Figure 4.23   Logic waveforms explaining square-wave inverter control.

However, at the transition of each half-cycle, a lock-out time interval is generated when the drive signal of both the upper and lower devices is inhibited. For a thyristor inverter, the commutation is initiated at the leading edge of the lock-out time so that at its trailing edge the outgoing thyristor has completely turned off before gating on the incoming thyristor. Similarly, for a GTO or transistor inverter, the lock-out time assures turn-off of the outgoing device to prevent any shoot-through fault. The lock-out time can be generated by a one-shot timer. After the lock-out interval the logic signals are separated to generate the respective logic drive signals shown in Fig. 4.23.

PWM Inverter

The control block diagram of a PWM inverter is shown in Fig. 4.24, which consists of the following functional elements:

- A VCO, which converts the analog frequency command v_f to a pulse train of frequency f_1.
- A three-phase sine-wave signal generator with variable amplitude and frequency as desired for the inverter output. The frequency f_s is a scaled-down value of f_1 and the amplitude is proportional to the analog voltage command v_s. The phase sequence of the generator is reversible by a forward/reverse logic signal.
- The phase-locked loops (PLL) (explained in Fig. 3.41) generate frequencies of different ratios. The ratios correspond (but do not equal) the carrier-to-signal frequency ratio in synchronized PWM mode.
- A multiplexer, which selects the appropriate frequency channel of the PLL depending on the range of f_1. A free-running frequency f_0 is also an input to the multiplexer.
- A triangular carrier wave generator of fixed amplitude but variable frequency (f_c), which is synthesized from the multiplexer-selected frequency. The frequency f_c is a submultiple of the input frequency and the waveform can be phase synchronized with the sine signal wave.
- An analog comparator for each phase, which compares the sine and triangular waves. The comparator has a small hysteresis band to prevent multiple crossing jitters.
- The lock-out and separator circuit for each phase, as explained in Fig. 4.22.

The elements of sine reference wave generator are shown in Fig. 4.25. The amplitudes of a unit sine wave at regular angular intervals are stored digitally in the form of a lookup table in a read-only memory (ROM). The frequency f_1 clocks the programmable up/down-counter, which generates the address of ROM in a normal UP counting mode. When the counter reaches the terminal count at the end of a cycle, it resets and a new cycle begins. The digital output from the ROM is fed to a multiplying digital-to-analog converter (DAC) to convert into an analog sine wave. The analog voltage v_s multiplies the DAC output to control the sine-

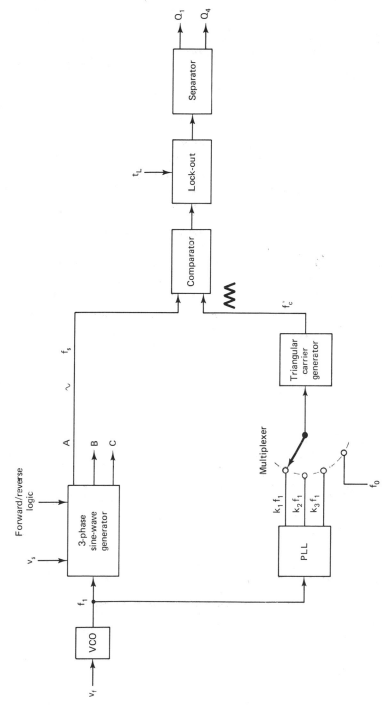

Figure 4.24 Control block diagram of PWM inverter.

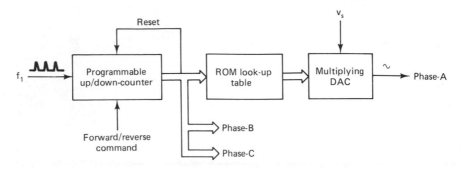

Figure 4.25 Block diagram showing elements of sine-wave generator.

wave amplitude. The phases B and C have identical ROMs and DACs except that the look-up tables are mutually phase shifted by 120°. The phase sequence of the sine wave can be reversed by the forward/reverse command as shown. At the reverse command, the counter changes into a down-counting mode and generates the sine waves in a backward direction, which is the reversal of the phase sequence.

The triangular carrier wave is generated using the same principle as above except that it has only one phase, its magnitude is fixed, and it does not involve phase sequence reversal.

REFERENCES

1. L. J. Penkowski and K. E. Pruzinsky, "Fundamentals of a Pulse Width Modulated Power Circuit," *IEEE Trans. Ind. Appl.*, Vol. IA-8, pp. 584–593, Sept.–Oct. 1972.

2. B. K. Bose and H. A. Sutherland, "A High Performance Pulse-Width Modulator for an Inverter-Fed Drive System Using a Microcomputer," *IEEE Trans. Ind. Appl.*, Vol. IA-19, pp. 235–243, Mar.–Apr. 1983.

3. A. B. Plunkett, "A Current-Controlled PWM Transistor Inverter Drive," *Conf. Rec. IEEE/IAS Annu. Meet.*, pp. 786–792, Oct. 1979.

4. H. S. Patel and R. G. Hoft, "Generalized Techniques of Harmonic Elimination and Voltage Control in Thyristor Inverters: Part I. Harmonic Elimination," *IEEE Trans. Ind. Appl.*, Vol. IA-9, pp. 310–317, May–June 1973.

5. H. S. Patel and R. G. Hoft, "Generalized Techniques of Harmonic Elimination and Voltage Control in Thyristor Inverters: Part II. Voltage Control Techniques," *IEEE Trans. Ind. Appl.*, Vol. IA-10, pp. 666–673, July–Aug. 1974.

6. P. C. Krause and T. A. Lipo, "Analysis and Simplified Representations of a Rectifier–Inverter Induction Motor Drive," *IEEE Trans. Power Appar. Syst.*, Vol. PAS-88, pp. 588–596, May 1969.

7. T. L. Grant and T. H. Barton, "A Highly Flexible Controller for a Pulse Width Modulated Inverter," *Conf. Rec. IEEE/IAS Annu. Meet.*, pp. 486–492, Oct. 1978.

8. G. S. Buja and G. B. Irdri, "Optimal Pulsewidth Modulation for Feeding AC Motors," *IEEE Trans. Ind. Appl.*, Vol. IA-13, pp. 38–44, Jan–Feb. 1977.

9. P. Bhagwat and V. Stefanovic, "Some New Aspects of the Design of PWM Inverters," *Conf. Rec. IEEE/IAS Annu. Meet.*, pp. 383–388, Oct. 1979.

10. J. G. Kassakian, "Simulating Power Electronic Systems—A New Approach," *Proc. IEEE*, Vol. 67, pp. 1428–1439, Oct. 1979.

11. H. A. Spang, "The Federated Computer-Aided Control Design System," *Proc. IEEE*, Vol. 72, pp. 1724–1731, 1984.

12. B. D. Bedford and R. G. Hoft, *Principles of Inverter Circuits*, Wiley, New York, 1964.

5

CURRENT-FED INVERTERS

5.0 INTRODUCTION

The basic principles of a current-fed inverter using phase control and line commutation techniques were discussed in Chapter 3. In this chapter the concepts are expanded and several more inverter circuits are introduced. A current-fed inverter likes to see a stiff dc current source (ideally, infinite Thévenin impedance) at the input, which is in contrast to the stiff voltage source desirable in a voltage-fed inverter. A variable-voltage source can be converted to a variable-current source by connecting an inductance in series and controlling the voltage within a current control loop. A variable dc voltage source can be obtained from a utility supply or from a rotating alternator by rectification, or from a battery-type power source through a chopper. With a stiff current source, the output current waves are not affected by the load (i.e., the current waves are dual to the voltage waves of a voltage-fed inverter). The power semiconductors in a current-fed inverter have to withstand reverse voltage, and therefore devices such as GTOs,* transistors, and power MOSs, are not suitable.

The application of current-fed inverters may include the following:

- General speed control of ac machines
- Synchronous motor starting in gas turbine, pumped hydro, and so on.
- Induction heating
- Lagging VAR generation

*GTOs are available with limited reverse voltage capability.

166

In this chapter we discuss different types of current-fed inverters, analyze their performances, discuss an inverter–machine model, and compare current-fed inverters with voltage-fed inverters. Again, the nonideal characteristics of power semiconductor devices will be ignored.

5.1 GENERAL OPERATION

A general power circuit for a current-fed inverter which is being supplied by a phase-controlled rectifier is shown in Fig. 5.1. A variable dc link voltage V_R is generated by phase control, which is converted to current source I_d by connecting the series inductance L_d. Although an infinite value of L_d is desirable for an ideal current source, the cost and size constraints limit L_d to a reasonable value. Ignoring the commutation circuit for the present, the inverter appears to be identical with the rectifier. The load of the inverter is an induction or synchronous machine which can be approximately represented by a counter emf in series with an equivalent leakage inductance, as shown in the figure. The power circuit thus appears symmetrical about the dc link. The dc current I_d is switched through the inverter thyristors so as to establish three-phase, six-stepped, symmetrical line current waves as shown in Fig. 5.2. Each thyristor conducts for 120° and at any instant one upper thyristor and one lower thyristor remain in conduction. The dc link current is considered harmonic free ($L_d \rightarrow \infty$) and the commutation effect is ignored. The waves in Fig. 5.2 are drawn for maximum power inversion (i.e., $\alpha = 180°$), when the fundamental-component of phase current is out of phase with the respective phase voltage. The inverter input voltage V_d can be constructed by the amplitude between the two phase voltage envelopes. The dc voltage $V_R = V_d$ if the resistance of the inductance is neglected. For variable-speed drive applications, the inverter can be operated at a variable frequency with an adjustable magnitude of current I_d.

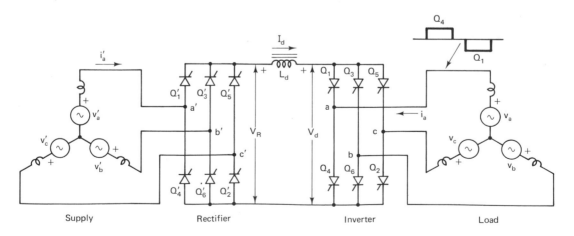

Figure 5.1   General power circuit of current-fed inverter.

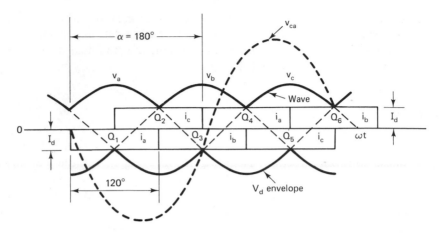

Figure 5.2 Idealized voltage and current waves.

Inverter Operation Modes

The inverter firing angle α can be varied in the range 0 to 360° with respect to the counter emf wave, and the following modes, as explained in Fig. 5.3, can be obtained.

Mode 1: Load-Commutated Rectifier, $0° \le \alpha \le 90°$. This mode corresponds to the familiar line-commutated rectifier mode of operation, except that here the commutation is performed by the load instead of the line. Figure 5.3(a) shows the phase voltage and current waves for $\alpha = 45°$. When the incoming thyristor Q_4 in Fig. 5.2 is fired, the outgoing thyristor Q_2 will be impressed with the negative anode voltage v_{ca} shown by the dashed line, and therefore will cause load commutation. The fundamental component of i_a will lag the voltage wave by 45°, and the corresponding phasor diagram is shown on the right side. The active power will flow from the load to the dc link, which will then be pumped back to the ac supply line by the rectifier operating in the line-commutated inverter mode. The dc link current I_d is always positive, but the voltages in general are given by

$$V_d = -V_{d0} \cos \alpha \qquad \text{for the inverter} \qquad (5.1)$$

$$V_R = V_{RO} \cos \alpha' = V_d = V_{d0} \cos (\pi - \alpha) \qquad \text{for the rectifier} \qquad (5.2)$$

where α' is the firing angle of the rectifier, and it is assumed that the supply voltages and load emfs are equal. For example, at $\alpha = 45°$, $\alpha' = 135°$, and V_d and V_R are negative, so that the active power will flow in the reverse direction. The load will supply lagging VAR to the inverter (i.e., the leading VAR will be supplied to the load from the inverter). Such a condition can be met by a synchronous machine operating with over-excitation. Therefore, this mode can be considered as a synchronous motor operating in a regenerative braking condition, as indicated in Fig. 5.4.

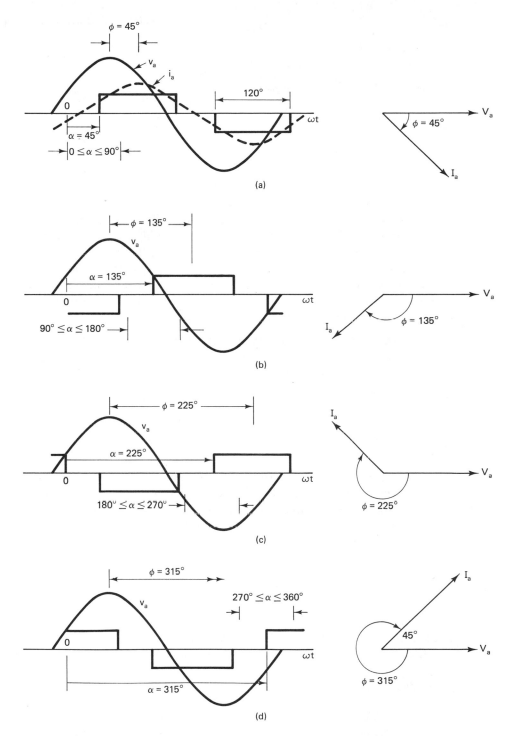

Figure 5.3 Modes of operation with counter emf load: (a) $0° \le \alpha \le 90°$: load-commutated rectifier; (b) $90° \le \alpha \le 180°$: load-commutated inverter; (c) $180° \le \alpha \le 270°$: force-commutated inverter; (d) $270° \le \alpha \le 360°$: force-commutated rectifier.

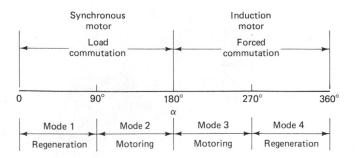

Figure 5.4 Modes of ac machine operation.

Mode 2: Load-Commutated Inverter, 90° ≤ α ≤ 180°.   This mode is explained in Fig. 5.3(b) for a typical angle α = 135°.   The outgoing thyristor Q_2 is commutated by the load since v_{ca} is negative in this range.   The active power flows to the load and the dc link voltage are positive, as determined by equation (5.1).   The load is required to operate at leading power factor as in the previous mode.   This mode can therefore be considered as the motoring mode of a synchronous motor operating at over-excitation.

Mode 3: Force-Commutated Inverter, 180° ≤ α ≤ 270°.   By delaying the inverter firing angle further beyond 180°, the advantage of load commutation is lost since the outgoing thyristor Q_2 is impressed with a positive v_{ca} voltage. Therefore, for successful operation in this range, some type of forced commutation is required.   The phasor diagram at the typical phase angle ϕ = 225° indicates that the active power flows to the load, causing motoring operation, and the lagging VAR is consumed by the load.   This mode therefore corresponds to induction motor operation.

Mode 4: Force-Commutated Rectifier, 270° ≤ α ≤ 360°.   In this mode, as in mode 3, the inverter requires forced commutation.   The phasor diagram indicates rectifier operation, with the load demanding lagging VAR.   This mode can therefore be identified as induction motor operation with regenerative braking, as indicated in Fig. 5.4.

5.2 LOAD-COMMUTATED INVERTERS

In the preceding section, we discussed different modes of current-fed inverter with a counter emf type of load.   The modes are also generally valid for a single-phase inverter and a passive-type load.   We expand this concept further in this section.

Single-Phase Inverter

Let us consider a single-phase bridge inverter with a passive RL load as shown in Fig. 5.5.   A capacitor C of sufficiently high value is connected across the load so that effectively the load has a leading power factor.   The purpose of the capacitor

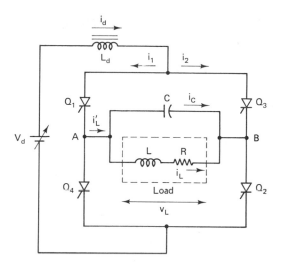

Figure 5.5 Single-phase bridge inverter with load commutation.

is to have load commutation of the thyristors.   The circuit is generally used for typical induction heating applications.   Figure 5.6(a) shows the inverter load voltage and current waves, again assuming that the dc link inductance is very high. The thyristor pairs Q_1, Q_2 and Q_3, Q_4 are switched alternately for 180° duration to impress a square current wave across the load.   The resulting load voltage wave is nearly sinusoidal and lags the current wave by $\beta°$.   When the thyristor pair Q_1, Q_2 is switched on, the outgoing pair Q_3, Q_4 is impressed with negative load voltage for duration $\beta°$, causing load commutation.   The minimum value of the commutating angle β should be sufficient to turn off the outgoing thyristors.   Figure 5.6(b) shows the fundamental frequency phasor diagram where the load rms voltage V_L is drawn as a reference phasor.   The load current I_L is at lagging power factor angle ϕ and can be resolved into a reactive component I_Q and an active component I_P.   The leading capacitor current I_C overcomes the lagging current I_Q so that the effective power factor cos β becomes leading.

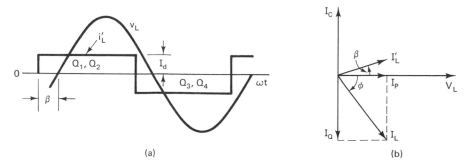

(a)   (b)

Figure 5.6   (a) Load voltage and current waves; (b) phasor diagram.

Circuit analysis.   The general circuit equations of the inverter can be written as follows:

$$v_L = i_L R + L \frac{di_L}{dt} \tag{5.3}$$

$$i_C = C \frac{dv_L}{dt} \tag{5.4}$$

$$i'_L = i_L + i_C \tag{5.5}$$

$$i'_L = i_1 - i_2 \tag{5.6}$$

$$i_d = i_1 + i_2 \tag{5.7}$$

$$V_d - v_L = L_d \frac{di_d}{dt} + R_d i_d \tag{5.8}$$

where R_d is the resistance of inductor L_d.

The equations can be expressed in state-variable form and solved for both transient- and steady-state conditions with the help of a computer program. Such a generalized analysis permits complete design of the inverter, including the effects of harmonics. We will attempt here an approximate steady-state analysis assuming that L_d is lossless and of infinite value and that the load is highly inductive ($\omega L \gg R$). Again, the operation near load resonance (small β) will be considered and the harmonic effect on load will be neglected. The series RL load can be resolved into parallel L_1 and R_1 components so that the reactive component I_Q flows through L_1 and the active component I_P flows through R_1. Therefore, the load Z_L impedance can be written in the form

$$Z_L = R + j\omega L = \frac{R_1 j\omega L_1}{R_1 + j\omega L_1}$$

$$= \frac{R_1 \omega^2 L_1^2}{R_1^2 + \omega^2 L_1^2} + j \frac{R_1^2 \omega L_1}{R_1^2 + \omega^2 L_1^2} \tag{5.9}$$

If the circuit is highly inductive (i.e., $I_Q \gg I_P$, or $R_1 \gg \omega L_1$). Then

$$R \simeq \frac{\omega^2 L_1^2}{R_1} \qquad L \simeq L_1$$

or

$$R_1 = \frac{\omega^2 L^2}{R} \qquad L_1 = L \tag{5.10}$$

The total load fundamental component of current (i.e., the fundamental of the square current wave) is given by

$$I'_L = \frac{2\sqrt{2}}{\pi} I_d \tag{5.11}$$

The active dc power supplied by the source is consumed by the load; that is,

$$V_d I_d = \frac{V_L^2}{R_1} \tag{5.12}$$

Again, the active and reactive components of total load current can be given as

$$I_P = I_L' \cos \beta = \frac{V_L}{R_1} \tag{5.13}$$

$$I_Q' = I_C - I_Q = I_L' \sin \beta = \frac{V_L}{X_C'} \tag{5.14}$$

where $X_C' = j\omega L_1 \parallel 1/j\omega C$. Combining equations (5.13) and (5.14) yields

$$V_L^2 = I_L'^2 \left(\frac{R_1^2 X_C'^2}{R_1^2 + X_C'^2} \right) \tag{5.15}$$

$$\cos \beta = \frac{I_P}{\sqrt{I_P^2 + I_Q'^2}} = \frac{X_C'}{\sqrt{R_1^2 + X_C'^2}} \tag{5.16}$$

Substituting equations (5.11) and (5.12) in (5.15) gives

$$V_L = \frac{\pi}{\sqrt{8}} V_d \frac{\sqrt{R_1^2 + X_C'^2}}{X_C'} \tag{5.17}$$

The load voltage, currents, and commutating angle can be calculated from the equations above for the given circuit parameters. Or else, for a specified angle β, the value of capacitance can be determined for a given load.

Variable load.   In practice, the load may be variable. Therefore, for satisfactory load commutation, the capacitance C may be varied such that the margin angle β is always maintained. Alternatively, for a fixed C value, the inverter frequency ω can be manipulated such that it is always slightly higher than the load resonance frequency. This condition will assure leading power factor operation of the load irrespective of its variation. The frequency variation is of no concern for an induction heating type of load. In fact, the inverter firing control can be synchronized with the load voltage wave so that the commutating margin angle β is always maintained. Instead of a constant margin angle, a constant margin time t_β is desirable in variable-frequency operation. With a constant β angle, as the frequency decreases, the margin time ($t_\beta = \beta/\omega$) becomes larger than necessary, causing unnecessary VAR loading of the inverter.

Three-Phase Inverter

The concept of load commutation as explained above with a single-phase inverter can be extended to polyphase inverters. Figure 5.7 shows a three-phase bridge inverter with lagging power factor load where the load commutation is being achieved with a leading VAR load connected at the load terminal. Again, as discussed

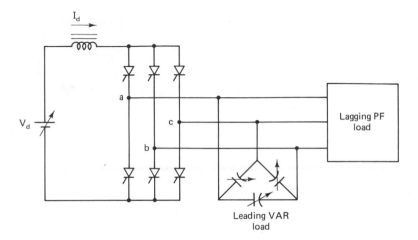

Figure 5.7   Three-phase bridge inverter with load commutation.

before, with a variable load a fixed capacitor bank can be connected at the terminal and the inverter frequency can be manipulated so that load commutation occurs with fixed advance angle β.   If, however, the load is an induction motor, the inverter frequency is dictated by the machine speed requirement.   For this condition, the leading VAR requirement can be met by a variable VAR generator at the machine terminal.   The leading VAR generator has to supply lagging VAR demand of the load and, in addition, supply leading VAR so that the effective load power factor is leading at angle β.   The leading VAR generator can be one of the following types:

- Switched capacitors
- Rotating synchronous condenser (no-load operation of synchronous motor with excitation control)
- Current-fed inverter leading VAR generator (Ref. 11)
- Voltage-fed PWM inverter leading VAR generator (Ref. 13)
- Cycloconverter-type leading VAR generator (Ref. 12)

This type of induction motor speed control system with a static leading VAR generator has been used, but the additional cost and complexity of the VAR generator is to be weighed carefully against the force-commutated inverter drive system.

Synchronous machine load.   The current-fed inverters with load commutation are extremely popular in large power synchronous motor drives, where it is easy to maintain the required leading power factor angle by adjustment of machine excitation.   The absence of a forced commutation requirement makes the inverter simple, reliable, and more efficient.   The thyristors of the inverter can be of the slow-speed rectifier type, reducing the inverter cost further.   The torque and speed of the machine are controlled by the dc link current and frequency of the inverter, respectively, and the machine is operated in the brushless and commutatorless motor (CLM) mode, which is explained later.

Figure 5.8 shows the phasor diagram of a salient-pole machine for load commutation under the condition of motoring, and Fig. 5.9 gives the phase voltage and current waves with the effect of overlapping. The phasor diagram has been drawn using standard notation, neglecting the winding resistance and commutation overlapping effect. A flux linkage phasor diagram has been added in the figure, where ψ_f is the field flux, ψ_a the armature reaction flux, and ψ_s the resultant stator flux. At steady state, current does not flow through the damper winding (neglecting commutation transient) and therefore the d^e and q^e components of ψ_a can be written as

$$\psi_{ds} = L_{ds}I_{ds} = L_{ds}I_s \sin(\delta + \phi) \tag{5.18}$$

$$\psi_{qs} = L_{qs}I_{qs} = L_{qs}I_s \cos(\delta + \phi) \tag{5.19}$$

Since $L_{ds} \neq L_{qs}$, the phasors ψ_a and I_s are not cophasal.

The voltage and current waves of the load-commutated inverter are identical to those of the phase-controlled inverter, and therefore Fig. 5.9 has already been discussed in Chapter 3. The machine terminal voltages will deviate from the nominal stator voltages v_a, v_b, and v_c by a resistance drop and commutation spikes. The commutating inductance L_c which causes the voltage spike and overlap angle μ should be small. An expression of the overlap angle that was derived in equation (3.38) can be given as

$$\cos(\beta - \mu) - \cos\beta = \frac{2\omega L_c I_d}{\sqrt{6}V_s} \tag{5.20}$$

Figure 5.10 shows the steady-state characteristics as a function of machine current that is proportional to I_d. The reactance $X_c = \omega L_c$ is the subtransient reactance

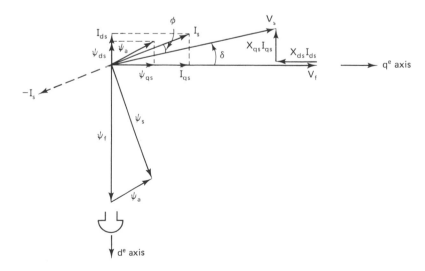

Figure 5.8 Phasor diagram of synchronous motor with load commutation.

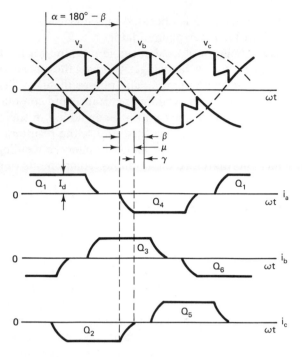

Figure 5.9 Phase voltage and current waves with load commutation showing overlapping effect.

of the machine and can be given approximately by the components of equivalent circuit of Fig. 2.29 as

$$X''_d = X_{1s} \parallel X_{dm} \parallel X_{1fr} \parallel X_{1dr} \tag{5.21}$$

$$X''_q = X_{1s} \parallel X_{qm} \parallel X_{1qr} \tag{5.22}$$

The exact value of X_c varies as a function of no-load advance angle β_0 and lies

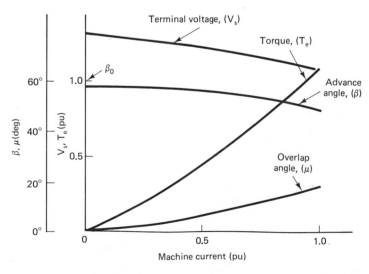

Figure 5.10 Steady-state characteristics with increase of machine current.

within X_d'' and X_q'' as shown in Fig. 5.11. Evidently, the damper winding plays an important part in determining the value of X_c and it should be designed to have small leakage reactance. During the dwell time, when the phase current magnitude does not change, current does not flow through the damper winding. But during each commutation, a burst of current flows through the damper winding. The distribution of typical rms damper currents I_{Dq} and I_{Dd} as a function of the β_0 angle is shown in Fig. 5.11. These curves help to design the damper winding so that it is not overheated.

The power circuit design criteria of a load-commutated inverter is essentially the same as that of the phase-controlled converter discussed in Chapter 3. Although load commutation is generally used with a wound field synchronous machine, in principle it can also be used with a permanent-magnet machine. Lack of controllability of field flux does not permit optimum adjustment of the β angle in the constant-torque region. The condition becomes much worse if the machine is operated above the base speed. With a constant field, the induced emf increases proportional to the speed, but because of the constant dc link voltage, the β angle becomes excessively large at higher speeds. This causes higher machine losses and the inverter rating becomes higher.

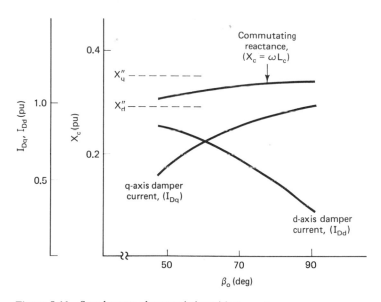

Figure 5.11   Steady-state characteristics with β_0 angle.

5.3 SYNCHRONOUS MOTOR STARTING

The load commutation of an inverter with a synchronous machine load as discussed above, depends on counter emf and therefore the machine must not operate below a critical speed. The counter emf becomes lower at lower speed and causes the overlap angle to be larger. Typically, below 10% of the base speed, the load commutation cannot be applied satisfactorily. In this speed range, the inverter requires some type of forced commutation, which is discussed later. A type of forced commutation that does not require any additional power circuit components is known as the pulsed or dc link current interruption method and is explained in Fig. 5.12. In this method, the inverter thyristors are commutated by periodically interrupting the dc link current as shown. The three-phase inverter firing pulses can be derived either from an independent voltage-controlled oscillator or from a rotor position encoder. The firing pulses through logic circuits also generate control signals of a line-side converter so as to generate dc current pulses at $60°$ intervals. Assume, for example, that commutation is desired from thyristors Q_2 to Q_4 in phase a. At point A, the line-side converter is blocked. The converter will temporarily go into the inverting mode, pumping the inductive energy into the line. During current interruption, the inverter thyristors Q_2 and Q_3 will turn off, but then at point B, Q_3 and Q_4 are fired and the line-side converter is enabled to establish the current I_d. Thus conduction is successfully transferred from Q_2 to Q_4. This mode of control is repeated at every $60°$ interval and the machine phase current becomes slightly less than $120°$ wide per half-cycle with a gap in the middle. The machine is started from zero speed with an initial field current and the developed torque gradually increases the speed until control is transferred to load commutation. Nearly fully developed torque can be obtained in this method and the machine can have sustained operation in either the motoring or regenerating mode. The advantage of triggering with a rotor position sensor instead of a free-running oscillator is the improvement of system stability, which is discussed later.

Although load-commutated inverters with synchronous machines are popular

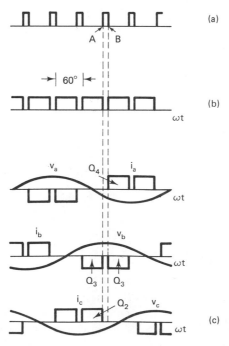

Figure 5.12 Pulse method of starting synchronous motor: (a) pulse train from rotor position encoder; (b) dc link current; (c) inverter phase voltage and current waves.

for large-horsepower, variable-speed applications, these are also used as solid-state starters for synchronous machines for such applications as gas turbine and pumped storage hydro turbine starting.   As a general-purpose starter for a synchronous machine connected with the load, the drive has the following three modes of operation: (1) at low speed, the motor is started with forced commutation of the inverter as explained above; (2) when adequate counter emf is developed, the drive enters into the load commutation mode; and (3) as the machine speeds up and the voltage, frequency, and phase conditions are matched with the utility line, the starter is bypassed and the machine is switched into the utility system.   The solid-state starter, although expensive, permits the drive to operate at variable speed with four-quadrant control capability.   This starting scheme becomes specially attractive in an installation where a number of synchronous machines can time share a common starter.

5.4 FORCE-COMMUTATED INVERTERS

It was mentioned before that a current-fed inverter requires forced commutation if it supplies a nonleading power factor load.   In this section, several types of force-commutated inverters are described.

Auto-Sequential-Commutated Inverter (ASCI)

This class of inverters is the most popular among all the types of force-commutated current-fed inverters, and therefore it will be given maximum emphasis.   The ASCI inverter can have various single-phase and polyphase circuit configurations.   The three-phase bridge configuration is used almost exclusively in practice.   However, we will discuss the single-phase bridge inverter first to explain the basic fundamentals and then describe the three-phase bridge circuit in detail with an ac machine load.

Single-phase bridge ASCI inverter.   A single-phase bridge inverter with auto-sequential commutation is shown in Fig. 5.13.   A constant dc current I_d is supplied from a variable-voltage source, typically from a phase-controlled rectifier operating in the feedback current control mode.   The current I_d, as is usual in any current-fed inverter, is not affected by the inverter load condition and therefore its ripple is neglected.   The thyristor pairs Q_1, Q_2 and Q_3, Q_4 are alternately switched on to establish a nearly square wave current through the series RL load. A commutating capacitor is connected across each of the upper and lower halves of the bridge as shown.   A diode is connected in series with each thyristor, which helps to isolate the capacitors from the load.   The explanatory voltage and current waves of the inverter are shown in Fig. 5.14.   Assume in the beginning that the thyristor pair Q_3, Q_4 is on and that a steady dc current I_d is flowing through the load in the path Q_3, D_3, R, L, D_4, and Q_4.   The commutating capacitors are assumed to be initially charged equally in the polarity shown (i.e., $v_{C_1} = v_{C_2} = -V_{CO}$).   At time $t = 0$, the thyristors Q_1, Q_2 are fired.   The thyristors Q_3 and

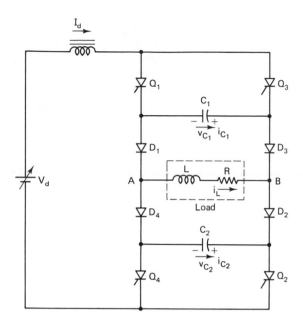

Figure 5.13 Single-phase bridge auto-sequential-commutated inverter (ASCI).

Q_4 will be turned off instantly by the reverse capacitor voltages, and Q_1 and Q_2 will conduct the current I_d. The diodes D_1 and D_2 will remain reverse biased initially and the current I_d will flow in the path Q_1, C_1, D_3, R, L, D_4, C_2, and Q_2, linearly charging the capacitors. The diode D_1 voltage is given as

$$v_{D_1} = -V_{CO} + \frac{1}{C} \int_0^t I_d \, dt + I_d R = -V_{CO} + \frac{I_d t}{C} + I_d R \qquad (5.23)$$

The voltage v_{D_1} rises linearly and eventually at $t = t_1$, $v_{D_1} = 0$, and then the diode will start conducting. The voltage across D_2 is given by the identical expression and both D_1 and D_2 will start conducting at the same instant. The constant-current charging interval t_1 of capacitors is given from equation (5.23) as

$$0 = -V_{CO} + \frac{I_d t_1}{C} + I_d R$$

that is

$$t_1 = \frac{C}{I_d}(V_{CO} - I_d R) \qquad (5.24)$$

The capacitor voltage $v_{C_1} = v_{C_2} = v_C$ and is the same as the thyristor reverse voltage (Q_3 or Q_4), which is given as

$$v_C = -V_{CO} + \frac{1}{C} \int_0^t I_d \, dt \qquad (5.25)$$

and therefore at $t = t_1$ using equation (5.24).

$$v_C(t_1) = -I_d R \qquad (5.26)$$

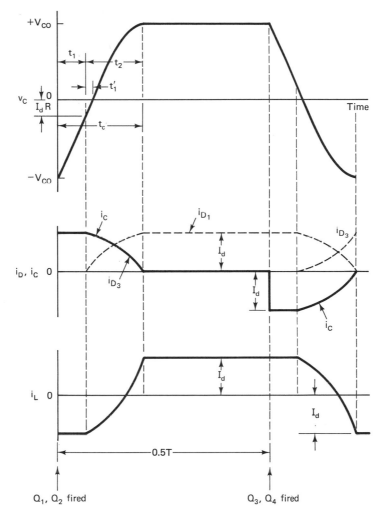

Figure 5.14   Voltage and current waves of single-phase bridge ASCI inverter.

At the end of time t_1, all four diodes D_1, D_2, D_3, and D_4 will conduct, connecting the commutating capacitors directly in parallel with the load.   The parallel circuits, consisting of equivalent capacitance $2C$ and the load circuit, will share the total input current I_d.   In this mode, the circuit equations can be written as

$$i_L + i'_C = I_d \tag{5.27}$$

$$v_C = L \frac{di_L}{dt} + i_L R \tag{5.28}$$

$$i'_C = C' \frac{dv_C}{dt} \tag{5.29}$$

where $i'_C = i_{C_1} + i_{C_2} = 2i_C$ and $C' = 2C$.   Substituting equations (5.27) and (5.29) in (5.28) and simplifying, we get

$$LC' \frac{d^2v_C}{dt^2} + RC' \frac{dv_C}{dt} + v_C = RI_d \tag{5.30}$$

The second-order differential equation of capacitor voltage can be solved for the initial conditions

$$v_C|_{t=0} = -I_d R \tag{5.31}$$

$$\frac{dv_C}{dt}\bigg|_{t=0} = \frac{I_d}{C} \tag{5.32}$$

where a new $t = 0$ is defined at t_1 (i.e., when D_1 and D_2 start conducting). The solution of equation (5.30) with the initial conditions above can be given in the form of a damped sinusoid as

$$v_C(t) = I_d R + K_1 K_2 I_d e^{-\alpha t} \sin(\omega_d t - \theta) \tag{5.33}$$

where

$$K_1 = \frac{R}{\lambda \omega_0}$$

$$K_2 = \sqrt{\frac{1 + 2\alpha\lambda + \lambda^2 \omega_0^2}{1 - (\alpha/\omega_0)^2}}$$

$$\omega_0 = \frac{1}{\sqrt{LC'}}$$

$$\alpha = \frac{R}{2L}$$

$$\omega_d^2 = \omega_0^2 - \alpha^2$$

$$\frac{1}{\lambda} = \frac{1}{RC'} - \frac{R}{L}$$

$$\theta = \tan^{-1} \frac{\lambda \omega_d}{1 + \alpha\lambda}$$

Equation (5.33) consists of a damped sinusoid with a constant term and indicates that the steady-state capacitor voltage is $I_d R$. For a purely inductive load, $R \to 0$ and $v_C(t)$ can be simplified as

$$v_C(t)_L = I_d \sqrt{\frac{L}{C'}} \sin \omega_0 t \tag{5.34}$$

Similarly, for a purely resistive load, $L \to 0$ and the first-order differential equation from equation (5.30) can be solved as

$$v_C(t)_R = RI_d(1 - e^{-t/RC'}) \tag{5.35}$$

The individual capacitor current i_C can be derived from equation (5.33) as

$$i_C = C\frac{dv_C}{dt} = CK_1K_2I_de^{-\alpha t}[\omega_d \cos(\omega_d t - \theta) - \alpha \sin(\omega_d t - \theta)] \qquad (5.36)$$

which verifies that $i_C = I_d$ at $t = 0$. The diodes D_3 and D_4 will stop conducting when $i_C = 0$ at $t = t_2$; that is, from equation (5.36),

$$\omega_d \cos(\omega_d t_2 - \theta) - \alpha \sin(\omega_d t_2 - \theta) = 0 \qquad (5.37)$$

The transcendental equation can be solved to determine t_2 for the given parameters. At the end of t_2 the commutation is complete and the current I_d flows in the path Q_1, D_1, L, R, D_2, and Q_2. The total commutation time t_c is given as

$$t_c = t_1 + t_2 \qquad (5.38)$$

and the corresponding angle is

$$\theta_c = \omega t_c = \omega t_1 + \omega t_2 \qquad (5.39)$$

where $\omega = 2\pi f = 2\pi/T$ and T is the time period of the inverter. At low frequency, the commutation angle may be small, but at high frequency it may occupy a significant portion of the half-cycle. The reverse half-cycle starts when the thyristor pair Q_3, Q_4 is fired and the commutation waveforms are symmetrical.

The outgoing thyristors remain reverse biased for the duration $t_R = t_1 + t_1'$, where t_1' can be evaluated from equation (5.33) by solving

$$I_dR + K_1K_2I_de^{-\alpha t_1'} \sin(\omega_d t_1' - \theta) = 0 \qquad (5.40)$$

The capacitor voltage is the same but of opposite polarity at the beginning and at the end of commutation in steady state. Therefore,

$$v_C(t_2) = I_dR + K_1K_2I_de^{-\alpha t_2} \sin(\omega_d t_2 - \theta) - -V_{CO} \qquad (5.41)$$

Equation (5.41) can be solved for V_{CO}. The capacitor voltage changes by $2V_{CO}$ during each commutation interval, as shown in Fig. 5.14. The equations derived so far are useful in designing the inverter.

Three-phase bridge ASCI inverter.   A three-phase bridge version of an auto-sequential-commutated inverter is shown in Fig. 5.15. The circuit is popularly used in medium- to large-power induction motor drives. However, since the induction motor operates at lagging power factor, the current-fed inverter requires forced commutation, as discussed earlier. The induction motor model can be approximately represented by a per phase equivalent circuit which consists of a sinusoidal counter emf in series with inductance L (Fig. 2.19) and where winding resistances have been neglected for simplicity. This equivalent circuit is satisfactory for explaining the commutation phenomena and design of inverter circuit components. At a stalled condition of the machine, the counter emf vanishes and the motor becomes ideally an inductive load.

The thyristors Q_1 to Q_6 are the principal switching devices of the inverter,

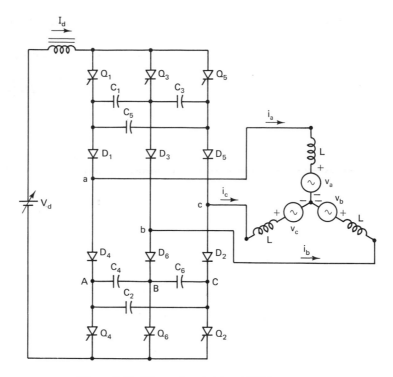

Figure 5.15 Three-phase bridge ASCI inverter.

where each of them conduct in sequence, ideally for 120°, to establish a six-stepped current wave in the output line. The series diodes and the delta capacitor bank of equal values which are connected to each of the upper and lower groups of thyristors constitute the commutating elements. The capacitors store charge for commutation and the series diodes tend to isolate them from the load as explained in Fig. 5.13. The snubber circuit, including di/dt inductance, is not shown as usual, for simplicity.

In normal inverter operation, the upper group and lower group operate independently and there are six commutations per cycle of fundamental frequency. We will discuss commutation phenomena in detail for commutation from Q_2 to Q_4, and the discussion will be valid for all other commutations. Again, the motoring mode of the machine is considered where $180° \leq \alpha \leq 270°$ (Fig. 5.4), but the discussion is valid for the regenerative mode also.

Figure 5.16 shows the equivalent conduction circuit during Q_2 to Q_4 commutation and Fig. 5.17 shows the corresponding voltage and current waves. Assume that initially $v_{CA} = -V_{CO}$, so that when Q_4 is fired, Q_2 is instantaneously turned off by the reverse voltage. The dc current I_d flows through Q_3 and D_3 in the upper group, the phases b and c of the machine, and the devices D_2, the delta capacitor bank, and Q_4 to the negative supply. The capacitor bank charges linearly

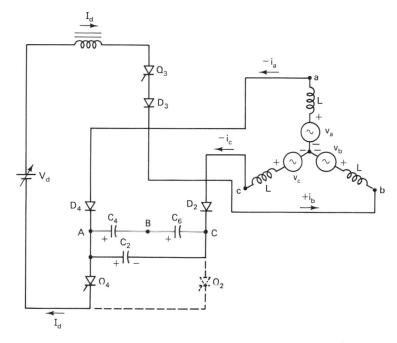

Figure 5.16 Conduction paths of ASCI inverter during commutation from Q_2 to Q_4.

with current I_d and the voltatge v_{CA} is given by

$$v_{CA} = -V_{CO} + \frac{1}{C'}\int_0^t I_d \, dt$$

$$= -V_{CO} + \frac{I_d t}{C'}$$

(5.42)

where $C' = 1.5C$ is the effective capacitance and C represents the individual capacitance. The distribution of capacitor currents will be in inverse ratio of their values; that is,

$$i_{C_2} = 0.67 I_d$$

(5.43)

$$i_{C_4} = i_{C_6} = 0.33 I_d$$

(5.44)

During the constant-current charging of capacitors, the diode D_4 will remain reverse biased by the voltage

$$v_{D_4} = v_{CA} - v_a + v_c$$

$$= -V_{CO} + \frac{I_d t}{C'} + v_{ca}$$

(5.45)

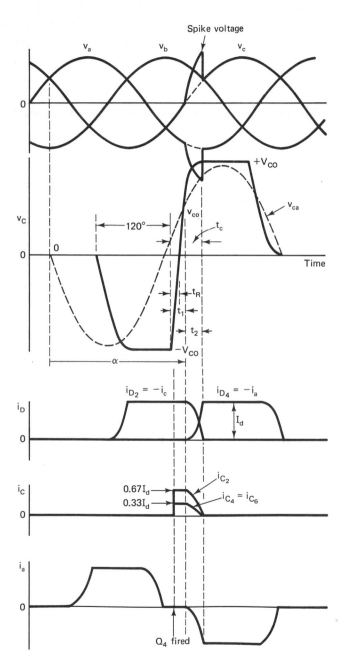

Figure 5.17 Voltage and current waves of three-phase bridge ASCI inverter during Q_2 to Q_4 commutation.

because inductance L does not cause any voltage drop.   The linear charging period will end at time t_1 when the capacitor bank voltage equals the line voltage, and the diode D_4 begins to conduct.   Then the capacitor charging will slow down as the current begins to bypass through D_4.   At time t_1, $v_{D4} = 0$; that is, from equation (5.45)

$$-V_{CO} + \frac{I_d t_1}{C'} - \sqrt{3}\ V_m \sin \alpha = 0 \qquad (5.46)$$

where $v_{ca} = -\sqrt{3}\ V_m \sin \alpha$ and α is defined at time t_1 as shown in Fig. 5.17. Therefore, t_1 is given by

$$t_1 = \frac{C'}{I_d}(V_{CO} - \sqrt{3}\ V_m \sin \alpha) \qquad (5.47)$$

The thyristor Q_2 will remain reverse biased for the duration t_R until $v_{CA} - 0$; that is, from equation (5.42),

$$t_R = \frac{C'V_{CO}}{I_d} \qquad (5.48)$$

Assuming an ideal six-stepped line current wave and neglecting harmonic and inverter losses, we can write

$$I_d = \frac{\pi}{2\sqrt{3}}\ I_m = \frac{\pi}{\sqrt{6}}\ I \qquad (5.49)$$

$$P = V_d I_d = \sqrt{3}\ V_L I \cos \phi \qquad (5.50)$$

where I is the fundamental rms line current and I_m its peak value, P the power delivered from dc-to-ac side, and $\phi = \pi - \alpha$.   Combining equations (5.49) and (5.50) yields

$$V_d = \frac{3\sqrt{2}}{\pi}\ V_L \cos \phi \qquad (5.51)$$

The equations above relate the basic power circuit quantities at steady state.

When the diode D_4 starts conducting, the current will gradually transfer from D_2 to D_4, and the commutation process will be complete at time t_2 when $i_{D2} = 0$. During the transfer interval, the loop voltage equation can be written as

$$L \frac{di_{D4}}{dt} + v_{ca} - L \frac{di_{D2}}{dt} - v_{CO} - \frac{1}{C'}\int_0^t i_{D2}\ dt = 0 \qquad (5.52)$$

where $t = 0$ is defined at the beginning of the transfer interval.   The initial conditions are

$$i_{D4}(0) = 0$$

$$i_{D2}(0) = I_d$$

$$v_{CO} = -\sqrt{3}\ V_m \sin \alpha$$

During this interval,

$$i_{D_4} + i_{D_2} = I_d \tag{5.53}$$

$$v_{ca} = -\sqrt{3}\, V_m \sin(\omega t + \alpha) \tag{5.54}$$

Substituting equations (5.53) and (5.54) in (5.52) gives us

$$2L\frac{di_{D_2}}{dt} + \frac{1}{C'}\int_0^t i_{D_2}\, dt = -\sqrt{3}\, V_m \sin(\omega t + \alpha) + \sqrt{3}\, V_m \sin\alpha \tag{5.55}$$

We consider next the following conditions of counter emf.

Case 1: Invariant counter emf during transfer.   At low-frequency operation of the inverter, the current transfer will occur at a small angular interval during which the counter emf may be considered as invariant and is equal to the initial value (i.e., $v_{ca} = -\sqrt{3}\, V_m \sin\alpha$).   This assumption simplifies the mathematical analysis.   Substituting the condition in equation (5.55) gives us

$$2L\frac{di_{D_2}}{dt} + \frac{1}{C'}\int_0^t i_{D_2}\, dt = 0 \tag{5.56}$$

The differential equation can be solved easily for i_{D_2} with the initial condition, which gives

$$i_{D_2} = I_d \cos\omega_0 t \tag{5.57}$$

where $\omega_0 = 1/\sqrt{2LC'}$.   Correspondingly, the current i_{D_4} can be given as

$$i_{D_4} = I_d(1 - \cos\omega_0 t) \tag{5.58}$$

At time t_2, $i_{D_2} = 0$.   Therefore, from equation (5.57),

$$t_2 = \frac{\pi}{2\omega_0} \tag{5.59}$$

which occupies a quarter-cycle time period of the commutation resonance frequency ω_0.   The total commutation time t_c is therefore

$$t_c = t_1 + t_2 \tag{5.60}$$

The capacitor bank charges to the peak voltage V_{CO}, which can be given as

$$
\begin{aligned}
V_{CO} &= \sqrt{3}\, V_m \sin\alpha + \frac{1}{C'}\int_0^{t_2} i_{D_2}\, dt \\
&= -\sqrt{3}\, V_m \sin\alpha + \frac{I_d}{\omega_0 C'}
\end{aligned}
\tag{5.61}
$$

which should be equal to $-V_{CO}$ at the beginning of t_1 period for steady-state operation.   The voltage V_{CO} appears as the peak voltage across each diode and thyristor.   Equation (5.61) can be substituted in equations (5.47) and (5.48) to determine t_1 and t_R intervals.   During the transfer interval, a voltage spike is induced across the leakage inductance L of each phase as shown in Fig. 5.17. The peak spike voltage can be derived from equation (5.57) as

$$V_{\text{spike}} = L \left. \frac{di_{D2}}{dt} \right|_{t=t_2} = I_d L \omega_0$$

$$= \frac{I_d}{2\omega_0 C'} \tag{5.62}$$

The spike voltage at a large current may be significant and a separate clamping circuit may be desirable. A simple but practical clamping circuit consists of a diode bridge rectifier at the machine terminal with a string of zener diodes as the load.

It may be noted that the parameters V_{CO}, t_1, and t_R are a function of the α angle, which may vary widely depending on the operating condition of the machine. For example, V_{CO} will tend to be maximum at no-load conditions, when ideally $\alpha = 270°$, and will tend to be lower in motoring and regenerating conditions, when α deviates from this position. Of course, the inverter is to be designed for the worst condition. This also suggests that if an ASCI inverter drives a synchronous machine load and the machine excitation is controlled so that it operates at unity power factor (i.e., $\alpha = 180°$ for motoring and $\alpha = 0°$ for regeneration), then V_{CO} will be of lowest amplitude. Since the voltage V_{CO} determines the voltage rating of the inverter components, the inverter cost can be reduced by limiting the V_{CO} amplitude.

Case 2: Varying counter emf during transfer. As the inverter frequency increases with increasing motor speed, the commutation angle ωt_c may occupy a significant angular interval and the assumption of constant counter emf during commutation is no longer valid. Again, as $\omega t_c > 60°$, partial overlap will occur between the upper group and the lower group of the bridge (i.e., the thyristor Q_5 will be fired and the upper group will be in a constant-current charging mode before D_2-to-D_4 current transfer is complete in the lower group). This mode of operation is permissible for inverter operation. During the partial overlap mode, the operation of the upper and lower groups of the bridge can be considered as independent and therefore the analytical derivations in this section remain valid. However, it should be noted that if the current transfer interval lasts more than 60° ($\omega t_2 > 60°$), full overlap between the upper group and the lower group will start. This means that before D_2-to-D_4 transfer is complete, the current transfer will start from D_3 to D_5 in the upper group. The conduction of diodes D_5 and D_2 tends to bypass power from motor phase c, and therefore this mode of operation should be avoided.

With varying counter emf during current transfer, equations (5.47) and (5.48) will remain valid and the current i_{D_2} from equation (5.55) can be solved by the Laplace transform method for varying counter emf. Considering $i_{D_2}(0) = I_d$, the solution can be given as

$$i_{D_2} = \frac{\sqrt{3}I_d}{2K_1(K_2^2 - 1)} \left\{ \left[\frac{2K_1(K_2^2 - 1)}{\sqrt{3}} + \cos\alpha \right] \cos K_2\omega t \right.$$

$$\left. - \frac{\sin\alpha}{K_2} \sin K_2\omega t - \cos(\omega t + \alpha) \right\} \qquad (5.63)$$

where

$$K_1 = \frac{\omega L I_d}{V_m} \qquad \text{and} \qquad K_2 = \frac{\omega_0}{\omega}$$

The current transfer will be complete when $i_{D_2} = 0$ at time t_2. This can be evaluated from the transcendental equation

$$\left[\frac{2K_1(K_2^2 - 1)}{\sqrt{3}} + \cos\alpha \right] \cos K_2\omega t_2 - \frac{\sin\alpha}{K_2} \sin K_2\omega t_2$$

$$- \cos(\omega t_2 + \alpha) = 0 \qquad (5.64)$$

The capacitor voltage v_{CA} can be determined by substituting equation (5.63) into

$$v_{CA} = -\sqrt{3}\,V_m \sin\alpha + \frac{1}{C'} \int_0^t i_{D_2}\,dt \qquad (5.65)$$

which gives, after simplification,

$$\frac{v_{CA}}{\sqrt{3}V_m} = K_2 \left(\frac{2K_1}{\sqrt{3}} + \frac{\cos\alpha}{K_2^2 - 1} \right) \sin K_2\omega t + \frac{\sin\alpha}{K_2^2 - 1} \cos K_2\,\omega t$$

$$- \frac{K_2^2}{K_2^2 - 1} \sin(K_2\,\omega t + \alpha) \qquad (5.66)$$

The normalized peak capacitor voltage $V_{CO}/\sqrt{3}\,V_m$ can be solved by substituting t_2 from equation (5.64) into equation (5.66). The angular intervals ωt_1 and ωt_R from equations (5.47) and (5.48), respectively, can be derived in the following forms after substituting K_1 and K_2:

$$\omega t_1 = \frac{\sqrt{3}}{2K_1K_2^2} \left(\frac{V_{CO}}{\sqrt{3}\,V_m} - \sin\alpha \right) \qquad (5.67)$$

$$\omega t_R = \frac{\sqrt{3}}{2K_1K_2^2} \frac{V_{CO}}{\sqrt{3}\,V_m} \qquad (5.68)$$

A computer program can be written to solve ωt_2 from equation (5.63) for different parameters K_1, K_2, and α, and correspondingly, $V_{CO}/\sqrt{3}\,V_m$, ωt_1, and ωt_2 can be solved. The results substantiate that $V_{CO}/\sqrt{3}\,V_m$ approaches the maximum as $\alpha = 270°$ and both K_1 and K_2 increase. This α angle also coincides for maximum values of commutating angle and turn-off angle. The solutions above will be valid only for $\omega t_2 < 60°$ (i.e., before the full overlap mode starts).

Selection of commutating capacitor.   Inverter thyristors and diodes can easily be designed on the basis of a 120° duty cycle of current and peak capacitor voltage, as discussed before.   However, the selection of commutating capacitor is not very obvious.   It has been indicated in the previous analysis that a larger capacitance will reduce the operating frequency range, but the voltage rating of the devices will be reduced.   On the other hand, a smaller capacitance increases the frequency range but increases the voltage spike.

A question naturally arises: For a given maximum frequency, what should be the capacitor size?   We will make a simplified analysis here based on invariant counter emf during the current transfer condition, and the maximum frequency will be limited until partial overlap mode starts; that is,

$$\omega_m(t_1 + t_2) = \frac{\pi}{3} \tag{5.69}$$

where ω_m is the maximum inverter frequency.   Combining equations (5.61) and (5.47), we obtain

$$t_1 = \frac{C'}{I_d}\left(-2\sqrt{3}\,V_m \sin\alpha + \frac{I_d}{\omega_0 C'}\right) \tag{5.70}$$

The value of t_1 will be maximum at minimum I_d and at $\alpha = 270°$.   Therefore,

$$t_{1max} = \frac{2\sqrt{3}\,C'V_m}{I_d} + \frac{1}{\omega_0} \tag{5.71}$$

Combining equations (5.59), (5.69), and (5.71), we have

$$\omega_m\left(\frac{\pi}{2\omega_0} + \frac{1}{\omega_0} + \frac{2\sqrt{3}\,C'V_m}{I_d}\right) = \frac{\pi}{3} \tag{5.72}$$

The limiting frequency ω_m can be expressed as a fraction of resonance frequency ω_0 as

$$\frac{\omega_m}{\omega_0} = \frac{\pi/3}{1 + \pi/2 + 2\sqrt{3}\,C'V_m\omega_0/I_d} \tag{5.73}$$

Or the frequency should be restricted below the value

$$\omega < \frac{(\pi/3)\omega_0}{1 + \pi/2 + 2\sqrt{3}\,C'V_m\omega_0/I_d} \tag{5.74}$$

so that overlapping does not occur in the existing system.   For selection of the capacitor so that the specified maximum frequency ω_m can be obtained, equation (5.72) can be written in quadratic equation form as

$$\frac{2\sqrt{3}\,V_m}{I_d}(\sqrt{C'})^2 + \left(1 + \frac{\pi}{2}\right)\sqrt{2L}\,\sqrt{C'} - \frac{\pi}{3\omega_m} = 0 \tag{5.75}$$

where ω_0 has been substituted for $1/\sqrt{2LC'}$.   The solution of equation (5.75) yields

$$\sqrt{C'} = \frac{\pi(1 + \pi/2)}{24}\frac{I_m}{V_m}\left[-\sqrt{2L} + \sqrt{2L + \frac{16}{\omega_m(1+\pi/2)^2}\frac{V_m}{I_m}}\right] \tag{5.76}$$

where $C' = 1.5C$, $L = L_{ls} + L_m L_{lr}/(L_m + L_{lr})$ in standard machine notation, and I_d has been substituted by I_m using equation (5.49). The ratio V_m/I_m is the maximum impedance of the induction motor equivalent circuit at frequency ω_m and at zero slip, that is,

$$\frac{V_m}{I_m} = \omega_m(L_{ls} + L_m) \tag{5.77}$$

Torque Pulsation

The six-stepped current waves of a current-fed inverter cause harmonic heating and torque pulsation problems of the machine as discussed in Chapter 4. The pulsating torque problem may be serious for inverter operation below a few hertz of fundamental frequency. Fortunately, the ASCI inverter can be operated to generate pulse-width-modulated phase current waves as shown in Fig. 5.18. The dc current I_d can be switched back and forth between the adjacent phases for both the upper and lower thyristor groups by forced commutation to create the desired notch angles. Consider, for example, that the thyristors Q_1 and Q_2 are conducting and that pulse width modulation is desired between the phase a and phase b currents. The upper capacitor bank will be correctly biased so that when Q_3 is fired at angle α_1, Q_1 will be turned off and I_d will be transferred to Q_3. When commutation is complete, C_1 voltage will reverse so that when Q_1 is fired again at α_2, Q_2 will be off and phase a current will be restored. Thus the desired number of notches in a phase current wave at specified α angles can be provided. The minimum pulse and notch widths will depend on the commutation delay t_c. In

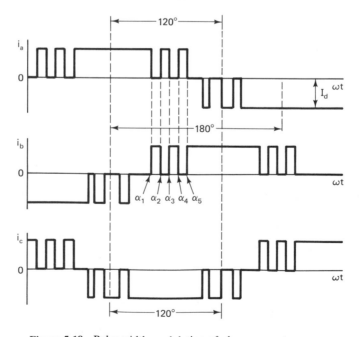

Figure 5.18 Pulse width modulation of phase current waves.

Fig. 5.18 there are two notches per quarter-cycle of wave, which according to selected harmonic elimination theory will help to eliminate two harmonics from the current wave. The lowest two harmonics (i.e., the 5th and 7th) can be eliminated to eliminate the 6th harmonic pulsating torque. As the inverter frequency decreases, higher number of notches can be incorporated and therefore higher harmonic torques (e.g., 6th, 12th, etc.) can be eliminated.

Another method of pulsating torque elimination, which is based on the modulation of dc current I_d by the front-end converter, is explained in Fig. 5.19. Figure 5.19(a) shows the torque wave with the ideal six-stepped phase current wave. The torque ripple follows the dc voltage ripple of a six-pulse phase-controlled bridge converter and therefore is given by 60° segments of a sine wave. The dc current I_d can be modulated by the front-end rectifier so that it follows the inverse profile of the torque ripple as indicated in Fig. 5.19(b). With such an ideal modulation, the torque ripple can be completely eliminated. In practice, the finite slope of I_d when falling will introduce some torque spikes.

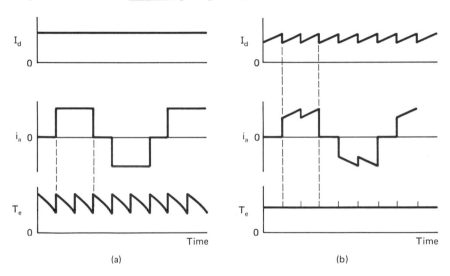

Figure 5.19   (a) Pulsating torque with smooth dc current; (b) smoothing of pulsating torque by modulation of dc current.

ASCI Inverter with Reset Circuits

The frequency range of an ASCI inverter can be extended to some extent by adding reset circuits as shown in Fig. 5.20. A reset circuit consisting of a reset inductance L_R in series with a diode has been connected across each thyristor in the inverse direction as shown. The reset circuits help to reverse or reset the capacitor voltages during commutation, and therefore the commutation time is shortened (i.e., the inverter frequency is raised). The reset circuits become particularly effective in light-load and high-speed motor conditions, where in the normal circuit the interval t_1 tends to be too high. Consider, for example, the commutation from thyristor Q_2 to Q_4 in Fig. 5.20. As the thyristor Q_4 is fired, Q_2 will turn off immediately

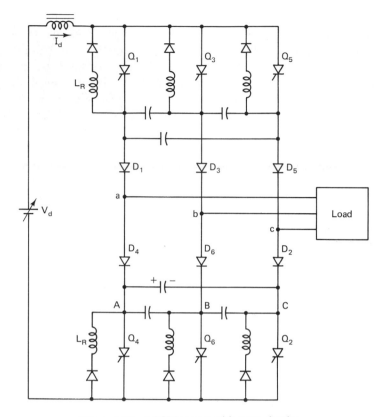

Figure 5.20   ASCI inverter with reset circuits.

by the reverse capacitor voltage, and the constant-current charging period of the capacitor bank will begin as usual.   But the reset circuits across Q_6 and Q_2 will offer resonant discharge paths for fast reversal of capacitor voltages.   This will shorten the interval t_1.   As the voltage v_{CA} exceeds the machine counter emf, conduction of diode D_4 starts and during this transfer interval the reset circuits may continue conduction.   Then at the end of commutation, the reset circuit currents are zero, and only the devices D_4 and Q_4 will conduct.

The addition of reset circuits permits boosting the frequency range by providing load-independent commutation paths, but at the expense of higher voltage rating of the components.   For the same voltage rating of the components, the frequency range of the present circuit will be higher than that of the normal circuit by simple reduction of capacitor size.   Of course, the additional components will add to the cost of the inverter and the efficiency will be poorer.

Auxiliary-Bridge-Commutated Inverter

A three-phase bridge inverter where forced commutation of thyristors is performed by an auxiliary bridge inverter is shown in Fig. 5.21. The circuit is also called an individually commutated inverter, because each thyristor in the main bridge has

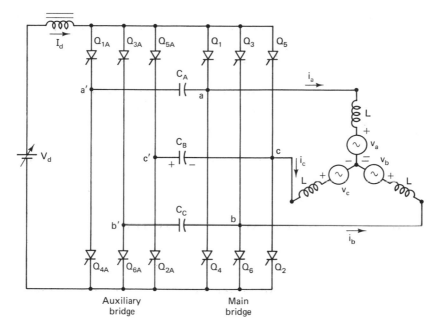

Figure 5.21 Individually commutated current-fed inverter with auxiliary bridge.

its corresponding commutating thyristor in the auxiliary bridge.   There are only three commutating capacitors, one for each phase, and the commutating circuit consists of the auxiliary bridge and the capacitors.   When a main thyristor is required to be commutated, its auxiliary thyristor and the incoming main thyristor are fired simultaneously.   With the correct polarity of capacitor voltage, the outgoing thyristor turns off and the incoming thyristor conducts after the lapse of commutation time.   At each commutation, the corresponding capacitor voltage is reversed so that at a 180° interval the opposite thyristor on the same leg can be commutated successfully.   While the main thyristors conduct for nearly 120°, the auxiliary thyristors conduct during commutation only and therefore the auxiliary bridge power rating is substantially small.

There is some analogy in the commutation process between this inverter and the ASCI inverter which will be evident as we describe the commutation from thyristors Q_2 to Q_4.   In the beginning, thyristors Q_3 and Q_2 are conducting and the capacitor C_B is charged to the polarity as shown.   Thyristors Q_{2A} and Q_4 are fired simultaneously.   The thyristor Q_2 will turn off instantaneously by the reverse capacitor voltage and its current $-i_c = I_d$ will be transferred to thyristor Q_{2A}.   The capacitor will charge at a constant current rate and the interval t_1 will end when $v_{cc'}$ approaches the counter emf $v_{ca} = -\sqrt{3}\,V_m \sin\alpha$.   At this point, Q_4 will be forward biased and the current will begin to transfer from Q_{2A} to Q_4.   Note that the gate of Q_4 should have a pulse train firing so that conduction can start when the device is forward biased.   The current transfer will be complete in interval t_2 when $v_{cc'}$ will be opposite and of equal magnitude.   This will permit Q_5 to be commutated successfully by Q_{5A} firing 180° later.   With the analogy of the ASCI inverter equations (5.55), (5.63), (5.66), (5.67), and (5.68), we can write the corresponding equations as

$$2L \frac{d(-i_c)}{dt} + \frac{1}{C} \int_0^t -i_C \, dt = -\sqrt{3} \, V_m \sin(\omega t + \alpha) + \sqrt{3} \, V_m \sin \alpha \qquad (5.78)$$

$$-i_C = \frac{\sqrt{3} \, I_d}{2K_1(K_2^2 - 1)} \left\{ \left[\frac{2K_1(K_2^2 - 1)}{\sqrt{3}} + \cos \alpha \right] \cos K_2\omega t \right.$$
$$\left. - \frac{\sin \alpha}{K_2} \sin K_2\omega t - \cos(\omega t + \alpha) \right\} \qquad (5.79)$$

$$v_{cc'} = K_2 \left(\frac{2K_1}{\sqrt{3}} + \frac{\cos \alpha}{K_2^2 - 1} \right) \sin K_2\omega t + \frac{\sin \alpha}{K_2^2 - 1} \cos K_2\omega t$$
$$- \frac{K_2^2}{K_2^2 - 1} \sin(K_2\omega t + \alpha) \qquad (5.80)$$

$$\omega t_1 = \frac{\sqrt{3}}{2K_1 K_2^2} \left(\frac{V_{CO}}{\sqrt{3} \, V_m} - \sin \alpha \right) \qquad (5.81)$$

$$\omega t_R = \frac{\sqrt{3}}{2K_1 K_2^2} \frac{V_{CO}}{\sqrt{3} \, V_m} \qquad (5.82)$$

The frequency range of the inverter should be restricted so that $\omega t_c < 60°$; otherwise, objectionable overlap will occur between the main and auxiliary bridges bypassing the load power. This inverter is definitely more expensive than the ASCI inverter.

Inverter with Fourth Leg Commutation

A somewhat simpler inverter circuit where forced commutation is performed by a fourth bridge leg and a single capacitor is shown in Fig. 5.22. The commutation circuit requires the load to be star connected and its neutral point be available. The auxiliary thyristor Q_A is responsible for commutation of the upper group of the main thyristors (Q_1, Q_3, Q_5) and thyristor Q_B commutates the main thyristors in the lower group (Q_4, Q_6, Q_2). Since the main thyristors in each group are commutated at 120° intervals, the auxiliary thyristors conduct alternately six times per cycle and the capacitor current i_C alternates three times per fundamental period. For this reason, the circuit is also called a 3rd-harmonic commutated inverter. The operation of the commutation circuit can be explained by the following three modes of operation:

Mode 1. Assume that in the beginning thyristors Q_3 and Q_2 are conducting and dc current I_d is flowing steadily through phases b and c. It is desired to commutate thyristor Q_2 to thyristor Q_4. Assume also that the capacitor C is charged to the polarity shown and its voltage V_{CO} is greater than the phase voltage v_c. Mode 1 starts when thyristor Q_B is fired to commutate thyristor Q_2. The dominating reverse capacitor voltage applied across phase c gradually transfers

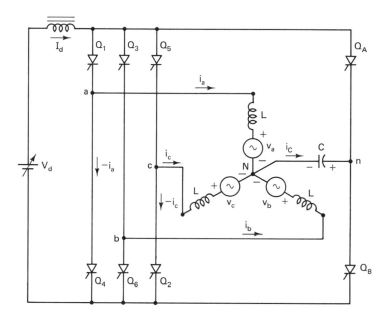

Figure 5.22   Current-fed inverter with fourth leg commutation.

phase c current to the capacitor and the following equations hold true:

$$L \frac{d(-i_c)}{dt} + V_{CO} - \frac{1}{C} \int_0^t i_C \, dt - v_c = 0 \tag{5.83}$$

$$i_C + (-i_c) = I_d \tag{5.84}$$

Mode 1 will end when phase c current is zero.

Mode 2.   In mode 2, the current I_d will flow through phase b, the capacitor, and thyristor Q_B to the negative of dc supply.   With no current through Q_2, the device will turn off.   Since the motor current in phase a and phase c is zero, it indicates that the actual phase current will be less than the usual 120° duration. The capacitor will charge linearly with current I_d by the following equation:

$$v_{Nn} = v_{C(0)} + \frac{I_d}{C} \tag{5.85}$$

where $v_{C(0)}$ is the capacitor voltage at the beginning of mode 2.   Mode 2 will end when the capacitor voltage equals the phase a counter emf [i.e., $V_{Nn} = v_a$ (v_a will be negative because $\alpha > 180°$)].

Mode 3.   The mode 3 will begin when current spills over to phase a due to excess capacitor voltage.   It is assumed that the thyristor Q_4 is fired by a pulse train when Q_B is fired.   The capacitor current will decrease and phase a current $-i_a$ will build up with the equation

$$L \frac{d(-i_a)}{dt} - v_a + v'_{C(0)} - \frac{1}{C} \int_0^t i_C \, dt = 0 \tag{5.86}$$

where $v'_{C(0)}$ is the capacitor voltage at the beginning of mode 3. Mode 3 (i.e., the commutation) will be complete when $-i_a = I_d$ and $i_c = 0$. At this point, the capacitor voltage will be equal but of opposite polarity in steady-state operation to the initial voltage V_{CO}. With this capacitor voltage, next Q_A can be fired to commutate Q_5.

Since the initial capacitor voltage has to overcome the phase voltage amplitude for successful commutation, the commutation circuit will be effective only within a range of α angles. One great demerit of the circuit is that at high frequency the machine torque is reduced substantially because dc link current is diverted through the motor neutral in mode 2. The circuit has been used successfully in starting synchronous motors.

5.5 TWELVE-STEP INVERTER

The torque pulsation and harmonic heating problems of a current-fed inverter can be substantially improved by connecting two bridge inverters in the 12-step mode using phase-shifting transformers as shown in Fig. 3.33. In a 12-step wave, the harmonic currents present are 11th, 13th, 23rd, 25th, and so on, and the lowest frequency of the pulsating torque is the 12th harmonic. The higher harmonic frequencies and their corresponding lower amplitudes reduce pulsating torque and harmonic heating effects considerably. However, to justify the additional complexity of 12-pulse operation the drive system power rating should be very large.

The circuit configuration of Fig. 3.33 can be directly used for a synchronous motor drive with load commutation, where the ac source is replaced by the three-phase machine. The upper and lower inverters operate symmetrically but with a 30° phase shift since the machine counter emfs at the inverter outputs are also shifted by 30°. A similar circuit configuration can be used to convert ac to dc for feeding the inverter. Of course, during regeneration, the circuits will exchange their functions. For an induction motor load, the inverters require forced commutation, and therefore the load-commutated inverters may be replaced by ASCI inverters, other circuit components remaining the same.

Twelve-pulse operation using the phase-shifting transformers becomes very expensive. The transformers can be eliminated by using a six-phase machine as shown in Fig. 5.23. The machine has an asymmetric six-phase winding in the sense that the xyz winding group is phase advanced by 30° with respect to the abc winding group. In a conventional or symmetric six-phase machine, the three-phase winding groups are displaced by 60°. The advantage of the asymmetric connection is that if the component three-phase windings are supplied by the inverters at a 30° phase shift, as shown in the figure, an equivalent 12-pulse operation of the machine results as far as the harmonic torque pulsation is concerned. The machine counter emfs are at 30° phase shift for each three-phase group, and therefore the firing pulse train of the lower bridge will be 30° phase shifted with respect to the upper bridge. The inverter input voltages V_{d_1} and V_{d_2} will be equal and their sum $V_d = V_{d_1} + V_{d_2}$ will have a 12-pulse ripple. The same circuit configuration can be used for a

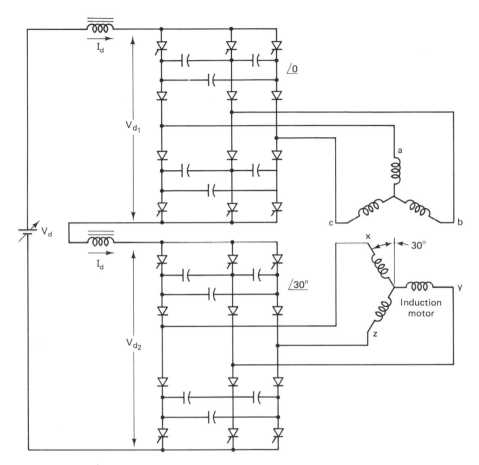

Figure 5.23 Dual six-pulse ASCI inverter for induction motor drive.

load-commutated inverter synchronous motor drive also except that the ASCI inverters are replaced by ordinary thyristor bridges.

5.6 INVERTER–MACHINE MODELING

The general discussion on modeling and simulation given in Section 4.7 is also applicable for a current-fed inverter drive and therefore will not be repeated here. In this section we will develop a simplified dynamic model of a six-step current-fed inverter induction motor drive in a synchronously rotating reference frame which will be adequate for system stability studies. Again, as before, the discrete-time effect of the inverter will be neglected and we will assume the inverter to be an ideal three-phase balanced current source. An equivalent stationary reference model can also be developed but will not be considered because of limited importance.

The motor line current waves can be expressed by the following Fourier series:

$$i_a = \frac{2\sqrt{3}}{\pi} I_d \left(\cos \omega_e t - \frac{1}{5} \cos 5\omega_e t + \frac{1}{7} \cos 7\omega_e t - \cdots \right) \tag{5.87}$$

$$i_b = \frac{2\sqrt{3}}{\pi} I_d \left[\cos(\omega_e t - 120°) - \frac{1}{5} \cos (5\omega_e t \right.$$
$$\left. + 120°) + \frac{1}{7} \cos (7\omega_e t - 120°) - \cdots \right] \tag{5.88}$$

$$i_c = \frac{2\sqrt{3}}{\pi} I_d \left[\cos(\omega_e t + 120°) - \frac{1}{5} \cos(5\omega_e t \right.$$
$$\left. - 120°) + \frac{1}{7} \cos(7\omega_e t + 120°) + \cdots \right] \tag{5.89}$$

The line currents can be converted into a d^s-q^s stationary reference frame by using the following relations:

$$i_{qs}^s = \frac{2}{3} i_a - \frac{1}{3} i_b - \frac{1}{3} i_c \tag{5.90}$$

$$i_{ds}^s = \frac{1}{\sqrt{3}} (i_c - i_b) \tag{5.91}$$

The resulting expressions are

$$i_{qs}^s = \frac{2\sqrt{3}}{\pi} I_d \left[\cos \omega_e t - \frac{1}{5} \cos 5\omega_e t + \frac{1}{7} \cos 7\omega_e t - \cdots \right] \tag{5.92}$$

$$i_{ds}^s = \frac{2\sqrt{3}}{\pi} I_d \left[-\sin \omega_e t - \frac{1}{5} \sin 5\omega_e t - \frac{1}{7} \sin 7\omega_e t - \cdots \right] \tag{5.93}$$

Equations (5.92) and (5.93) can be converted to the corresponding rotating frame relations as follows:

$$i_{qs} = i_{qs}^s \cos \omega_e t - i_{ds}^s \sin \omega_e t$$
$$= \frac{2\sqrt{3}}{\pi} I_d \left(1 - \frac{2}{35} \cos 6\omega_e t - \frac{2}{143} \cos 12\omega_e t - \cdots \right) \tag{5.94}$$

$$i_{ds} = i_{qs}^s \sin \omega_e t + i_{ds}^s \cos \omega_e t$$
$$= \frac{2\sqrt{3}}{\pi} I_d \left(-\frac{12}{35} \sin 6\omega_e t - \frac{24}{143} \sin 12\omega_e t - \cdots \right) \tag{5.95}$$

assuming that at time $t = 0$, the q^e axis and phase a axis are aligned. Equations (5.94) and (5.95) relate the dc link current I_d to the machine currents i_{qs} and i_{ds}, and the switching functions are expressed in Fourier series.

The relation between inverter dc voltage and machine voltages can be derived by the instantaneous power balance equation between the input and output of the inverter; that is,

$$V_d I_d = v_{an} i_a + v_{bn} i_b + v_{cn} i_c \tag{4.75}$$

The corresponding rotating frame relation is

$$V_d I_d = \frac{3}{2} (v_{ds} i_{ds} + v_{qs} i_{qs}) \tag{4.77}$$

Substituting equations (5.94) and (5.95) in (4.77), the inverter dc voltage can be related with the machine voltages as

$$V_d = \frac{3\sqrt{3}}{\pi} v_{qs} \left(1 - \frac{2}{35} \cos 6\omega_e t - \frac{2}{143} \cos 12\omega_e t - \cdots \right)$$

$$+ \frac{3\sqrt{3}}{\pi} v_{ds} \left(-\frac{12}{35} \sin 6\omega_e t - \frac{24}{143} \sin 12\omega_e t - \cdots \right) \tag{5.96}$$

The inverter switching functions can be defined as

$$G_{qs} = 1 - \frac{2}{35} \cos 6\omega_e t - \frac{2}{143} \cos 12\omega_e t - \cdots \tag{5.97}$$

$$G_{ds} = \frac{12}{35} \sin 6\omega_e t - \frac{24}{143} \sin 12\omega_e t - \cdots \tag{5.98}$$

so that

$$i_{qs} = \frac{2\sqrt{3}}{\pi} I_d G_{qs} \tag{5.99}$$

$$i_{ds} = \frac{2\sqrt{3}}{\pi} I_d G_{ds} \tag{5.100}$$

$$V_d = \frac{3\sqrt{3}}{\pi} (v_{qs} G_{qs} + v_{ds} G_{ds}) \tag{5.101}$$

We have considered so far that the inverter is fed by an ideal current source. The inverter in general will be supplied by a phase-controlled rectifier in the front end and the dc voltage will be adjusted by feedback control to establish the desired current I_d. There will be an inductance filter to smoothen the dc link current. The dc link equation can be given as

$$V_R = L_d \frac{dI_d}{dt} + I_d R_d + V_d \tag{5.102}$$

where L_d and R_d are filter parameters and V_R is the rectifier voltage. The equation can be modified by multiplying with constants as

$$\frac{\pi}{3\sqrt{3}} V_R = \left(\frac{\pi^2}{18} L_d \right) \frac{d}{dt} \left(\frac{2\sqrt{3}}{\pi} I_d \right)$$

$$+ \left(\frac{\pi^2}{18} R_d \right) \left(\frac{2\sqrt{3}}{\pi} I_d \right) + \frac{\pi}{3\sqrt{3}} V_R \tag{5.103}$$

or

$$V'_R = L'_d \frac{d}{dt} I'_d + I'_d R'_d + V'_d \qquad (4.88)$$

where the primed quantities replace the coefficient quantities as shown. In terms of the primed variables, equations (5.99), (5.100), and (5.101) can be written as

$$i_{qs} = I'_d G_{qs} \qquad (5.104)$$

$$i_{ds} = I'_d G_{ds} \qquad (5.105)$$

$$V'_d = v_{qs} G_{qs} + v_{ds} G_{ds} \qquad (5.106)$$

The equivalent circuit in a synchronously rotating reference frame in terms of these equations is shown in Fig. 5.24. The derivation of rotor speed ω_r using the following additional equations is also shown in the figure:

$$T_e = \frac{3}{2}\left(\frac{P}{2}\right)(\psi_{dr} i_{qs} - \psi_{qr} i_{ds}) \qquad (2.82)$$

$$T_e - T_L = \frac{2}{P} J \frac{d\omega_r}{dt} \qquad (2.78)$$

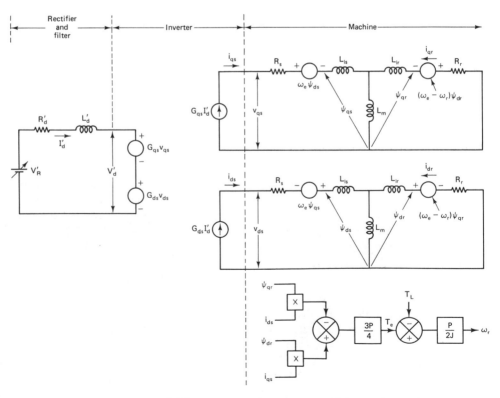

Figure 5.24 Current-fed inverter–induction motor model in synchronously rotating reference frame.

In the model, the parameters V'_R, ω_e, and T_L are the impressed variables.   It can be shown that the drive system stability is determined primarily by the fundamental components of variables.   The model then can be further simplified by neglecting the harmonics of line currents; that is,

$$G_{qs} = 1$$

$$G_{ds} = 0$$

$$i_{qs} = I'_d$$

$$i_{ds} = 0$$

$$V'_d = v_{qs}$$

In terms of these simplifications, the equivalent circuit can be further simplified. The dc link voltage V'_d is now directly connected to the q^e-axis equivalent circuit, and the d^e-axis equivalent circuit is open at the input.   This is because of the alignment of the q^e axis with the a axis at zero time.   Note that for the same reason, the input voltage $v_{ds} = 0$ in the voltage-fed inverter drive equivalent circuit shown in Fig. 4.21.   The simplified equivalent circuit can be modeled by the following set of equations:

$$\begin{bmatrix} V'_R \\ 0 \\ 0 \end{bmatrix} = \begin{bmatrix} R_s + R'_d + S(L'_d + L_s) & SL_m & \omega_e L_m \\ SL_m & R_r + SL_r & (\omega_e - \omega_r)L_r \\ -(\omega_e - \omega_r)L_m & -(\omega_e - \omega_r)L_r & R_r + SL_r \end{bmatrix} \begin{bmatrix} i_{qs} \\ i_{qr} \\ i_{ds} \end{bmatrix}$$

$$(5.107)$$

$$T_e = \frac{3}{2} \left(\frac{P}{2}\right) L_m (i_{qs} i_{dr} - i_{ds} i_{qr}) \tag{2.83}$$

$$T_e - T_L = \frac{2}{P} J \frac{d\omega_r}{dt} \tag{2.78}$$

where $L_s = L_m + L_{ls}$ and $L_r = L_m + L_{lr}$.   Equations (5.107), (2.83), and (2.78) can be the basis if small-signal analysis of the system is desired at steady-state operating points.

The equivalent circuit of Fig. 5.24 remains valid for a current-fed inverter irrespective of the nature of commutation.   The model of a synchronous machine with a load-commutated inverter remains the same expect that the induction machine model is replaced by the synchronous machine model.

5.7 CURRENT-FED VERSUS VOLTAGE-FED INVERTERS

We have discussed in the previous sections the performance characteristics of different types of current-fed inverters.   At this point it is useful to compare the features of current-fed inverters with those of voltage-fed inverters.   Such a comparison helps the designer to select the particular type of inverter depending on the application. Note that a voltage-fed inverter can be operated in the current control mode by adding a current control feedback loop. Similarly, a current-fed inverter can be operated in the voltage control mode, if desired, by adding a voltage control loop. But our definition of voltage-fed and current-fed inverters is on the basis of whether the dc supply is a voltage source or a current source. The two classes of inverters have duality in many aspects and the general discussion on the inverter–machine interface given in Chapter 4 holds true here also. A summary comparison of the two classes of inverters can be given as follows:

- In a current-fed inverter, the inverter is more interactive with the load, and therefore a close match between the inverter and machine is desirable. For example, the inverter likes to see a low leakage inductance, unlike that of voltage-fed inverter, because this parameter directly influences the inverter commutation process. A large leakage inductance of machine filters harmonics in a voltage-fed inverter, but in a current-fed inverter it lengthens the current transfer interval and worsens the voltage spike problem. The thyristors and diodes of the inverter may require series connection to combat the spike voltage, which may be several times the peak counter emf.

- Because of the load-dependent commutation process, the commutation is time consuming. As a result, the commutation angle of force-commutated inverters becomes large at light load and high frequency, and thereby restricts the highest frequency of operation.

- The current-fed inverter has inherent four-quadrant operation capability and does not require any extra power circuit component. On the other hand, a voltage-fed inverter requires a line-commutated inverter connected in inverse-parallel with the rectifier for regeneration. In case of 60-Hz power failure, the regenerated power in the current-fed inverter cannot be absorbed in the line and therefore the machine speed can be reduced only by a mechanical brake. For a voltage-fed inverter, however, the dynamic braking is applicable for line-power-failure conditions.

- A current-fed inverter is more rugged and reliable, and problems such as shoot-through fault do not exist. A momentary short circuit in load and misfiring of thyristors are acceptable. Fault interruption by gate circuit suppression is simple and straightforward.

- Inverter thyristors have to withstand reverse voltage during part of the cycle and therefore such devices as GTOs, power MOSs, and transistors cannot be used. Due to the large turn-off time, the thyristors can be the inexpensive rectifier-grade instead of the expensive inverter-grade (low turn-off time) thyristors required for voltage-fed inverters.

- The control of current-fed inverters, especially for commonly used ASCI and load-commutated inverters, is simple and similar to the phase-controlled line-commutated converters discussed in Chapter 3.

- Multi-machine load on a single inverter or multi-inverter load on a single rectifier is very difficult with current-fed inverters. An industrial drive usually consists of one rectifier, one inverter, and one machine system. In applications where multi-machine or multi-inverter capability is required, a voltage-fed inverter may prove very economical.

- Current-fed inverters have a sluggish dynamic performance compared to PWM voltage-fed inverters. The stability problem is more severe at light load and high-frequency conditions. On the other hand, stability problems are minimal in voltage-fed inverters and often the drive systems can be operated in open loop. The control of the drive systems is discussed in Chapters 7 and 8.

- The torque pulsation and harmonic heating problems are more severe in low-frequency operation. Because of the 120° conduction mode, PWM operation is not very efficient and can only be applied at very low frequency. Similarly, dc current modulation by pulsating torque control loop becomes difficult due to the large bandwidth requirement. The simplicity of the control principle is lost when such additional control requirements are considered.

- The successful operation of current-fed inverters requires that a minimum load should always remain connected. The inability to operate at no-load invalidates its application in general-purpose power supply applications such as UPS systems.

REFERENCES

1. T. Okuyama et al., "Effects of Machine Constants on Steady State and Transient Characteristics of Commutatorless Motors," *Conf. Rec. IEEE/IAS Annu. Meet.*, pp. 272–279, 1977.

2. B. Mueller, T. Spinanger, and D. Wallstein, "Static Variable Frequency Starting and Drive System for Large Synchronous Motors," *Conf. Rec. IEEE/IAS Annu. Meet*, pp. 429–438, 1979.

3. M. K. Parasuram and B. Ramaswami, "Analysis and Design of a Current-Fed Inverter," *Conf. Rec. 2nd IFAC Symp. Control in Power Electron. Electr. Drives*, pp. 235–245, 1977.

4. M. Showleh, W. A. Maslowski, and V. R. Stefanovic, "An Exact Modeling and Design of Current Source Inverters," *Conf. Rec. IEEE/IAS Annu. Meet.*, pp. 439–449, 1978.

5. L. H. Walker and P. M. Espelage, "A High Performance Controlled Current Inverter Drive," *IEEE Trans. Ind. Appl.*, Vol. IA-16, pp. 193–202, Mar–Apr. 1980.

6. M. B. Brennen, "A Comparative Analysis of Two Commutation Circuits for Adjustable Current Input Inverters Feeding Induction Motors," *Conf. Rec. IEEE Power Electr. Spec. Conf.*, pp. 201–212, 1973.

7. R. L. Steigerwald, "Characteristics of a Current-Fed Inverter with Commutation Applied through Load Neutral Point," *IEEE Trans. Ind. Appl.*, Vol. IA-15, pp. 538–553, Sept.–Oct. 1979.

8. E. P. Cornell and T. A. Lipo, "Modeling and Design of Controlled Current Induction Motor Drive System," *IEEE Trans. Ind. Appl.*, Vol. IA-13, pp. 321–330, July–Aug. 1977.

9. T. A. Lipo and L. H. Walker, "Design and Control Techniques for Extending High

Frequency Operation of a CSI Induction Motor Drive," *IEEE Trans. Ind. Appl.*, Vol. IA-19, pp. 744–753, 1983.

10. B. K. Bose and T. A. Lipo, "Control and Simulation of a Current-Fed Linear Inductor Machine," *IEEE Trans. Ind. Appl.*, Vol. IA-15, pp. 591–600, Nov.–Dec. 1979.

11. N. Sato, Y. Hayashi, and H. Umida, "A System of Induction Generator with Static Exciter and Paralleled to AC Power Lines," *Conf. Rec. IEEE/IAS Int. Sem. Power Conv. Conf.*, pp. 295–305, 1982.

12. P. M. Espelage and B. K. Bose, "High-Frequency Link Power Conversion," *IEEE Trans. Ind. Appl.*, Vol. IA-13, pp. 387–394, Sept.–Oct. 1977.

13. H. Akagi, Y. Kanazawa, and A. Nabae, "Instantaneous Reactive Power Compensators Comprising Switching Devices without Energy Storage Components," *IEEE Trans. Ind. Appl.*, Vol. IA-20, pp. 625–630, May–June 1984.

6

SLIP-POWER-CONTROLLED DRIVES

6.0 INTRODUCTION

So far we have discussed induction motor drives where the power is controlled in the stator circuit only. In such a drive system, the converters are to be designed to handle the full power flowing to the machine. Between the two classes of induction machines—squirrel cage and wound rotor—the former is always preferred, because the wound-rotor machine is more bulky and expensive and has the disadvantages of a dc machine due to the presence of slip rings and brushes. The wound-rotor machines had long been used for inexpensive speed control by mechanically varying the rotor circuit rheostats. One advantage of this type of machine is that slip power becomes available which can be controlled to control the speed of the machine. For limited-range speed-control applications, where the slip power is only a fraction of the total power of the machine, the converter cost reduction may be substantial. This advantage offsets the demerits of the wound-rotor machine to some extent, and the drive system becomes viable for large-horsepower pump and compressor-type applications, where the speed does not usually deviate more than 50% from the synchronous speed. Another advantage is that the slip power can be controlled to flow either out of the rotor or to the rotor, and thus speed can be controlled in both subsynchronous and supersynchronous regions with motoring and regeneration. Slip-power-controlled machines have been used successfully in VSCF (variable-speed, constant-frequency) systems, such as wind energy generators and shipboard power supplies, where the mechanical energy from a variable-speed shaft is converted to fixed-frequency (60 Hz), fixed-voltage power supply. It will be shown later that slip-power-controlled drives have dc-machine-like characteristics both in steady-state and dynamic conditions, and therefore control is simple with minimal stability problems.

In this chapter we discuss the principle and theory of slip power control, review slip power recovery schemes by static Kramer and static Scherbius drives, and discuss VSCF generation principles.

6.1 STATIC ROTOR RESISTANCE CONTROL

The speed of a wound-rotor machine can be varied by mechanically varying the rotor circuit resistance, as indicated earlier. This method of speed control is very inefficient because the slip energy is wasted in the rotor circuit resistance. However, the advantages are that high starting torque is available at low starting current and improved power factor is possible with a wide range of speed control. Instead of mechanically varying the resistance, the equivalent resistance in the rotor circuit can be varied statically by using a diode bridge rectifier and chopper as shown in Fig. 6.1. The stator of the machine is connected directly to the line power supply, but in the rotor circuit the slip voltage is rectified to dc by the bridge rectifier. The dc voltage is converted to a current source by connecting a large series inductor L_d and is then fed to a GTO chopper with an external shunt resistor R as shown. The chopper periodically connects and disconnects the resistance R. When the chopper is off, the resistance is connected in the circuit and dc link current I_d flows through it. On the other hand, if the chopper is on, the resistance is short circuited and the current I_d is bypassed. The chopper operates with a duty cycle $\delta = t_{on}/T$, where t_{on} is the on-time and T the time period. Apparently, the equivalent resistance between terminals A and B can be given by $R_0 = (1 - \delta)R$, where δ can be varied to vary the equivalent resistance. The derivation of this relation is given

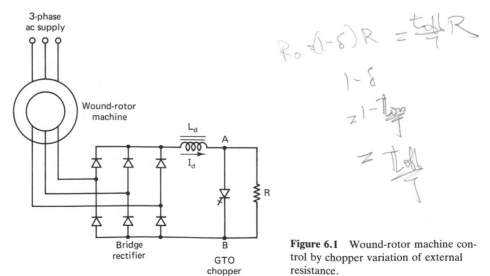

Figure 6.1 Wound-rotor machine control by chopper variation of external resistance.

later. Therefore, the speed or torque of the system can be varied by variation of duty cycle δ of the chopper.

DC Equivalent Circuit

To analyze the performance of the machine, the rotor circuit can be represented either by dc or by ac per phase equivalent circuit. We will attempt here to develop an approximate dc equivalent circuit, and an ac equivalent circuit will be derived in the next section. Figure 6.2(a) shows a hybrid equivalent circuit where the machine has been represented by a per phase equivalent circuit with respect to the rotor. For simplicity, the magnetizing branch is assumed to be connected at the input, where the slip voltage SV'_s is connected at the equivalent stator terminal.

The rotor current of the machine ideally has a six-stepped wave if the commutation effect of the rectifier is neglected and the dc link current I_d is assumed

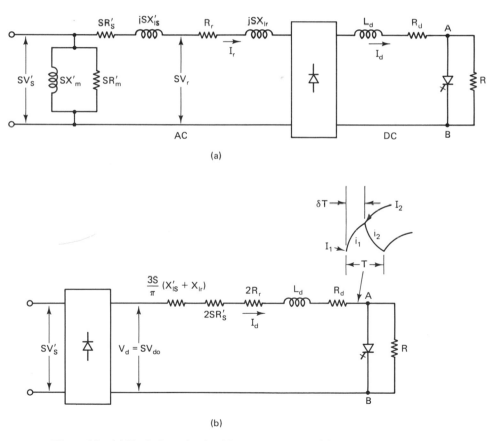

Figure 6.2 (a) Equivalent circuit with respect to rotor; (b) equivalent circuit with respect to dc link.

to be perfectly filtered. The rotor current, consisting of fundamental and distortion components, is reflected to the stator, and the fundamental component remains in phase with the rotor-induced voltage. The fundamental rms and total rms currents, respectively, are given by

$$I_{rf} = \frac{\sqrt{6}}{\pi} I_d \tag{6.1}$$

$$I_r = \frac{\sqrt{2}}{\sqrt{3}} I_d \tag{6.2}$$

The ac side resistances can be translated to the dc side by balancing the power loss expression, that is,

$$3I_r^2(SR_s' + R_r) = I_d^2 R_e \tag{6.3}$$

Substituting equation (6.2) yields

$$R_e = 2SR_s' + 2R_r \tag{6.4}$$

The presence of leakage inductances SX_{ls}' and SX_{lr} will cause commutation delay of the rectifier, and as a result the dc output voltage will droop with current as given by equation (3.44). The commutation distortion effect has of course been ignored in the rotor current wave. The expression for commutation voltage drop V_x' is given as

$$V_x' = \frac{3S}{\pi}(X_{ls}' + X_{lr})I_d$$
$$= R_e' I_d \tag{6.5}$$

where R_e' is the equivalent resistance reflected to the dc circuit. The dc equivalent circuit is shown in Fig. 6.2(b), where

$$V_d = 1.35(\sqrt{3}SV_s') = SV_{d0} \tag{6.6}$$

Torque Relation

Let us consider the steady-state operating condition of the system when the chopper is operating at duty cycle δ and time period T. During the on-time of the chopper, the dc link current i_1 will rise exponentially, which is given by the equation

$$L_d \frac{di_1}{dt} + R_1 i_1 = V_d \qquad \text{for } 0 < t \leq \delta T \tag{6.7}$$

where $R_1 = (3S/\pi)(X_{ls}' + X_{lr}) + 2SR_s' + 2R_r + R_d$. During the off-time of the chopper, the resistance R will be inserted in the circuit and the dc link current i_2 will fall exponentially and is given by the equation

$$L_d \frac{di_2}{dt} + R_2 i_2 = V_d \qquad \text{for } \delta T \leq t \leq T \tag{6.8}$$

where $R_2 = R_1 + R$. At steady-state operation, Δi_1 during time δT must be equal to Δi_2 during the time $(1 - \delta)T$, as indicated in Fig. 6.2(b). Equations (6.7) and (6.8) can be solved as

$$i_1 = I_{d_1} + (I_1 - I_{d_1}) \exp\left(\frac{-t}{\tau_1}\right) \qquad \text{for } 0 \le t \le \delta T \qquad (6.9)$$

$$i_2 = I_{d_2} + (I_2 - I_{d_2}) \exp\left(-\frac{t - \delta T}{\tau_2}\right) \qquad \text{for } \delta T \le t \le T \qquad (6.10)$$

where I_{d_1} and I_{d_2} are the short-circuit and open-circuit current, respectively, of the chopper given by

$$I_{d_1} = \frac{V_d}{R_1} \qquad (6.11)$$

$$I_{d_2} = \frac{V_d}{R_2} \qquad (6.12)$$

and τ_1 and τ_2 are the respective time constants, given by

$$\tau_1 = \frac{L_d}{R_1} \qquad (6.13)$$

$$\tau_2 = \frac{L_d}{R_2} \qquad (6.14)$$

The currents I_1 and I_2 are the crest and trough, respectively, of the current wave shown in Fig. 6.2(b).

In steady-state conditions,

$$i_1(\delta T) = i_2(\delta T) \qquad (6.15)$$

$$i_1(0) = i_2(T) \qquad (6.16)$$

These conditions can be substituted in equations (6.9) and (6.10) and simplified to derive expressions of I_1 and I_2 as

$$I_1 = \frac{I_{d_1}(1 - x)y + I_{d_2}(1 - x)}{1 - xy} \qquad (6.17)$$

$$I_2 = \frac{I_{d_1}(1 - x) + I_{d_2}(1 - x)y}{1 - xy} \qquad (6.18)$$

where

$$x = \exp\left(-\frac{\delta T}{\tau_1}\right) \qquad (6.19)$$

$$y = \exp\left[-\frac{(1 - \delta)T}{\tau_2}\right] \qquad (6.20)$$

The general expressions of average and rms values of the dc link current can be given as

$$I_d = \frac{1}{T} \left(\int_0^{\delta T} i_1 \, dt + \int_{\delta T}^{T} i_2 \, dt \right) \tag{6.21}$$

$$I_{rms} = \left[\frac{1}{T} \left(\int_0^{\delta T} i_1^2 \, dt + \int_{\delta T}^{T} i_2^2 \, dt \right) \right]^{1/2} \tag{6.22}$$

If the chopping frequency is high (i.e., the time period T is small compared to the time constants τ_1 and τ_2), x and y can be approximated as

$$x \simeq 1 - \frac{\delta T}{\tau_1} \tag{6.23}$$

$$y \simeq 1 - \frac{(1 - \delta)T}{\tau_2} \tag{6.24}$$

This means that I_1 and I_2 are very close to the mean current I_d. In practice, the chopper duty cycle δ and period T will be controlled to regulate the mean dc link current I_d, and I_1 and I_2 values will be limited within a very small hysteresis band. Substituting the approximate values of x and y in equations (6.17) and (6.18) and simplifying give us

$$I_d \simeq I_1 \simeq I_2 = \frac{I_{d_1} \delta \tau_2 + I_{d_2}(1 - \delta)\tau_1}{\delta \tau_2 + (1 - \delta)\tau_1} \tag{6.25}$$

Substituting equations (6.11) to (6.14) in (6.25), we have

$$I_d = \frac{V_d}{R_1 + (1 - \delta)R} \tag{6.26}$$

Equation (6.26) shows that the equivalent resistance reflected to the chopper input is $(1 - \delta)R$ as originally assumed. This equation indicates that the current I_d is a function of slip and chopper duty cycle for a given value of R. This means that at a constant motor speed, I_d can be regulated by controlling δ.

The rotor circuit copper loss of the machine can be given as

$$P_{lr} = V_d I_d - I_d^2 \frac{3S}{\pi} (X'_{ls} + X_{lr}) - I_{rms}^2(2SR'_s) \tag{6.27}$$

If peak-to-peak ripple $I_2 - I_1$ is neglected, then I_{rms} approaches the value of I_d. Then equation (6.27) is given as

$$P_{lr} = S\left\{ V_{d0}I_d - \left[\frac{3}{\pi}(X'_{ls} + X_{lr}) + 2 R'_S \right] I_d^2 \right\} \tag{6.28}$$

where V_d has been substituted for by SV_{d0}. The developed torque T_e at slip S can be given as

$$T_e = \frac{P_{cl}}{S\omega_e}$$

$$= \frac{1}{\omega_e} \left\{ V_{d0}I_d - \left[\frac{3}{\pi} (X_{ls}' + X_{lr}) + 2R_s' \right] I_d^2 \right\} \tag{6.29}$$

Equation (6.29) shows the relation between torque and dc link current and it is independent of speed. The torque becomes proportional to current I_d if contribution by the second term is neglected.

The torque can also be expressed as a function of slip and duty cycle δ by eliminating I_d from equation (6.29) with the help of equation (6.26). This gives

$$T_e = \frac{1}{\omega_e} \left\{ \frac{SV_{d0}^2}{R_1 + (1 - \delta)R} - \frac{S^2 V_{d0}^2 R_x}{[R_1 + (1 - \delta)R]^2} \right\} \tag{6.30}$$

where $R_x = (3/\pi) (X_{ls}' + X_{lr}) + 2R_s'$.

The torque–slip curves for different values of δ are shown in Fig. 6.3. The chopper control of resistance indeed reflects as a variable resistance in the per phase ac equivalent circuit, and therefore the torque–slip curves are similar to these with mechanically varying rotor rheostats. At $\delta = 1$, the rotor resistance is short circuited and as its value decreases, the equivalent rotor resistance increases and the torque curve becomes flatter as δ approaches zero. The torque curve at $\delta = 1$ appears flatter than that of a normal machine because of the additional equivalent resistance introduced by the diode and GTO drops and filter resistance R_d.

A desired torque or dc link current I_d can be obtained within a slip range of the machine as illustrated in Fig. 6.3 and can easily be explained from the equivalent circuit of Fig. 6.2(b). For $T_e = 1.0$ pu, the minimum and maximum slips corre-

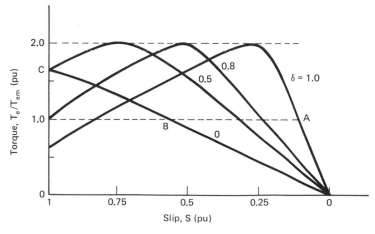

Figure 6.3 Torque–slip curves with chopper control.

spond to points A and B, respectively. The critical lower slip at $\delta = 1.0$ is determined by the adequate slip voltage V_d induced in the equivalent circuit. In the slip range between A and B, as the slip voltage increases the chopper equivalent resistance $(1 - \delta)R$ is controlled so as to maintain the constant current I_d. If the slip is increased beyond point B, the torque increases along curve BC as shown in the figure. The range AB can be increased by a higher value of resistance R.

The rectifier and chopper are to be designed to handle the current I_d and their voltage ratings depend on the desired speed range. This is with the assumption that the machine is not started with the chopper control, and a separate starting scheme is available so that the rectifier and chopper do not withstand the full voltage at unity slip. Therefore, for a limited-speed range application, the converter handles only a limited amount of power. This point is discussed further in the next section. The chopper method of speed control is not popular because of poor efficiency, as indicated earlier. However, the simple, low-cost converters and the simplicity of control are the attractive features. The scheme has been used in intermittent speed control applications where the efficiency penalty is not of great concern.

6.2 STATIC KRAMER DRIVE

Instead of wasting the slip power in the rotor circuit resistance, it can be converted to 60-Hz ac and pumped back to the line as shown in Fig. 6.4. This subsynchronous region speed control principle where the slip power is recovered back to the line through a converter cascade is known as a static Kramer drive. The original Kramer drive system used a rotory converter instead of a diode rectifier and fed power to a dc motor coupled to the same induction machine shaft. The slip power

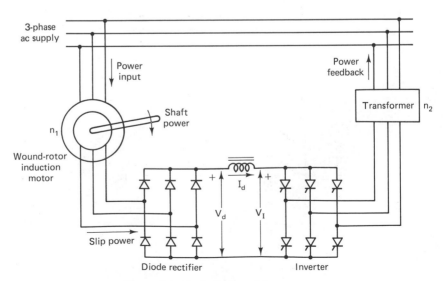

Figure 6.4   Static Kramer drive system.

in this principle is converted to mechanical power, which contributes partially to the mechanical power output of the induction machine shaft. The static Kramer system is popular in large power pump and compressor-type drives where the range of speed variation is usually limited. The drive system is not only efficient but the converter power rating is low, because it has to handle only the slip power. This power rating becomes lower for a more restricted speed range near the synchronous speed. The additional advantages are that the drive system has dc-machine-like characteristics and the control circuit is simple. These advantages offset to some extent the disadvantages of the wound-rotor machine and the poor power factor characteristics, discussed later.

The air gap flux of the machine is established by the stator supply and it remains practically constant if stator drops and supply voltage fluctuation are neglected. Ideally, the rotor current is a six-stepped wave in phase with the rotor phase voltage if the dc link current I_d is considered harmonic free, and commutation overlap angle of the diode rectifier is neglected. Thus the machine torque becomes directly proportional to the fundamental component of rotor current. Instead of static resistance control as discussed in the preceding section, the scheme here can be considered as counter emf control where a variable counter emf V_I is being presented by a phase-controlled line-commutated inverter. In steady-state operation, the rectified slip voltage V_d and the inverter voltage V_I will balance for a certain dc current I_d. The voltage V_d will be proportional to slip and the current I_d will be proportional to torque. The simplified speed and torque expressions can be derived as follows. Neglecting the stator and rotor drops, the voltage V_d is given by

$$V_d = \frac{1.35}{n_1} S V_L \tag{6.31}$$

where n_1 is the stator-to-rotor turns ratio of the machine, V_L the stator line voltage, and S the per unit slip. Again, the inverter terminal voltage V_I is given as

$$V_I = \frac{1.35}{n_2} V_L \,|\cos \alpha\,| \tag{6.32}$$

where n_2 is the transformer line side-to-inverter ac side turns ratio and α is the inverter firing angle, which is in the range $90°$ to $180°$. Since V_d and V_I must balance in the ideal case, equations (6.31) and (6.32) give

$$S = \frac{n_1}{n_2} \,|\cos \alpha\,| \tag{6.33}$$

that is,

$$\omega_r = \omega_e \,(1 - |\cos \alpha\,|) \tag{6.34}$$

assuming that $n_1/n_2 = 1$. Equation (6.34) indicates that ideally speed can be controlled between zero and synchronous speed ω_e by controlling the inverter firing angle α. At zero speed, the voltage V_d is maximum, which corresponds to angle $\alpha = 180°$, and at synchronous speed $V_d = 0$ when $\alpha = 90°$.

Again neglecting losses, the following power equations can be written:

$$SP_g = V_I I_d \tag{6.35}$$

$$P_m = (1 - S)P_g = T_e \omega_r = T_e \omega_e (1 - S) \tag{6.36}$$

where P_g is the air gap power and P_m the mechanical output power. Combining the equations above yields

$$T_e = \frac{V_I I_d}{S \omega_e} \tag{6.37}$$

Substituting equations (6.32) and (6.33) gives

$$T_e = \frac{1.35 V_L}{\omega_e n_1} I_d \tag{6.38}$$

which indicates that the torque is proportional to current I_d. The drive system has nearly the characteristics of a separately excited dc motor. The air gap flux is constant and the torque is proportional to current I_d. With a higher load torque, I_d will increase and for a fixed V_I, V_d should slightly increase to overcome the dc link drop, indicating a speed droop like a dc machine. A more accurate torque–speed relation will be derived later.

Phasor Diagram

A fundamental-frequency phasor diagram can be drawn to explain the performance of the drive system. The rotor current displacement factor will deviate from unity in practice because of the commutation overlap angle shown in Fig. 6.5. The overlap angle μ will introduce a lagging angle ϕ_r to the fundamental current. This angle will increase as the current I_d increases and slip S decreases. In fact, near zero slip, when the rotor voltage is very small, the large current I_d may cause overlap angle μ to exceed 60°, causing a short circuit between the upper and lower diodes.

Figure 6.6 shows the approximate phasor diagram of the drive system at the rated torque condition, where all the phasors are referred to the line or stator side. The stator draws a magnetizing current I_m which is lagging 90° with respect to the stator phase voltage V_s. The rotor fundamental current I_r is reflected to the stator as I_r' which lags the voltage V_s. The total stator current I_s lags the stator voltage by angle ϕ_s as shown.

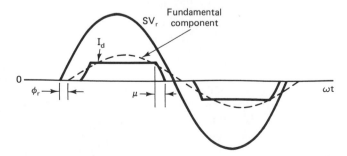

Figure 6.5 Rotor phase voltage and current waves.

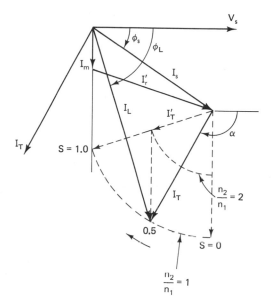

Figure 6.6 Phasor diagram of static Kramer system at rated torque.

On the inverter side, although the active power is fed back to the line, it will also consume reactive power because of phase control. This additional reactive power demand by the inverter reduces the overall power factor of the system. Assuming continuous conduction of inverter and ripple-free current I_d, the inverter input power factor is $|\cos \alpha|$ (i.e., the power factor varies linearly with dc voltage V_I). This, of course, neglects the inverter commutation overlap effect. The phasor diagram shows the inverter line current I_T at slip $S = 0.5$ when no transformer is used. The phasors I_T and I'_r are nearly equal in magnitude because of the nearly identical waveshape of the currents. The active component $I_T \cos \alpha$ opposes the stator active current, whereas the reactive component $I_T \sin \alpha$ adds to the magnetizing current. The total line current I_L is the phasor sum of I_s and I_T and it lags at angle ϕ_L, which is larger than the stator power factor angle ϕ_s. With constant torque, the magnitude of I_T remains constant, but as slip varies between 0 to 1, the phasor I_I rotates from $\alpha = 90$ to $160°$ as shown in Fig. 6.6. At zero speed, the machine acts as a transformer and all the active power is transferred back to the line through the inverter. The result is that both the machine and inverter consume only reactive power. The margin angle of $20°$ for the inverter provides commutation and turn-off angles. From the phasor diagram it will be evident that at $S = 0$ the system power factor will be lagging and it will deteriorate as the slip increases. Therefore, to avoid very poor power factor, the speed range should be restricted close to the synchronous speed. The phasor diagram can easily be modified for different torque conditions and is left as an exercise for the reader. For a restricted speed range, the system power factor can be further improved by using a transformer. The transformer turns ratio can be adjusted so that at maximum slip $\alpha \simeq 180°$. Substituting this in equation (6.33) gives us

$$S_{\max} = \frac{n_1}{n_2} \tag{6.39}$$

For example, if $S_{max} = 0.5$ and $n_1 = 1$, then n_2 should be 2 ideally. In a phasor diagram, this condition corresponds to the transformer line current phasor I'_T, which clearly indicates the power factor improvement of line current I_L. As the speed is increased to change the slip from 0.5 to 0, the phasor I'_T describes a concentric arc as shown.

The transformer also helps in reducing the power rating of the converters. Both the rectifier and inverter are to be designed to handle the same current I_d as determined by the torque requirement. The rectifier is to be designed for the slip voltage given by SV_L/n_1, whereas the inverter is to be designed for the line voltage V_L in the absence of the transformer. The rectifier voltage or power rating goes down with the smaller speed range, but the inverter has to be designed for full power. Installation of a transformer reduces the voltage or power rating of the inverter and the criteria for turns ratio n_2 design is the same as that of equation (6.39). For the same example (i.e., $S_{max} = 0.5$, $n_1 = 1$, and $n_2 = 2$) both the rectifier and the inverter have an equal power rating, which is 50% of the full power. It can be shown that the converter power rating is reduced proportionally as S_{max} is reduced. This is one of the important merits of the slip power recovery scheme as mentioned earlier. The discussion above assumes that the machine is not started with the converters in the circuit.

A typical starting method with resistor switching is shown in Fig. 6.7. The motor is started with switch 1 on and switches 2 and 3 off. As the speed builds up, the resistances R_1 and R_2 are shorted sequentially until at the designed S_{max} value switch 1 is opened and the controller is brought into operation.

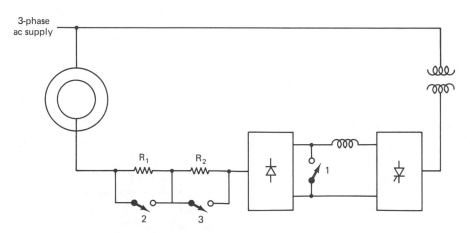

Figure 6.7 Motor starting method.

AC Equivalent Circuit

The drive system performances can be analyzed with the help of dc as well as ac equivalent circuits. The dc equivalent circuit described in the preceding section can easily be extended for the static Kramer drive. We will attempt, however, an approximate ac equivalent circuit with respect to the rotor.

Neglecting drops in semiconductor devices, the slip output power is partly dissipated in the dc link inductor and is partly fed back to the ac line. The respective power components can be given as

$$P_1 = I_d^2 R_d \tag{6.40}$$

$$P_f = \frac{1.35 V_L I_d}{n_2} \mid \cos \alpha \mid \tag{6.41}$$

The equivalent rotor circuit power per phase is given as

$$P' = P_1' + P_f' = \frac{1}{3} I_d^2 R_d + \frac{1}{3} \frac{1.35 V_L I_d}{n_2} \mid \cos \alpha \mid \tag{6.42}$$

Therefore, the machine air gap power per phase, which includes the rotor copper loss, is given as

$$P_g = I_r^2 R_r + P' + P_m \tag{6.43}$$

where P_m is the mechanical output per phase. The torque and the corresponding mechanical power P_m can be assumed to be contributed by the fundamental component of rotor current I_{rf} only. The expression for rotor circuit copper loss per phase is

$$\begin{aligned} P_{rl} &= I_r^2 R_r + \frac{1}{3} I_d^2 R_d \\ &= I_r^2 (R_r + 0.5 R_d) \end{aligned} \tag{6.44}$$

where $I_d = \sqrt{3}/\sqrt{2}\, I_r$ has been substituted. Therefore, the expression for P_m can be given as

$$\begin{aligned} P_m &= (\text{slip power with } I_{rf}) \frac{1 - S}{S} \\ &= \left[I_{rf}^2 (R_r + 0.5 R_d) + \frac{\pi}{3\sqrt{6}} \frac{1.35 V_L}{n_2} I_{rf} \mid \cos \alpha \mid \right] \left(\frac{1 - S}{S} \right) \end{aligned} \tag{6.45}$$

In equation (6.45), the current I_d in the second term has been replaced by I_{rf} from equation (6.1). The air gap power in equation (6.43) can then be written by substituting equation (6.45) as

$$\begin{aligned} P_g &= (I_r^2 - I_{rf}^2)(R_r + 0.5 R_d) \\ &\quad + \frac{1}{S} \left[I_{rf}^2 (R_r + 0.5 R_d) + \frac{\pi}{3\sqrt{6}} \frac{1.35 V_L}{n_2} I_{rf} \mid \cos \alpha \mid \right] \tag{6.46} \\ &= I_{rf}^2 R_X + I_{rf}^2 \frac{R_A}{S} \end{aligned}$$

where

$$R_X = \left(\frac{\pi^2}{9} - 1 \right)(R_r + 0.5 R_d) \tag{6.47}$$

$$R_A = (R_r + 0.5R_d) + \frac{\pi}{3\sqrt{6}} \frac{1.35V_L}{n_2 I_{rf}} |\cos \alpha|$$

$$= (R_r + 0.5R_d) + \frac{V_s}{n_2 I_{rf}} |\cos \alpha| \tag{6.48}$$

Equation (6.46) indicates that the rotor circuit which absorbs the active power can be represented by a per phase passive ac equivalent circuit where R_A is the equivalent resistance given by equation (6.48). The resistance R_X represents an additional resistance which consumes harmonic power. It is more convenient to represent the equivalent circuit in terms of counter emf presented by the inverter. Equation (6.46) can also be written in the form

$$P_g = I_{rf}^2(R_X + R_B) + V_C I_{rf} \tag{6.49}$$

where

$$R_B = \frac{1}{S}(R_r + 0.5R_d) \tag{6.50}$$

$$V_C = \frac{1}{S}\frac{V_s}{n_2}|\cos \alpha| \tag{6.51}$$

A complete rotor-referred per phase equivalent circuit is shown in Fig. 6.8, which includes leakage reactances and stator resistance. The equivalent circuit indicates that the speed and torque of the machine can be controlled by controlling the counter emf with the α angle.

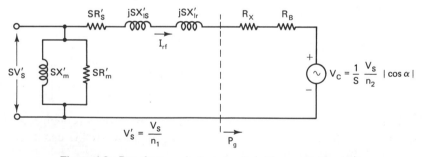

Figure 6.8   Per phase equivalent circuit (with respect to rotor).

Torque Expression

The average torque developed by the machine is given by the fundamental air gap power divided by the synchronous speed. The expression in terms of passive ac equivalent circuit is

$$T_e = \frac{1}{\omega_e}\left(3I_{rf}^2\frac{R_A}{S}\right) \tag{6.52}$$

where R_A is given by equation (6.48). The equation can be solved in terms of circuit parameters as

$$T_e = \frac{3SV_s'^2 R_A}{\omega_e[(SR_s' + R_A)^2 + (SX_{ls}' + SX_{lr})^2]}$$ (6.53)

where

$$I_{rf} = \frac{SV_s'}{\sqrt{(SR_s' + R_A)^2 + (SX_{ls}' + SX_{lr})^2}}$$ (6.54)

has been substituted. Equation (6.53) relates torque as a function of slip, rotor current, and inverter firing angle. An approximate torque equation relating slip and α angle can be derived more conveniently from the equivalent circuit of Fig. 6.8. The torque in terms of fundamental air gap power is given from equation (6.49) as

$$T_e = \frac{3}{\omega_e} (V_C I_{rf} + I_{rf}^2 R_B)$$ (6.55)

The current I_{rf} can be solved from the equivalent circuit by neglecting the reactances and the stator resistance, which are small at a small value of slip. Therefore,

$$I_{rf} \simeq \frac{SV_s' - V_C}{R_B} = \frac{SV_s/n_1 - (V_s/Sn_2) \mid \cos \alpha \mid}{R_B}$$ (6.56)

Substituting equations (6.56) and (6.51) in (6.55) yields

$$T_e \simeq \frac{3V_s^2}{\omega_e R_r'} \left[\frac{1}{Sn_2} \mid \cos \alpha \mid \left(\frac{S^2}{n_1} - \frac{\mid \cos \alpha \mid}{n_2} \right) + S \left(\frac{S}{n_1} - \frac{\mid \cos \alpha \mid}{Sn_2} \right)^2 \right]$$ (6.57)

Equation (6.57) relates torque as a function of slip and α angle approximately. A more accurate torque expression has been solved numerically from the equivalent circuit and plotted in Fig. 6.9.

Harmonics

The rectification of slip power causes harmonic currents in the rotor, and these harmonics are reflected to the stator by transformer action. The harmonic currents are also injected to the ac line by the inverter. As a result, the machine losses are increased and harmonic torques are produced. The rotor harmonic current components can be determined by Fourier analysis of the rotor current wave. The rotor current has a six-stepped wave shape which is given by the Fourier series

$$i_r = \frac{2\sqrt{3}}{\pi} I_d \left(\cos \omega_{sl} t - \frac{1}{5} \cos 5\omega_{sl} t + \frac{1}{7} \cos 7\omega_{sl} t \right.$$

$$\left. - \frac{1}{11} \cos 11 \omega_{sl} t + \cdots \right)$$ (6.58)

The lower harmonic components, such as the 5th and 7th, will have a dominating

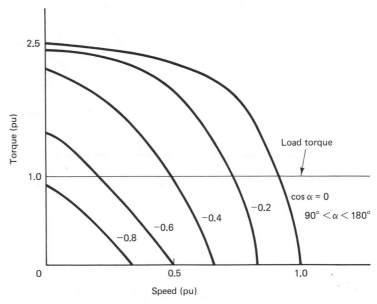

Figure 6.9 Torque–speed curves at different firing angles of inverter.

effect.   Each harmonic component in the rotor will create a rotating magnetic field
and its direction of rotation will depend on the order of the harmonic.   The 5th
harmonic current, for example, at frequency $5S\omega_e$ will rotate opposite to the direction of rotor, whereas the 7th harmonic at frequency $7S\omega_e$ will rotate in the
same direction.   Let us consider only the effect of the 5th harmonic.   Since the
rotor is moving at a speed $\omega_r = \omega_e(1 - S)$ in a direction opposite to that of the
rotating harmonic field, the harmonic current induced in the stator circuit will have
a frequency $\omega_e (1 - 6S)$.   The stator harmonic current will circulate through stator
reactance, resistance, and any other impedance in the stator circuit.   The stator
harmonic current equation can be given as

$$I_5 = \frac{V_{r5}(1 - 6S)}{R'_s + j(1 - 6S)X'_{ls}} \tag{6.59}$$

where I_5 is the harmonic current at frequency $\omega_e(1 - 6S)$ injected into the stator
by the 5th harmonic rotor current, and $V_{r5}(1 - 6S)$ is the corresponding stator
induced voltage.   In equation (6.59), it is convenient to consider the rotor as a
primary which is injecting voltage at slip $(1 - 6S)$ to the stator circuit.   Equation
(6.59) can be written as

$$I_5 = \frac{5SV_{r5}}{[5S/(1 - 6S)]\,R'_s + j5SX'_{ls}} \tag{6.60}$$

$$= \frac{5SV_{r5}}{(R'_s /S_5) + j5SX'_{ls}} \tag{6.61}$$

where

$$S_5 = \frac{1 - 6S}{5S} = \frac{\text{stator harmonic frequency}}{\text{rotor harmonic frequency}} \qquad (6.62)$$

is defined as the 5th harmonic slip. Equation (6.61) has the structure of the rotor current equation of a conventional machine. A per phase harmonic equivalent circuit can now be drawn with respect to the rotor and this is shown in Fig. 6.10. The equivalent circuit is valid only for the 5th harmonic but can be modified easily for the 7th harmonic. Actually, the rotor-injected harmonic current will be distributed in parallel between the stator and the magnetic circuit. Therefore,

$$
\begin{aligned}
I_5 &= I_{r5} \frac{5SX'_m}{\sqrt{(5SX'_m + 5SX'_{ls})^2 + \left(\dfrac{5S}{1 - 6S} R'_s\right)^2}} \\[2mm]
&= \frac{I_{rf}}{5} \frac{X'_m}{\sqrt{(X'_m + X'_{ls})^2 + \left(\dfrac{R'_s}{1 - 6S}\right)^2}}
\end{aligned}
\qquad (6.63)
$$

where I_{rf} is the fundamental rotor current. The current I_5 is maximum at unity slip, zero at slip $S = \frac{1}{6}$, and increases as the slip is further reduced. Equation (6.63) indicates that the harmonic current in the stator is smaller than that of the rotor, and the frequency, which is unrelated to the stator frequency, creates a beating effect.

A 5th-harmonic-related pulsating torque can be derived by considering the harmonic power injected to the stator. The equations are

$$P_{g5} = I_5^2 \frac{R'_s}{S_5} \qquad (6.64)$$

$$T_5 = \frac{P_{g5}}{5S\omega_e} \qquad (6.65)$$

where $5S\omega_e$ is the synchronous speed of the harmonic field and S_5 is the harmonic

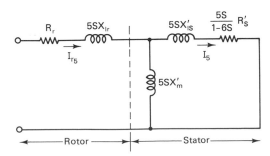

Figure 6.10   Harmonic equivalent circuit (with respect to rotor).

slip.   Combining equations (6.64), (6.65), and (6.63) yields

$$T_5 = \left(\frac{I_{rf}}{5}\right)^2 \frac{X_m'^2}{(X_m' + X_{ls}')^2 + \left(\dfrac{R_s'}{1-6S}\right)^2} \frac{R_s'}{(1-6S)\omega_e} \tag{6.66}$$

The equation gives the harmonic torque as a function of slip.   It can be shown that harmonic torque is small compared to the average torque and therefore can be neglected for practical purposes.

Power Factor Improvement

Various methods have been suggested for power factor improvement of the static Kramer system.   The phase-controlled current-fed inverter can be replaced by a voltage-fed PWM inverter which can operate at unity displacement factor and generate a small amount of line harmonics.   In fact, the inverter can be programmed to operate at leading power factor to compensate for the lagging VAR at the stator input.   Of course, a PWM inverter is more expensive and less reliable than a phase-controlled converter.

Another scheme for power factor improvement for very high-power applications is known as the commutatorless Kramer system, shown in Fig. 6.11.   The scheme is somewhat analogous to the conventional Kramer system, where the dc motor coupled to the induction machine shaft is replaced by a synchronous motor with a load-commutated inverter.   Such a machine–inverter system can be controlled to behave like a dc machine and is discussed in Chapter 8.   The power flow diagram in the commutatorless Kramer system is shown in Fig. 6.12.   The air gap power P_g flowing from the stator is split into shaft input power and slip power as in the static Kramer system.   But the slip power drives a synchronous motor and adds to the shaft input to constitute the total mechanical power.   The synchronous motor field is supplied from the line through a controlled rectifier.   The speed and

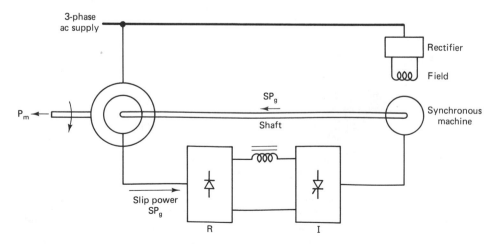

Figure 6.11 Commutatorless Kramer drive system.

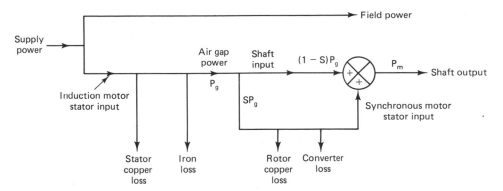

Figure 6.12 Power flow diagram in commutatorless Kramer drive system.

torque of the drive system are controlled by the inverter α angle and field current, so that the load commutation of the inverter is possible at an optimum α angle in a different speed.   As a characteristic of the load-commutated inverter drive, speed control is not possible at a low value because of insufficient counter emf.   Besides having improved power factor, the system will operate reliably with momentary line voltage dip, which will cause commutation failure in the static Kramer system.

As discussed above, the static Kramer systems do not have the regenerative mode of operation.   This feature requires that the slip power in the rotor should flow in the reverse direction, as explained in the next section.   If the diode rectifier is replaced by a thyristor bridge, the slip power flow can be controlled in either direction.   Such a static Kramer system with bidirectional slip power flow can be controlled for motoring and regeneration in both subsynchronous and supersynchronous ranges of speed.   The line commutation of the machine side converter is difficult near synchronous speed when the ac voltage is very small.   This problem can be solved by using a force-commutated ASCI-type converter.   Another limitation of the Kramer system is that speed reversal is not possible.   Speed reversal is possible by installing a phase sequence reversing contactor on the stator side.   This, of course, requires the drive system to go through zero speed, where the converters are not designed to operate.

6.3 STATIC SCHERBIUS DRIVE

The dual-converter system in a static Kramer drive can be replaced by a single phase-controlled line-commutated cycloconverter, as shown in Fig. 6.13.   The scheme is known as a static Scherbius drive and has found applications in very large horsepower pump and blower-type drives.   The cycloconverter permits the slip power to flow in either direction, and therefore the machine speed can be controlled in both subsynchronous and supersynchronous ranges with motoring and regeneration features.   The various modes of operation shown in Fig. 6.14 can be explained as follows.   It is assumed that the machine shaft torque is constant and that losses in the machine and cycloconverter are negligible.

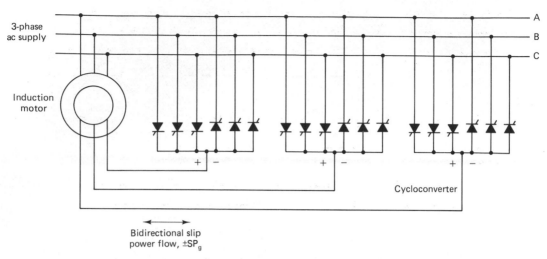

Figure 6.13 Static Scherbius drive system.

Mode 1: Subsynchronous Motoring. This mode is identical to that of the static Kramer system. The stator input or air gap power P_g remains constant and the slip power SP_g, which is proportional to the slip, is returned back to the line. Therefore, the line supplies the net mechanical power P_m consumed by the shaft. The slip frequency power in the rotor creates a rotating field in the same direction as in the stator and the rotor speed corresponds to the difference ($\omega_r = \omega_e - \omega_{sl}$) between these two frequencies. At true synchronous speed ($S = 0$), the cycloconverter supplies dc excitation to the rotor and the machine behaves like a synchronous motor.

Mode 2: Supersynchronous Motoring. As the shaft speed increases beyond the synchronous speed, the slip becomes negative and the slip power is absorbed by the rotor. The slip power supplements the stator power for total mechanical power output. The line therefore supplies slip power in addition to the stator power input. In this condition, the phase sequence of the slip voltage is reverse, so that the slip-frequency-induced rotating field is opposite to that of the stator.

Mode 3: Subsynchronous Regeneration. In a regenerative braking condition, the shaft is driven by the load and the mechanical energy is converted into electrical energy. With constant negative shaft torque, the mechanical power input to the shaft increases with speed and this equals the electrical power fed to the line. In the subsynchronous speed range, the slip power is fed to the rotor so that the total stator power output is constant. The slip voltage has a positive phase sequence (i.e., the direction of magnetic field is the same as that of stator field). At synchronous speed, the cycloconverter supplies dc excitation current to the rotor and the machine behaves as a synchronous generator.

Mode 4: Supersynchronous Regeneration. In this mode, the stator output power remains constant but the additional mechanical power input is reflected as slip power output.   The cycloconverter phase sequence is now reversed so that the rotor field rotates in the opposite direction.

The use of a cycloconverter instead of a dual converter system means additional cost and complexity, but the resulting advantages are obvious.   The problem of commutation near synchronous speed disappears and the near-sinusoidal current wave in the rotor substantially improves harmonic heating and torque pulsation effects.   The line current waveform is improved correspondingly.   The cycloconverter is to be controlled so that its output frequency tracks precisely with the slip frequency.

A step-up transformer can be connected to the line side of the cycloconverter with an appropriate voltage ratio for the desired range of speed control, as discussed

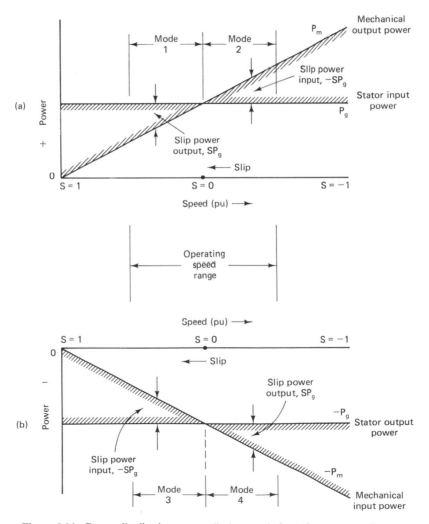

Figure 6.14   Power distribution versus slip in super/subsynchronous speed range; (a) motoring at constant torque; (b) generation at constant torque.

earlier. This reduces the voltage rating of the thyristors and the line power factor is improved. A separate starting scheme with resistance switching can be used. The drive system can be designed to operate within a fractional slip range about the synchronous speed. For example, if the operating speed range is $\pm 50\%$ about the synchronous speed is indicated in Fig. 6.14, the cycloconverter is to be designed to handle only 50% of the machine power rating. This does not, of course, consider the additional reactive power requirement of the cycloconverter. The Scherbius drive system does not permit speed reversal and requires a reversing contactor in the stator side for this function.

6.4 VSCF GENERATION

Variable-speed, constant-frequency generation involves generation of electrical power at fixed frequency and fixed voltage from a variable-speed prime mover coupled to the machine shaft. This type of generator is used on aircraft and naval ships where service power is generated from the propulsion turbine moving at a variable speed. Both synchronous and induction machines can be used for VSCF generation. In general, an adjustable-speed ac drive system which has a regeneration capability can be used as a VSCF generator. The various VSCF generation schemes can be summarized as follows:

- A variable-speed synchronous generator with variable-frequency output can be converted to fixed-frequency, fixed-voltage output through a cycloconverter. The cycloconverter can be replaced by a rectifier–inverter system, if desired.

- A squirrel-cage induction machine can operate as a VSCF generator if stator power is fed to a 60-Hz bus through a cycloconverter. The cycloconverter is commutated from the line side and its input frequency tracks the rotor speed so that the rotor operates at supersynchronous speed with a small slip. The small slip power output SP_g induced to the rotor is dissipated in the rotor circuit. The lagging VAR excitation requirement of the machine is supplied by the cycloconverter at stator frequency. The cycloconverter can be replaced by a rectifier–inverter system. The rectifier has to be force commutated so that the motor lagging excitation VAR can be supplied. If the output power is not fed to a 60-Hz bus, the inverter also requires forced commutation.

- In the scheme above, the rectifier can be a simple diode bridge. Then a separate VAR generator has to be connected at the stator terminal.

- A wound-rotor induction machine in the configuration of the static Scherbius drive system can be used.

- The static Scherbius system can be modified by feeding the slip power to a shaft-mounted synchronous machine as shown in Fig. 6.15. The scheme has some analogy with the commutatorless Kramer system shown in Fig. 6.11.

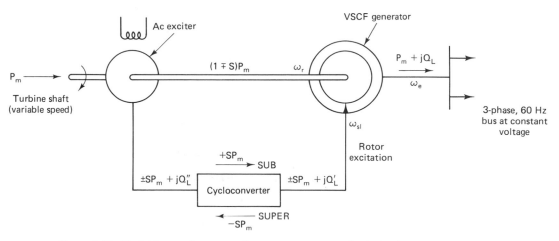

Figure 6.15 Variable-speed, constant-frequency generator using induction machine and separate exciter.

Modified Scherbius System

The modified Scherbius system of Fig. 6.15 has some special features and requires further discussion. The stator output power of the machine is connected to the independent 60-Hz constant-voltage bus which supplies the load. The distribution of active and reactive powers in supersynchronous and subsynchronous ranges is shown in the figure. The stator active power output P_m is equal to the turbine shaft power input neglecting the system losses. The stator reactive power output Q_L is reflected to the rotor as SQ_L, which is added by the machine magnetizing power requirement to constitute the total reactive power Q'_L of the cycloconverter. The power Q'_L is further increased to Q''_L at the cycloconverter input, which is supplied by the shaft-mounted synchronous exciter. The slip frequency magnitude and its phase sequence are adjusted with a different shaft speed so that the resultant air gap flux rotates at synchronous speed, as explained earlier. In the subsynchronous speed range, the slip power SP_m is supplied to the rotor by the exciter and therefore the remaining output power $(1 - S)P_m$ is supplied by the shaft. In the supersynchronous range, the rotor output power SP_m runs the exciter as a synchronous motor and therefore the total shaft power increases to $(1 + S)P_m$. The rotor voltage varies linearly with deviation from synchronous speed as shown in Fig. 6.16. This assumes negligible resistance and leakage inductance of the rotor winding so that the dc voltage absorbed by the rotor at synchronous speed is zero. The rotor reactive current decreases with speed because the magnetizing component of the current decreases at higher speeds. The exciter frequency increases linearly with speed as shown. For example, if the shaft speed varies in the range 800 to 1600 rpm, with 1200 rpm as the synchronous speed ($S = \pm 0.33$), the corresponding range of slip frequency is 0 to 20 Hz. Assuming a minimum frequency ratio of 4, the exciter frequency varies from 80 to 160 Hz, as shown.

The modified Scherbius system as a VSCF generator has several advantages

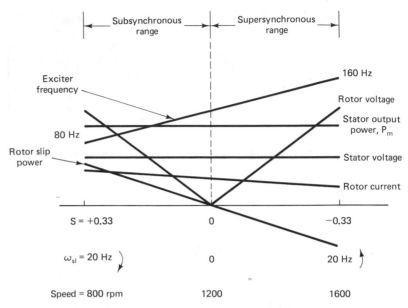

Figure 6.16 Idealized performance curves of VSCF generator at rated stator power.

over the conventional Scherbius system. One principal difference here is that the wraparound or circulating kVA demanded by the rotor is supplied from a separate exciter instead of being supplied from the machine stator terminals. As a result, the main machine is much smaller in size, although in this case, the power is distributed between the two machines. The VSCF bus is much cleaner with regard to harmonics, because the cycloconverter input harmonics are reflected to the exciter. The rotor excitation circuit can be designed with a higher voltage and the necessity of an input transformer is eliminated. In case of a supply brown-out or temporary short-circuit fault, the system has improved controllability and reliability of power supply than those of the standard Scherbius system.

REFERENCES

1. P. C. Sen and K. H. J. Ma, "Rotor Chopper Control for Induction Motor Drive: TRC Strategy," *IEEE Trans. Ind. Appl.*, Vol. IA-11, pp. 43–49, Jan.–Feb. 1975.
2. A. Lavi and R. J. Polge, "Induction Motor Speed Control with Static Inverter in the Rotor," *IEEE Trans. Power Appar. Syst.*, Vol. PAS-85, pp. 76–84, Jan. 1966.
3. A. Schonung, "Varying the Speed of Three-Phase Motors by Means of Static Frequency Changers," *Brown Boveri Rev.*, Vol. 51, pp. 540–554, Aug.–Sept. 1964.
4. A. Smith, "Static Scherbius System of Induction Motor Speed Control," *Proc. Inst. Electr. Eng.*, Vol. 124, pp. 557–565, 1977.

5. P. Zimmermann, "Super Synchronous Static Converter Cascade," *Conf. Rec. IFAC Symp. on Control in Power Elec. and Electrical Drives*, pp. 559–574, 1977.

6. T. Wakabayashi, T. Hori, K. Shimzu, and T. Yoshioka, "Commutatorless Kramer Control System for Large Capacity Induction Motors for Driving Water Service Pumps," *Conf. Rec. 1976 IEEE/IAS Annu. Meet.*, pp. 822–828, 1976.

7. H. W. Weiss, "Adjustable Speed AC Drive Systems for Pump and Compressor Applications," *IEEE Trans. Ind. Appl.*, Vol. IA-10, pp. 162–167, Jan.–Feb. 1975.

7

CONTROL OF INDUCTION MACHINES

7.0 INTRODUCTION

The control of ac machines is considerably more complex than that of dc machines and this complexity increases if stringent performance specifications are demanded. The complexity arises because of the variable-frequency power supply, ac signals processing, and complex dynamics of the ac machine. The induction machines can have various methods of control, and the particular method to be adopted depends on the nature of the application. The decision on control strategy should be based on the following general questions:

- What types of power converters should be used?
- Should the control be open loop or closed loop?
- Is it a position-, speed-, or torque-controlled system?
- Should it be a one-quadrant, two-quadrant, or four-quadrant drive system?
- What are the accuracy and response-time requirements?
- Is it a single-machine or a multimachine drive?
- What is the range of speed? Does it include zero speed and field weakening region?
- Is the drive required to give robust or parameter-insensitive response?
- Do pulsating torque, harmonics, and power factor need control?

We first review the classical and state variable control principles, and then discuss the control characteristics of induction machines. The various scalar control methods by voltage-fed and current-fed inverters are described. The principle of field-

232

oriented or vector control is discussed in detail, including the associated signal processing.   Finally, several adaptive control principles that can be applied to induction machines are reviewed.

7.1 REVIEW OF CLASSICAL CONTROL PRINCIPLES

Linear constant-coefficient systems are generally dealt within classical control theory.   In practice, very few physical systems will satisfy these criteria.   However, a large number of systems can be assumed to be linear and the coefficients can be considered as constant within a reasonable operating range.   The dynamics of a linear, continuous-time, constant-coefficient system can be analyzed by writing the system differential equations.   The differential equations in the time domain can be converted into algebraic equations by using Laplace transformation.   The response between an output and an input can be represented by a transfer function, which is defined as the ratio of the Laplace transform of the output to the Laplace transform of the input, with all other inputs as zero and assuming zero initial conditions.   Classical theory usually deals with a single-input single-output system compared with a multiple-input multiple-output system, which will be discussed in the next section.   The advantages of Laplace transform-domain analysis are that the manipulation of algebraic equations is easy and the transform-domain and frequency-domain analysis have a one-to-one correspondence.

Feedback System

A typical single-input single-output feedback control system in the transform domain can be represented by the block diagram shown in Fig. 7.1.   The input signal is $R(S)$, the output signal or response is $C(S)$, the feedback signal is $R'(S)$, and the error signal is $E(S)$.   The forward transfer function is $G(S)$ and the feedback transfer function is $H(S)$.   The open-loop transfer function can be defined as

$$A(S) = \frac{R'(S)}{R(S)} = G(S)H(S) \tag{7.1}$$

where the feedback loop is considered open at the feedback input.   The error and the closed-loop transfer functions can be derived as

$$\frac{E(S)}{R(S)} = \frac{1}{1 + G(S)H(S)} \tag{7.2}$$

$$\frac{C(S)}{R(S)} = \frac{G(S)}{1 + G(S)H(S)} \tag{7.3}$$

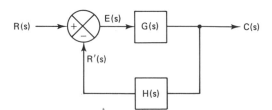

Figure 7.1   Block diagram of single-input single-output feedback control system.

For a unity feedback system, $H(S) = 1$ and then $E(S)$ represents the true error of the system. In general, any transfer function can be represented by a ratio of two polynomial functions of S. The roots of the numerator polynomial equation are defined as zeros and the roots of the denominator polynomial equation are defined as poles. Representing $G(S) = N_G(S)/D_G(S)$, $H(S) = N_H(S)/D_H(S)$, and substituting in equations (7.1) and (7.3) gives

$$A(S) = \frac{N_G(S)N_H(S)}{D_G(S)D_H(S)} \tag{7.4}$$

$$\frac{C(S)}{R(S)} = \frac{N_G(S)D_H(S)}{D_G(S)D_H(S) + N_G(S)N_H(S)} \tag{7.5}$$

which indicate that the zeros of the closed-loop transfer function are the same as the zeros of the open-loop transfer function. But the closed-loop poles are given by the roots of

$$D_G(S)D_H(S) + N_G(S)N_H(S) = 0 \tag{7.6}$$

or equivalently the roots of

$$1 + G(S)H(S) = 0 \tag{7.7}$$

Equation (7.6) or (7.7) is defined as the characteristic equation of the system and its roots determine the transient response of the closed-loop system.

It should be noted that a system need not necessarily always be controlled by feedback (i.e., the system may have open-loop control). A closed-loop system has the advantages that the output tracks the commanded input, the system response is less sensitive to parameter variation effect, and unwanted noise and disturbance effects are attenuated. However, the disadvantages include stability problems and the complex design criteria needed to make the system stable, the loss of system gain, and the requirement for precision feedback signals.

Analysis and Design

The time-domain response $C(t)$ can be expressed by inverting equation (7.3) as

$$C(t) = L^{-1}\left[\frac{R(S)G(S)}{1 + G(S)H(S)}\right] \tag{7.8}$$

For a simple system, the denominator can be factored and partial fraction expansion can be used to obtain a sum of easily invertable terms. Generally, the equation $1 + G(S)H(S) = 0$ is of higher degree and the explicit solution of $C(t)$ becomes quite difficult for every conceivable input. The initial and final value theorems, defined as follows, become very useful in determining the terminal behavior of the system:

Initial value theorem:

$$\lim_{t \to 0} f(t) = \lim_{s \to \infty} [SF(S)] \tag{7.9}$$

Final value theorem:

$$\lim_{t \to \infty} f(t) = \lim_{S \to 0} [SF(S)] \qquad (7.10)$$

where $f(t)$ is the time-domain function and $F(S)$ is the corresponding Laplace transfer function.

Classical control theory does not aim for explicit solution of $C(t)$, but rather deals with the closed-loop transfer function analysis to determine indirectly the system response for an arbitrary input.   The theory was developed before the advent of digital computers and remains a powerful tool for system analysis and design because of the intuitive understanding of the system it provides.

The quality of the feedback control system is judged by the criteria of stability, response time (or bandwidth), and steady-state error.   The stability means that the response of a system should be bounded for bounded input, initial condition, or disturbance.   The stability of a system is determined by the closed-loop characteristic equation $1 + G(S)H(S) = 0$ and it is necessary that all the roots lie in the left half of the S-plane.   The system stability is analyzed by graphical methods which can be summarized as follows:

Nyquist Method: This is a polar plot of the function $G(j\omega)H(j\omega)$, where the amplitude and phase are plotted as a function of frequency ω.   The plot checks the encirclement of the critical point -1 and determines the stability from the general criteria $n_p = N$, where n_p is the number of open-loop poles of equation (7.1) in the right half-plane and N the number of net counter-clockwise encirclements of the critical point -1 made by the plot.   If the system is open-loop stable, [i.e., poles of $A(S)$ lie in the left half-plane (which is usually the case], then $n_p = 0$.   The presence of closed-loop poles in the right half-plane is detected by the condition $n_p > N$.

Root-locus method: This is a plot of roots of the characteristic equation in the S-plane as the gain of the loop varies.   Since $G(S)H(S)$ is a complex number and is associated with a magnitude and phase, the solution of $G(S)H(S) = -1$ for any positive loop gain requires $\underline{/G(S)H(S)} = \pm K\pi$ and $|G(S)H(S)| = 1$, where K = any odd integer.   Thus the closed-loop roots are determined with the help of open-loop transfer function $G(S) H(S)$, which is normally available in factored form.

Bode method: This is a frequency-response plot of the open-loop transfer function with a minimum phase (poles and zeros in the left half-plane) as in the Nyquist method.   The magnitude and phase are plotted separately as a function of frequency.   The frequency and magnitude are plotted on log-log graph paper and asymptotic approximation permits rapid plot of magnitude in decibels.   The critical point for stability is the open-loop gain of -1 (i.e., 0 dB magnitude at a phase shift of $-180°$).   The frequency response of a physical system can be measured experimentally to identify its transfer function.

The satisfactory transient response of a system requires that there should be an acceptable value of response or settling time within a limited overshoot with a step

function input.   These criteria are met by the location of closed-loop poles in the
S-plane.   In frequency-domain analysis, the two important design criteria are gain
margin and phase margin.   The gain margin is the additional gain a system can
tolerate without a phase change to render it unstable and is defined at the frequency
where the phase reaches $-180°$.   The phase margin is the additional phase shift
that a system can tolerate with no gain change to make it unstable and is defined
at the frequency where the gain is unity (i.e., 0 dB).   Experience has shown that
for satisfactory transient response, a system should typically have phase margin
$> 35°$ and gain margin $> 6dB$.   These frequency-domain stability criteria can be
translated into time-domain characteristics by the following approximate relations
(Ref. 4):

$$\text{rise time} \times \text{closed-loop bandwidth} \simeq 2.83$$

$$\% \text{ overshoot} \simeq 75° - \text{phase margin}$$

$$\text{damping ratio} = 0.01 \text{ phase margin}$$

where the bandwidth is in rad/s, the phase margin is in degrees, and the time is in
seconds.

The performance of a feedback control system can be made satisfactory by
the following methods:

- The system loop gain can be increased.   Increase in the loop gain decreases
  the steady-state error and rise time, but the system is pushed toward instability.

- A compensating transfer function can be added within the feedback loop.
  In general, the undesirable poles and zeros of the system transfer function
  can be eliminated by the compensator.   A compensator can be designed so
  that the closed-loop transfer function of the system has a desirable structure.
  The following types of compensator are commonly used:

  —*PI compensator*: The proportional integral (PI) compensator is a sum of
  proportional and integral terms which has the structure

  $$F(S) = K_1 + \frac{K_2}{S}$$
  $$= K_2 \frac{1 + \tau_1 S}{S} \tag{7.11}$$

  where $\tau_1 = K_1/K_2$.   The compensator makes the system response sluggish
  but has the advantage that the steady-state error is zero due to the presence
  of the integration term.   This can be shown by applying the final value
  theorem.

  —*Lag–lead compensator*: This compensator is used for a system with acceptable stability margin but poor steady-state accuracy and has the structure

  $$F(S) = K \frac{1 + \tau_1 S}{1 + \tau_2 S} \tag{7.12}$$

what is phase margin?

where $\tau_2 > \tau_1$. The lag compensation lowers the loop gain at high frequency but keeps the phase unchanged for $\omega \gg 1/\tau_1$. As a result, the dc gain can be increased to improve the steady-state error without significantly affecting the phase margin. This assumes that the corner frequencies $1/\tau_1$ and $1/\tau_2$ are chosen well below the crossover frequency.

freq before

— *Lead–lag compensator*: This compensator has the same structure as equation (7.12) except that $\tau_1/\tau_2 > 1$. It can be used to improve the stability margin without affecting the error. The τ_1 and τ_2 values can be selected properly to increase the phase margin, which also increases the bandwidth (i.e., improves the response time). The low-frequency gain which affects the steady-state error remains unchanged.

- Minor loops can be added within the feedback system. A hierarchy of minor loops can considerably improve system performance, but the price paid is the requirement of additional feedback signals. Feedforward control can also be considered together with feedback control.

7.2 REVIEW OF STATE VARIABLE PRINCIPLES

The classical control theory based on Laplace transform and transfer function concepts is best suited for design of single-input single-output linear time-invariant systems. Multivariable systems which have multiple inputs and multiple outputs and which are either linear, nonlinear, or time varying can best be analyzed by modern control theory based on state space concepts. The performance specifications in terms of bandwidth, settling time, overshoot, and so on, have in many cases been replaced by optimal performance indexes, such as minimum time, minimum cost, and maximum efficiency. The complex optimal and adaptive control laws are based on modern control theory. Modern control theory analyzes and designs systems in the time domain and heavily uses digital computers, which also work in time domain. The state variable approach to modern control theory provides a uniform and powerful method of representing systems of arbitrary order—linear, nonlinear, or time varying—with the initial conditions taken into consideration, and provides an ideal formulation for the computer solution of systems. Modern control theory, although much more powerful than classical control theory, will not displace the latter unless in the future, modern control theory becomes better than the classical theory in every respect.

The concept of state is very important in modern control theory. The state means the status of a system at any particular time t. The state of a system can be described by a set of variables, called state variables. The state vector $X(t)$ is an array of these state variables. A state space can be defined as an n-dimensional space in which the n state variables $X_1, X_2, \ldots, X_n$ are coordinates. In general, the state space equations of a linear constant-coefficient multivariable continuous-time dynamical system are described in vector form as

$$\dot{X} = AX + BU \tag{7.13}$$

$$Y = CX + DU \tag{7.14}$$

where X is the state vector, U the input vector, Y the output vector, $\dot{X} = dX/dt$

= derivative of the state vector, A the system matrix, B the input matrix, C the output matrix, and D the feedforward matrix. The block diagram presentation of equations (7.13) and (7.14) is shown in Fig. 7.2. In general, for an n-dimensional system, the coefficient matrices A, B, C, D have dimension $n \times n$, $n \times r$, $m \times n$, and $m \times r$, respectively. More specifically, the component matrix structures of state space equations are given as

$$X = [X_1 \quad X_2 \quad \cdots \quad X_n]^T$$

$$\dot{X} = [\dot{X}_1 \quad \dot{X}_2 \quad \cdots \quad \dot{X}_n]^T$$

$$A = \begin{bmatrix} a_{11} & a_{12} & \cdots & a_{1n} \\ a_{21} & a_{22} & \cdots & a_{2n} \\ \vdots & & & \vdots \\ a_{n1} & a_{n2} & \cdots & a_{nn} \end{bmatrix}$$

$$B = \begin{bmatrix} b_{11} & b_{12} & \cdots & b_{1r} \\ b_{21} & b_{22} & \cdots & b_{2r} \\ \vdots & & & \vdots \\ b_{n1} & b_{n2} & \cdots & b_{nr} \end{bmatrix}$$

$$C = \begin{bmatrix} c_{11} & c_{12} & \cdots & c_{1n} \\ c_{21} & c_{22} & \cdots & c_{2n} \\ \vdots & & & \vdots \\ c_{m1} & c_{m2} & \cdots & c_{mn} \end{bmatrix}$$

$$D = \begin{bmatrix} d_{11} & d_{12} & \cdots & d_{1r} \\ d_{21} & d_{22} & \cdots & d_{2r} \\ \vdots & & & \vdots \\ d_{m1} & d_{m2} & \cdots & d_{mr} \end{bmatrix}$$

$$U = [U_1 \quad U_2 \quad \cdots \quad U_r]^T$$

$$Y = [Y_1 \quad Y_2 \quad \cdots \quad Y_m]^T$$

State space equation (7.13) is essentially a collection of first-order scalar differential equations where the first equation has the structure

$$\dot{X}_1 = (a_{11}X_1 + a_{12}X_2 + \cdots + a_{1n}X_n)$$

$$+ (b_{11}U_1 + b_{12}U_2 + \cdots + b_{1r}U_r) \qquad (7.15)$$

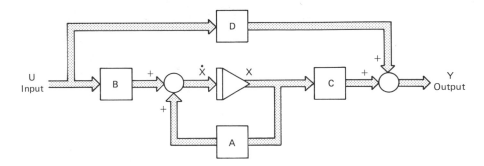

Figure 7.2   State space block diagram of continuous time linear system.

The output equation (7.14) is simply a collection of scalar algebraic equations and can easily be synthesized by a linear combination of input and state variables.   If a single-input single-output system is described by an nth-order differential equation, it can be transformed into the corresponding nth-order state space equations, where the state variables are the dependent variable and its higher-order derivatives.   The multiple-input multiple-output systems described by several coupled differential equations can similarly be transformed into state space equations.

Solving the State Space Equation

The solution of state space equation (7.13) can be given in the form

$$X = e^{At}X(0) + \int_0^t e^{A(t-\tau)}BU(\tau) \, d\tau \tag{7.16}$$

where $X(0)$ is the initial condition of the X vector and e^{At} is called the state transition matrix.   The solution is analogous with that of a first-order (scalar) differential equation in the form $dx/dt = ax + bu$.   In equation (7.16), the first term is known as the complementary function and corresponds to the unforced response of the system.   The second term is known as the particular integral and is in the form of a convolution integral.   The equation can be solved numerically with the help of a digital computer.   In special cases, if U vector is a constant, the solution is given as

$$X = e^{At}X(0) + A^{-1}(e^{At} - I)BU \tag{7.17}$$

where I is the unit vector and

$$A^{-1} = \frac{\text{Adj } (A)}{|A|}$$

$$e^{At} = I + \frac{At}{1!} + \frac{A^2t^2}{2!} + \frac{A^3t^3}{3!} + \cdots$$

The state space equation can also be solved by the eigenvalue–eigenvector method.   Consider an unforced system ($U = 0$) and assume that the general solution is in the form

$$X = Ve^{\lambda t} \tag{7.18}$$

Substituting the solution in equation (7.13) and simplifying yields

$$(\lambda I - A)V = 0 \tag{7.19}$$

For the elements of the vector V to be nonzero,

$$|\lambda I - A| = 0$$

that is,   (7.20)

$$|A - \lambda I| = 0$$

which in expanded form is

$$
\begin{vmatrix}
a_{11} - \lambda & a_{12} & \cdots & a_{1r} \\
a_{21} & a_{22} - \lambda & \cdots & a_{2n} \\
\cdot & \cdot & & \cdot \\
\cdot & \cdot & & \cdot \\
\cdot & \cdot & & \cdot \\
a_{n1} & a_{n2} & \cdots & a_{nn} - \lambda
\end{vmatrix} = 0
$$

The expression $|\lambda I - A|$ is called the characteristic polynomial, the roots of equation (7.20) are known as eigenvalues, and the vector V is known as an eigenvector. The eigenvalues are identical to poles of the transfer function and determine the stability of the system. Therefore, a system is open-loop stable if the eigenvalues have negative real roots. For an nth-order system, there are n number of eigenvalues and eigenvectors. The eigenvalues may appear in real or complex conjugate form. The solution of the unforced system with initial condition $X(0)$ can be given in eigenvalue–eigenvector form as

$$X = e^{At}X(0)$$

$$= \beta_1 e^{\lambda_1 t}V_1 + \beta_2 e^{\lambda_2 t}V_2 + \cdots + \beta_n e^{\lambda_n t}V_n \tag{7.21}$$

where

$$X(0) = \sum_{i=1}^{n} \beta_i V_i$$

and λ_1, λ_2, and so on, are the eigenvalues corresponding to eigenvectors, V_1, V_2, and so on, respectively.

Transfer Function

The state space equations can be converted into the Laplace transform domain and transfer functions can be derived. Assuming that $X(0) = 0$, the Laplace transform of the differential equation (7.13) is

$$SX(S) = AX(S) + BU(S)$$

that is,

$$X(S) = (SI - A)^{-1} BU(S) \tag{7.22}$$

Similarily, for the output equation (7.14),

$$Y(S) = CX(S) + DU(S) \tag{7.23}$$

Combining equations (7.22) and (7.23) gives

$$Y(S) = [C(SI - A)^{-1}B + D]U(S) = G(S)U(S) \tag{7.24}$$

where $G(S)$ is the transfer matrix. A transfer function is defined between a single input and a single output. Therefore, the transfer function $F(S)$ is given as

$$F(S) = c(SI - A)^{-1}b + d \tag{7.25}$$

where, c, b, and d are scalar parameters. Assuming that $d = 0$,

$$F(S) = \frac{c \, \text{Adj}(SI - A)b}{|SI - A|} \tag{7.26}$$

The poles of the transfer function are roots of $|SI - A| = 0$ since the eigenvalues of A are roots of $|\lambda I - A| = 0$, indicating that the eigenvalues are identical to poles.

Controllability and Observability

Controllability and observability are the two most important properties of a dynamical system. In testing the controllability criteria of a system, the question is asked: Is it possible to have a control function $U(t)$ which will transition the initial state $X(t_0)$ to a state $X(t)$ in the finite time $t - t_0$? If a system is not fully controllable, then for some initial states, no input exists which can drive the system to states at time t. An extreme example of an uncontrollable system is that the B matrix is zero, which causes complete decoupling of input from the state. The system of equation (7.13) is controllable or the matrix pair $[A, B]$ is controllable if and only if

$$\text{rank} \left[B \vdots AB \vdots A^2B \vdots \cdots \vdots A^{n-1}B \right] = n \tag{7.27}$$

which means that the foregoing $n \times (nr)$ matrix G has n linearly independent column vectors. Note that A and B matrices have dimensions $n \times n$ and $n \times r$, respectively.

The observability property permits us to observe or estimate the state variables of a system by knowing the input and output functions. In testing the observability criteria the question is asked: Can we calculate the initial state $X(t_0)$ from the measured values of output function $Y(t)$ and the measured values of input function $U(t)$? If the initial state is calculated, we can also calculate $X(t)$ using the system equation. Again, an extreme example of an unobservable system is that the C matrix is zero. Mathematically, a system is observable if and only if

$$\text{rank} \left[C^T \vdots A^T C^T \vdots A^{2^T} C^T \vdots \cdots \vdots A^{n-1^T} C^T \right] = n \tag{7.28}$$

which means that the foregoing $n \times (nm)$ matrix H has n linearly independent column vectors.   The C matrix has the dimension $n \times m$.

Control System with State Feedback

So far we have discussed the open-loop properties of multivariable system.   We now consider the feedback systems and their properties.   Figure 7.3 shows a block diagram of a state variable feedback system.   Here the state vector X is multiplied by the gain matrix G to construct the feedback vector.   The input vector W, assumed to be of the order $j \times 1$, is multiplied by the forward matrix E and its output is added to the feedback vector to construct the input equation

$$U = EW - GX \tag{7.29}$$

The matrices G and E are of the order $r \times n$ and $r \times j$, respectively.   The addition of feedback will tend to affect the controllability, observability, and stability of the system, and this is the subject of our discussion.   The question arises how practical is the state-variable feedback system?   For most practical systems, the state is inaccessible by measurement, and it is at least true for ac drive system.   However, for simple systems, such as the dc drive system, the state can be explicitly measured. If the state is inaccessible, it can be estimated or reconstructed from the available output and input signals by an observer technique that will be discussed later.   The reason for the emphasis on state variable feedback is that the state contains all the information about the system and by using it as feedback the system performances can be manipulated as desired.   Implementation of several optimal control laws is possible only by using state variable feedback.

The state space equations can be constructed by combining equations (7.13),

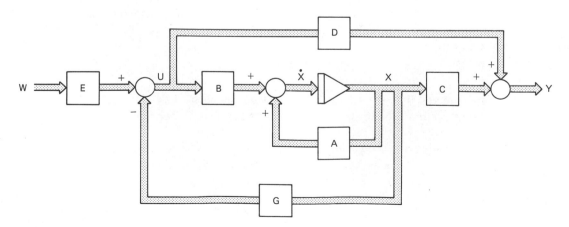

Figure 7.3 Block diagram of state variable feedback system.

(7.14), and (7.29) as

$$\dot{X} = [A - BG]X + [BE]W \qquad (7.30)$$

$$Y = [C - DG]X + [DE]W \qquad (7.31)$$

The properties of the closed-loop system are dictated by the matrices $[A - BG]$, $[BE]$, $[C - DG]$, and $[DE]$, of which G and E are unknown elements. The controllability of the system is determined by the criteria

$$\text{rank} \left[BE \vdots (A - BG)BE \vdots (A - BG)^2 BE \vdots \cdots \vdots (A - BG)^{n-1} BE \right] = n \qquad (7.32)$$

If the forward matrix E satisfies rank $(E) = r$ (i.e., the number of independent input components after adding feedback is the same as the number of inputs in U), then E will not affect the rank of the controllability matrix above. The manipulation of this matrix indicates that its rank is n, $(A \neq BG)$ (i.e., the controllability property of the open-loop system is not affected by state feedback).

The observability property of the closed-loop system is determined by the criteria

$$\text{rank} \left[(C - DG)^T \vdots (A - BG)^T (C - DG)^T \vdots \cdots \vdots \right.$$
$$\left. (A - BG)^{n-1^T} (C - DG)^T \right] = n \qquad (7.33)$$

State feedback in general can cause a loss of observability. For example, observability will be lost if $C = DG$.

The stability of the closed-loop system is determined by the eigenvalues of the characteristic equation

$$|\lambda I - A + BG| = 0 \qquad (7.34)$$

In fact, the eigenvalues or poles of the state feedback system can be assigned as desired by the design of gain matrix G. This capability of setting arbitrary pole positions (i.e., influencing the complete transient response) is a great power of state variable feedback system.

Control System with Output Feedback

An output feedback system is very practical because the output signals can be obtained by measurement. The block diagram of an output feedback system is shown in Fig. 7.4. Here the output vector Y is multiplied by the gain matrix G' to construct the feedback vector. The input vector W' of order $j \times 1$ is again multiplied by E' and its output is added to the feedback vector to construct the input equation

$$U = E'W' - G'Y \qquad (7.35)$$

where the matrices G' and E' have dimensions $r \times m$ and $r \times j$, respectively. The state space equations of output feedback system are obtained by combining

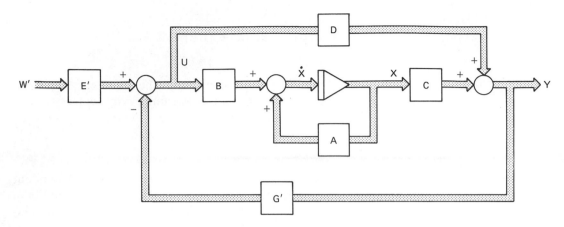

Figure 7.4 Block diagram of output feedback system.

equations (7.13), (7.14), and (7.35) as

$$\dot{X} = \left\{A - BG'[I_m + DG']^{-1}C\right\}X + B[I_r + G'D]^{-1}E'W' \qquad (7.36)$$

$$Y = [I_m + DG']^{-1}CX + [I_m + DG']^{-1}DE'W' \qquad (7.37)$$

The properties of controllability, observability, and stability are influenced by the three matrices

$$\{A - BG'[I_m + DG']^{-1}C\} \qquad B[I_r + G'D]^{-1}E' \qquad [I_m + DG']^{-1}C$$

It can be shown that the controllability of the system remains unaffected by output feedback if the matrix $[I_r + G'D]$ is nonsingular and rank $(E') = r$. By constructing the observability matrix of the output feedback system and manipulating, it can be shown that the open-loop observability condition is unaltered irrespective of G'. The system is observable with output feedback because of the presence of matrix C in the system matrix

$$\left\{A - BG'[I_m + DG']^{-1}C\right\}$$

With output feedback the eigenvalues are the roots of the characteristic equation

$$|\lambda I - A + BG'[I_m + DG']^{-1}C| = 0 \qquad (7.38)$$

The equation has n eigenvalues of which m eigenvalues can be assigned by design of the gain matrix G'.

State Estimation by Observer Technique

We will now discuss the method of estimating the state variables of a system by the observer technique. For simplicity, the D matrix will be assumed to be zero. If the model of the original open-loop system is known, which means if we have

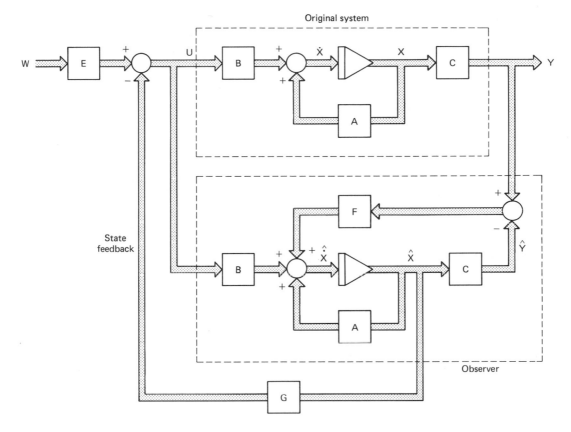

Figure 7.5 State estimation block diagram by observer technique.

accurate knowledge of the matrices, *A, B,* and *C* in Fig. 7.2, then the model can be solved to reconstruct the exact state of the system.   This method is somewhat unrealistic in practice because of the uncertainty and parameter variation of the model.   A superior arrangement of state estimation with the help of observer is shown in Fig. 7.5.   The observer is essentially a duplicate of the original system where the state is reconstructed by solving the original system.   If the estimated output $\hat{Y}$ does not match the actual output *Y* due to discrepancy of the model, the error $Y - \hat{Y}$ provides an auxiliary input to the model so that the estimated state $\hat{X}$ approaches the actual state *X*.   The matrix *F* can be designed to achieve this condition.   The estimated state $\hat{X}$ is fed back to the input through a gain matrix *G* as shown in Fig. 7.5.   The observer is *n*-dimensional because it estimates an *n*-dimensional state.   Note that the composite system is 2*n*-dimensional and has one set of *n* eigenvalues of $A - BG$ (the eigenvalues of a closed-loop system with state feedback) and another set of *n* eigenvalues of $A - FC$ (the eigenvalues of the observer).   The observer poles are assigned by the design of *F*.   The observer can be simulated on a computer but the real-time estimation of state always remains a challenge. In many systems (e.g., ac drive system), a partial state can be estimated by using a partial or reduced order observer and a system similar to Fig. 7.5 can be used. This may provide elimination or redundancy of some of the sensors in the system.

7.3 CHARACTERISTICS OF INDUCTION MOTOR CONTROL

The induction motor drive system is basically a multivariable control system, and therefore in principle the state variable control theory should be applicable. Here the voltage and frequency are the control inputs and the outputs may be speed, position, torque, air gap flux, stator current, or any combination of them. The machine model given by equations (2.77), (2.78), and (2.83) is nonlinear because of the presence of the ω_r term in the impedance matrix of equation (2.77) and product terms in equation (2.83). In addition, the parameters of the machine may vary with saturation, temperature, and skin effect, adding further nonlinearity to the system. The system is also discrete time because of the sampling nature of the converters. If a microcomputer or other digital circuits are used in the control system, then additional sampling characteristics are added. The discrete-time effect of the converters and controller can, of course, be neglected if the machine response is sluggish, which is normally the case.

Figure 7.6 shows the general block diagram of the control system. Here the stator current, air gap flux, speed, and developed torque are considered as outputs and the primary control signals are air gap flux, ψ_m^* and speed ω_r^*. All the outputs and control loop signals are dc voltages proportional to the respective variables. The control system is characterized by the hierarchy of control loops as shown. For example, the air gap flux control loop controls the stator current in the inner loop, which in turn controls the stator voltage. Similarly, in the speed control loop, the torque is controlled in the inner loop. The torque loop generates the slip signal ω_{sl}^*, which is added with the feedback speed to generate the frequency command ω_e^*. The controller elements G_1, G_2, G_3, and G_4 may be simple gain

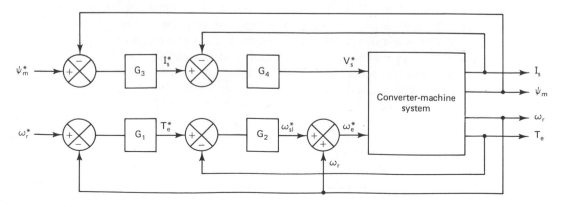

Figure 7.6 General control block diagram of induction motor.

or compensator functions and normally contain a limiter so that the excursion of the respective variable is limited.   It is also desirable that the progressively inner control loops have a faster response time.   Some of the output signals of an ac drive system are difficult to measure with the help of sensors but can be estimated with the help of a partial observer.   The number of output signals can be reduced if the number of inner control loops are reduced, but this may cause degradation of performance, which is discussed later.   Fig. 7.6 can be transformed into the standard multivariable structure as shown in Fig. 7.7.   If the control elements in Fig. 7.6 use pure gain, the feedback controller G in Fig. 7.7 is a pure-gain matrix. On the other hand, with compensator functions, G is characterized by a state space equation.

Since the ac drive system is multivariable, nonlinear, and discrete time in nature, stability analysis is very difficult.   Computer simulation or computer-aided analysis becomes very useful for stability study and performance analysis of a new control strategy as discussed before.   Once the control structure and the parameters of the controller are determined on the simulation, the prototype system is designed and laboratory tested with further iteration of the controller parameters.

Small-Signal Model

Neglecting the discrete-time nature of the converter, the converter–machine system in Fig. 7.6 can be linearized on a small-signal perturbation basis at a steady-state operating point and a transfer function can be described between a pair of output and input signals.   The advantage of such a transfer function model is that the stability analysis of the drive system is now possible using classical control theory, such as Bode, Nyquist, or root-locus techniques.   Since the system is nonlinear, the poles, zeros, and gain of the transfer functions will vary as the steady-state operating point shifts.   The closed-loop control system should then be designed such that at the worst operating point the system is stable and the performances are acceptable.

The electromechanical dynamics of an induction machine is described by a fifth-order nonlinear state-space equation which can be formed by equations (2.77), (2.78), and (2.83).   Combining the equations above and applying a small-signal

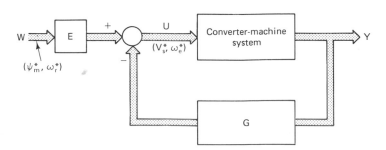

Figure 7.7 Multivariable structure of control system.

perturbation about a steady-state operating point, we get

$$
\begin{bmatrix}
v_{qso} + \Delta v_{qs} \\
v_{dso} + \Delta v_{ds} \\
v_{qro} + \Delta v_{qr} \\
v_{dro} + \Delta v_{dr} \\
T_{L0} + \Delta T_L
\end{bmatrix}
=
\begin{bmatrix}
R_s + SL_s & (\omega_{eo} + \Delta \omega_e)L_s & SL_m \\
-(\omega_{eo} + \Delta \omega_e)L_s & R_s + SL_s & -(\omega_{eo} + \Delta \omega_e)L_m \\
SL_m & (\omega_{eo} + \Delta \omega_e)L_m & R_r + SL_r \\
-(\omega_{eo} + \Delta \omega_e)L_m & SL_m & -(\omega_{eo} + \Delta \omega_e)L_r \\
\dfrac{3}{2}\dfrac{P}{2}L_m(i_{dro} + \Delta i_{dr}) & \dfrac{-3}{2}\dfrac{P}{2}L_m(i_{qro} + \Delta i_{qr}) & 0
\end{bmatrix}
$$

(7.39)

$$
\begin{bmatrix}
(\omega_{eo} + \Delta \omega_e)L_m & 0 \\
SL_m & 0 \\
(\omega_{eo} + \Delta \omega_e)L_r & -L_m(i_{dso} + \Delta i_{ds}) - L_r(i_{dro} + \Delta i_{dr}) \\
R_r + SL_r & L_m(i_{qso} + \Delta i_{qs}) + L_r(i_{qro} + \Delta i_{qr}) \\
0 & \dfrac{-2}{P}JS
\end{bmatrix}
\begin{bmatrix}
i_{qso} + \Delta i_{qs} \\
i_{dso} + \Delta i_{ds} \\
i_{qro} + \Delta i_{qr} \\
i_{dro} + \Delta i_{dr} \\
\omega_{ro} + \Delta \omega_r
\end{bmatrix}
$$

The quantities v_{qso}, v_{dso}, v_{qro}, v_{dro}, T_{Lo}, ω_{eo}, i_{qso}, i_{dso}, i_{qro}, i_{dro}, and ω_{ro} describe the steady-state operating point and can be found by solving the system equations with all time derivatives set equal to zero. Linearizing equation (7.39) by neglecting Δ^2 terms and eliminating the steady-state terms, we get the small-signal linear state space equation in the form

$$X = AX + BU \tag{7.40}$$

where

$$X = [\Delta i_{qs} \quad \Delta i_{ds} \quad \Delta i_{qr} \quad \Delta i_{dr} \quad \Delta \omega_r]^T$$

$$U = [\Delta V_s \quad 0 \quad 0 \quad 0 \quad \Delta \omega_e \quad \Delta T_L]^T$$

$$
A = \frac{-1}{L_sL_r - L_m^2}
\begin{bmatrix}
R_sL_r & (L_sL_r - L_m^2)\omega_{eo} + L_m^2\omega_{ro} \\
-(L_sL_r - L_m^2)\omega_{eo} - L_m^2\omega_{ro} & R_sL_r \\
-R_sL_m & -L_mL_s\omega_{r0} \\
L_mL_s\omega_{ro} & -R_sL_m \\
\dfrac{-3}{8}\dfrac{P^2}{J}L_m(L_sL_r - L_m^2)i_{dro} & \dfrac{3}{8}\dfrac{P^2}{J}L_m(L_sL_r - L_m^2)i_{qro}
\end{bmatrix}
$$

$$
\begin{bmatrix}
-R_rL_m & L_mL_r\omega_{ro} & L_m^2i_{dso} + L_mL_ri_{dro} \\
-L_mL_r\omega_{ro} & -R_rL_m & -L_m^2i_{qso} - L_mL_ri_{qro} \\
R_rL_s & (L_sL_r - L_m^2)\omega_{eo} - L_sL_r\omega_{ro} & -L_mL_si_{dso} - L_sL_ri_{dro} \\
-(L_sL_r - L_m^2)\omega_{eo} + L_sL_r\omega_{ro} & R_rL_s & L_mL_si_{qso} + L_sL_ri_{qro} \\
\dfrac{3}{8}\dfrac{P^2}{J}L_m(L_sL_r - L_m^2)i_{dso} & \dfrac{-3}{8}\dfrac{P^2}{J}L_m(L_sL_r - L_m^2)i_{qso} & 0
\end{bmatrix}
$$

$$
B = \frac{1}{L_sL_r - L_m^2}
\begin{bmatrix}
L_r & 0 & -L_m & 0 & -(L_sL_r - L_m^2)i_{dso} & 0 \\
0 & L_r & 0 & -L_m & (L_sL_r - L_m^2)i_{qso} & 0 \\
-L_m & 0 & L_s & 0 & -(L_sL_r - L_m^2)i_{dro} & 0 \\
0 & -L_m & 0 & L_s & (L_sL_r - L_m^2)i_{qro} & 0 \\
0 & 0 & 0 & 0 & 0 & \dfrac{-P}{2J}(L_sL_r - L_m^2)
\end{bmatrix}
$$

In the equation above, the machine is considered stator fed only if ($\Delta v_{qr} = \Delta v_{dr} = 0$) and the stator voltage phasor is aligned with the stator q-axis so that $\Delta v_{qs} = \Delta v_s$ and $\Delta v_{ds} = 0$, leaving Δv_s, $\Delta \omega_e$, and ΔT_L as the input variables. The state space equation will also represent the converter–machine system if converter dynamics are neglected. Any converter gain can be merged with the controller gain. The small-signal block diagram is shown in Fig. 7.8(a), where the electrical and mechanical responses have been separated. The converter–machine model generates the currents from the input control signals ΔV_s and $\Delta \omega_e$, and the feedback speed signal $\Delta \omega_r$ acts to generate the counter emf. The developed torque ΔT_e is synthesized from the currents by the equation

$$\Delta T_e = \frac{3}{2} \left(\frac{P}{2} \right) L_m[(i_{dro} \Delta i_{qs} + i_{qso} \Delta i_{dr}) - (i_{dso} \Delta i_{qr} + i_{qro} \Delta i_{ds})] \qquad (7.41)$$

The other small-signal outputs ΔI_s and $\Delta \psi_m$ in Fig. 7.6 can be synthesized from the current signals as follows:

The stator current $\hat{I}_s$ is given as

$$|\hat{I}_s| = \sqrt{i_{qs}^2 + i_{ds}^2} \qquad (7.42)$$

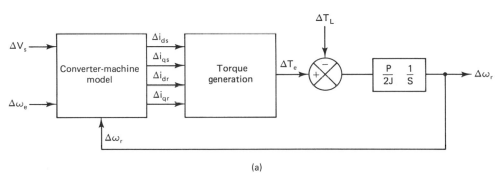

(a)

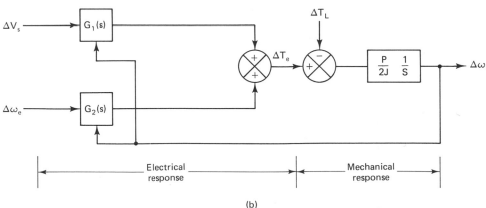

(b)

Figure 7.9 (a) Open loop volts/hertz control; (b) acceleration and deceleration characteristics.

which can be small-signal linearized as

$$|\Delta \hat{I}_s| = \frac{i_{qso}}{\sqrt{i_{qso}^2 + i_{dso}^2}} \Delta i_{qs} + \frac{i_{dso}}{\sqrt{i_{qso}^2 + i_{dso}^2}} \Delta i_{ds} \qquad (7.43)$$

Similarly, the air gap flux is given as

$$|\hat{\psi}_m| = \sqrt{\psi_{qm}^2 + \psi_{dm}^2} \qquad (7.44)$$

where

$$\psi_{qm} = L_m(i_{qs} + i_{qr})$$

$$\psi_{dm} = L_m(i_{ds} + i_{dr})$$

The linearized equation for the air gap flux is then

$$|\Delta \hat{\psi}_m| = \frac{L_m(i_{qso} + i_{qro})}{\sqrt{(i_{qso} + i_{qro})^2 + (i_{dso} + i_{dro})^2}} \Delta i_{qs}$$

$$+ \frac{L_m(i_{dso} + i_{dro})}{\sqrt{(i_{qso} + i_{qro})^2 + (i_{dso} + i_{dro})^2}} \Delta i_{ds}$$

$$+ \frac{L_m(i_{qso} + i_{qro})}{\sqrt{(i_{qso} + i_{qro})^2 + (i_{dso} + i_{dro})^2}} \Delta i_{qr} \qquad (7.45)$$

$$+ \frac{L_m(i_{dso} + i_{dro})}{\sqrt{(i_{qso} + i_{qro})^2 + (i_{dso} + i_{dro})^2}} \Delta i_{dr}$$

The small-signal transfer function block diagram derived from Fig. 7.8(a) is shown in Fig. 7.8(b). Here the transfer functions $G_1(S)$ and $G_2(S)$ are defined as

$$G_1(S) = \frac{\Delta T_e}{\Delta V_s} \left|\begin{array}{c} \Delta \omega_r = 0 \\ \Delta \omega_e = 0 \end{array}\right. \qquad (7.46)$$

$$G_2(S) = \frac{\Delta T_e}{\Delta \omega_e} \left|\begin{array}{c} \Delta \omega_r = 0 \\ \Delta V_s = 0 \end{array}\right. \qquad (7.47)$$

In the transfer functions above, the speed can be treated as a constant parameter ($\Delta \omega_r \rightarrow 0$) if the system inertia J is very large (i.e., $\tau_m \gg \tau_e$, where τ_m is the equivalent mechanical time constant and τ_e is the equivalent electrical time constant). In such a case, the speed response is dictated mainly by the mechanical time constant. For any finite inertia system, the transfer function $\Delta \omega_r/\Delta V_s$, $\Delta \omega_r/\Delta \omega_e$, and $\Delta \omega_r/\Delta T_L$ can be derived from the state space equation (7.40).

7.4 SCALAR CONTROL METHODS

In this section the selected scalar control techniques of an induction motor using voltage-fed inverters, current-fed inverters, and slip power recovery control are described. The cycloconverter-fed drives will follow as a natural consequence and will not be discussed.   Scalar control relates to the magnitude control of a variable only, and the command and feedback signals are dc quantities which are proportional to the respective variables.   This is in contrast to vector control, where both magnitude and phase of a vector variable are controlled, and this is described in the next section.

Voltage-Fed Inverter Control

Volts/hertz control.   A simple and popular open-loop volts/hertz speed control method for an induction motor is shown in Fig. 7.9(a).   The power circuit consists of a phase-controlled rectifier with a single- or three-phase ac supply, LC filter, and six-stepped inverter.   The frequency ω_e^* is the command variable and it is close to the motor speed, neglecting the small slip frequency.   The scheme is defined as volts/hertz control because the rectifier voltage command V_s^* is generated directly from the frequency signal through a volts/hertz gain constant G.   In steady-state operation, the machine air gap flux ψ_m is approximately related to the ratio V_s/ω_e.   Therefore maintaining the rated air gap flux will provide maximum torque sensitivity with stator current which is similar to that of a dc machine.   As the frequency approaches zero near zero speed, the stator voltage will tend to be zero and it will essentially be absorbed by the stator resistance.   Therefore, an auxiliary voltage V_o is injected to overcome the effects of stator resistance so that rated air

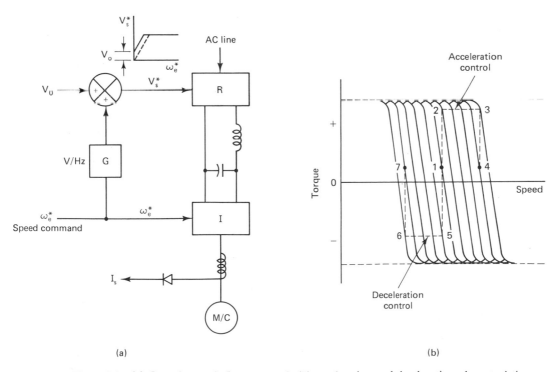

Figure 7.9 (a) Open loop volts/hertz control; (b) acceleration and deceleration characteristics.

gap flux and full torque become available up to zero speed. At steady-state operation, if the load torque is increased, the slip will increase within the stability limit and a balance will be maintained between the developed torque and the load torque. If the command frequency exceeds the base frequency of the machine, the rectifier voltage will reach saturation and the machine will transition from constant torque to the field weakening region. In this region, the flux ψ_m will be less and therefore the developed torque will be reduced for the same stator current limit. With open-loop voltage control, the ac line voltage fluctuation and impedance drop will cause fluctuation of the air gap flux. This fluctuation can be prevented by providing closed-loop voltage control of the rectifier. The acceleration/deceleration performance characteristics of the drive system in the constant-torque region is shown in Fig. 7.9(b). During steady-state operation (point 1), if the command frequency ω_e^* is increased by a step, the slip will exceed that of breakdown torque and the machine will become unstable. Similar instability will occur if the frequency is decreased by a step. Therefore, it is necessary that during both acceleration and deceleration, the frequency should track the speed so that slip does not exceed that of the breakdown torque. The stable operation with prescribed acceleration and deceleration control is possible as indicated in Fig. 7.9(b) with adjustable current limit control. For a step-up frequency command, the slip increases and the stator current I_s comes up to a limit which correspondingly transitions the operating point from 1 to 2 in the torque–speed curves. Thereafter the frequency is ramped under the current limit control for the constant-torque-acceleration region 2 to 3. Then between points 3 and 4, the current falls below the limit and the steady-state operating point 4 is reached, with decrease of slip frequency. For a step-down frequency command, the deceleration trajectory is 1-5-6-7, as shown in Fig. 7.9(b). During deceleration, the stator frequency falls below the speed and the machine energy in electrical form is pumped to the dc link, raising its voltage. With a unipolar rectifier, a dynamic braking resistor with a chopping mode switch absorbs the energy while controlling the dc link voltage.

In this scheme, the speed will tend to drift with variation in load torque and fluctuation of supply voltage. If open-loop speed fluctuation is not permissible, a closed-loop speed control can be provided, as shown in Fig. 7.10. The speed loop error signal controls the PWM inverter frequency and voltage through the current-limit controller as shown in the figure.

An alternative volts/hertz control scheme with slip regulation is shown in Fig. 7.11. Here the error of the speed control loop generates the slip command ω_{sl}^* through a proportional-integral (PI) controller and limiter. The slip is added with the speed signal to generate the frequency command. The frequency command also generates the voltage command through a volts/hertz function generator which incorporates the low-frequency stator drop compensation. Since the slip is proportional to developed torque, the scheme can be considered as torque control within a speed control loop. It differs from the previous scheme, where torque limit control was provided indirectly by stator current limit control. In addition to torque control, the advantage here is that the expensive current sensor is avoided and the same speed signal has been used in both the loops. With a step-up speed command, the machine accelerates freely with a slip limit that corresponds to

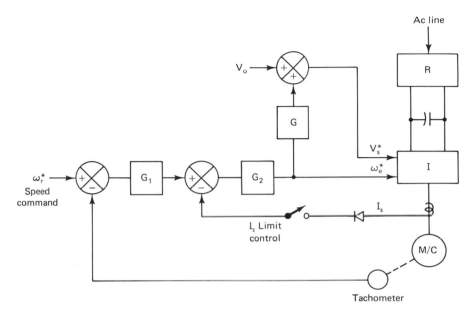

Figure 7.10   Closed-loop volts/hertz speed control.

maximum torque and then settles down to the slip value at steady state as dictated by the load torque.   If the command speed ω_r^* is reduced, the slip becomes negative and the machine goes into the dynamic or regenerative braking mode, as explained earlier.

Instead of controlling the slip, it can be held constant and the speed loop error may control only the inverter output voltage.   Variation in the volts/hertz

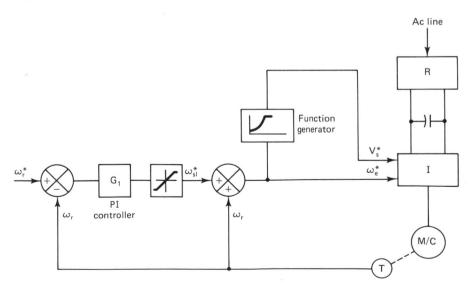

Figure 7.11   Constant volts/hertz speed control with slip regulation.

ratio causes variation of air gap flux and correspondingly the developed torque ($T_e \propto |\hat{\psi}_m|^2 \omega_{sl}$) is varied. The reduced air gap flux operation at light-load condition may result in some improvement of efficiency by trading off core loss with copper loss. But the disadvantage is that due to sluggishness of flux response, the machine may become unstable with a sudden increase of torque load.

Torque and flux control. As described earlier, the volts/hertz control scheme has the disadvantage that the air gap flux may drift, and as a result the torque sensitivity with slip or stator current will vary. If the correct volts/hertz ratio is not maintained, the flux may be weak or may saturate. The stator circuit parameters may also vary due to temperature and saturation, causing drift in the air gap flux. In Fig. 7.11, if the air gap flux decreases, the slip ω_{sl} has to increase for the same torque demand. As a result, the machine's maximum torque capability will decrease and the transient response will deteriorate.

A speed control scheme with independent torque and flux control loops is shown in Fig. 7.12. An additional torque loop has been added within the speed loop to make the speed loop response faster and more stable. The speed regulator G_1 may be a PI compensator, so that the steady-state speed error is zero. The torque regulator G_2 may be a pure gain or PI compensator but must have a limiter as shown in Fig. 7.11. The flux control loop compares the command flux and feedback flux and generates the voltage command of the PWM inverter as shown. The direct flux control loop solves the flux drift problem as discussed before, but getting an accurate feedback flux signal is difficult.

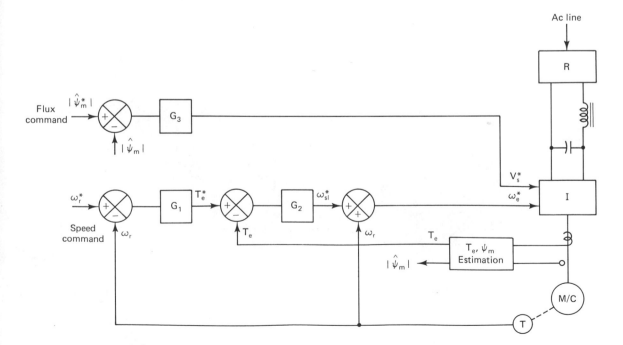

Figure 7.12 Speed control with independent torque and flux control.

The machine terminal voltages and currents can be sensed and torque and flux signals can be estimated by a partial observer. A computation flow diagram for estimation of torque and flux feedback signals is shown in Fig. 7.13. A simple method of flux measurement is the mounting of Hall effect sensors in the machine air gap. A problem here is that the Hall sensor outputs drift with temperature, which is difficult to compensate. Alternatively, flux coils may be mounted in the air gap and the correspondingly induced voltages may be integrated to get the flux information. The mounting of external devices, such as Hall sensors or flux coils, in the air gap is not favored by machine designers. In Fig. 7.13, the torque and flux signals are synthesized from the machine phase voltage and current signals by the equations

$$T_e = \frac{3}{2}\left(\frac{P}{2}\right)\left(\psi^s_{dm}i^s_{qs} \quad \psi^s_{qm}i^s_{ds}\right) \tag{7.48}$$

$$|\hat{\psi}_m| = \sqrt{(\psi^s_{dm})^2 + (\psi^s_{qm})^2} \tag{7.49}$$

The terminal phase voltages of the machine are sensed and stator drops are sub-tracted to generate the air gap voltage signals. These signals are then integrated and combined to generate the desired signals as shown. Typically, if the machine speed does not fall below 10%, the stator drop compensation can be ignored to

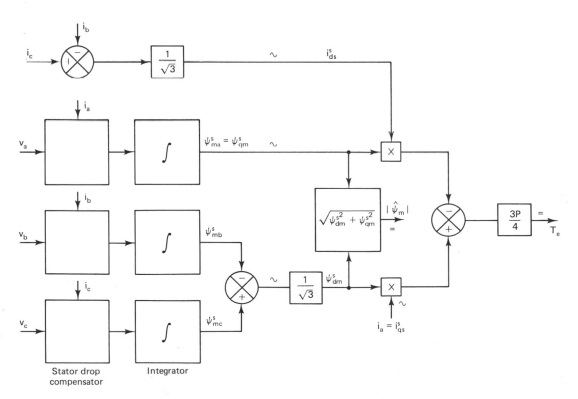

Figure 7.13   Estimation of torque and air gap flux signals.

give the computed accuracy within reasonable limits. Of course, this problem does not arise if flux coil voltages are available. The synthesis of the signals is accurate if the voltage and current waves are balanced and are assumed to be sinusoidal. In practice, the latter assumption is far from true. The $|\hat{\psi}_m|$ and T_e signals will contain harmonic ripple and therefore will require some filtering. Flux computation is discussed further in the next section.

The control scheme in Fig. 7.12 can operate satisfactorily in both the constant-torque and field weakening regions, and in the motoring and regeneration modes. The machine also has the capability to start with full-load torque. As the machine accelerates with maximum torque, the slip ω_{sl}^* remains clamped at a positive value and the frequency ω_e increases proportional to speed. The increase in ω_e tends to weaken the flux, but the flux control loop proportionately increases the voltage to maintain the flux constant. Beyond the base speed, the inverter operates in the square-wave mode, weakening the field flux, and therefore the flux loop loses its control. For excursion into the constant-power and "series motor" regions, the limiting slip can be programmed as shown in Fig. 2.11.

Current-controlled PWM inverter drive.   Instead of controlling the inverter voltage by the flux control loop, the stator current can also be controlled. The current control feature is important in a GTO or thyristor inverter because transient current surge may cause commutation failure. This is more important in a transistor inverter because of its limited surge current capability. In fact, the current control loop command can be generated from the slip command through a function generator to control the air gap flux indirectly. This scheme eliminates the feedback flux signal and is discussed further later. A current-controlled PWM inverter using the bang-bang current control method is shown in Fig. 7.14. The drive system can be used in electric-vehicle-type applications where torque instead of speed control is required and the supply battery can freely absorb energy during regeneration. The flux control loop generates the magnitude of the stator current command and the frequency command is generated by the torque control loop. With these as inputs, a three-phase sine-wave generator generates the balanced reference current waves for the three phases. Individual phase stator current waves are sensed, compared with the reference waves within a hysteresis band, and transistor base drive signals are generated through the lock-out circuits, as discussed in Chapter 4. The reference waves can be generated by ROM look-up tables and D/A converters. The analog frequency signal can drive a voltage-controlled oscillator and then, through a counter, a ROM-based three-phase look-up table can be retrieved. The output is converted to an analog signal through a multiplying-type D/A converter where the amplitude is modulated by the current signal $|I_s^*|$. Alternatively, a microcomputer can generate the waveforms with the help of A/D and D/A converters.

The performance of the drive system in both the acceleration and braking modes is explained in Fig. 7.15. This figure is the same as Fig. 2.11 except that the braking region has been added. The performance is somewhat symmetrical in both modes and therefore an explanation will be given for the motoring mode only. The drive system has three operating regions: the constant-torque region,

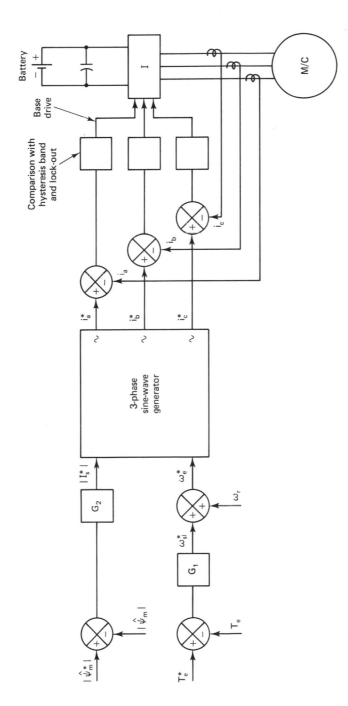

Figure 7.14 Bang-bang current-controlled PWM inverter drive.

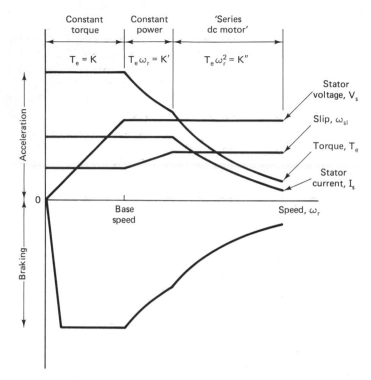

Figure 7.15 Drive characteristics in acceleration and braking modes.

the constant-power region, and the equivalent series dc motor region, where the product of torque and speed squared is constant. In the constant-torque region, the inverter operates in the PWM current control mode and has the characteristics of a current-fed inverter. But beyond the base speed, the inverter operates in the square-wave mode and the current control by PWM is lost. Assume that the machine accelerates from zero speed with full torque and with limit slip and stator current. It is assumed that the maximum stator current limits the slip below that of the breakdown torque. The stator voltage increases proportional to speed until the base speed is reached. In the transition PWM region below the base speed, there will be fewer chops near the edges of the square-wave until the inverter smoothly transitions into the square-wave region, where direct current control is lost. In the constant-power region, the slip is increased to the maximum value in a preprogrammed manner so that the stator current remains constant. Beyond the constant-power region, the slip remains constant but the stator current decreases as shown. The drive system may operate at reduced torque at any speed by reduction of slip. The available braking torque diminishes as the speed is reduced and vanishes at zero speed, as shown in the figure.

Operation of multiple machines. In a voltage-fed inverter scheme, multiple inverters can operate on a single rectifier, or multiple machines can be operated on a single inverter. In many applications, such as conveyer lines, extruder mills, and subway transit cars, several identical induction motors are required to operate in parallel on one inverter. If the machines have matched torque–speed characteristics, these will offer identical equivalent impedance on the variable-frequency supply line, and the developed torque and speed of each machine will be identical. In practice, there will be a small amount of mismatch and as a result, the torque–speed curves will not be identical. If the machines have very low slip characteristics, they will behave almost like synchronous machines and the speeds will be nearly matched. On the other hand, the speed matching may be worse for high-slip machines. If the machines are constrained to have identical speed, the slip will be the same but the machine with the lower slip characteristic will share the higher torque. If the machines' drive independent loads and closed-loop speed control are desired, the mean value of the speed can be used as a feedback signal, with the inverter operating in the voltage control mode.

In subway transit car or electric locomotive applications, the parallel operation of machines with unequal wheel diameters creates a special problem, which is explained in Fig. 7.16. It is assumed that an inverter has two machines in parallel and that each is driving an axle of the locomotive. The axle with the slightly smaller wheel diameter is driven by machine 1 and it is slightly mismatched from machine 2, as shown. During motoring operation, the control system will command the mean slip, but machine 1 will be constrained to move at a higher speed than machine 2, and as a result the torque distribution will be unequal (*A* and *B*), as shown in the figure. The mismatched torque distribution will be valid even if the machines have matched characteristics. Machine 1 shares the smaller torque in motoring, but in regeneration its torque share becomes larger. For the same speed differential, the torque mismatch becomes worse if the machines have lower slip characteristics, and it is quite possible that one machine may develop motoring torque while the other generates braking torque when both of them are supposed to operate in either the motoring or the regeneration mode. Again, the torque sharing of machine 2 may be excessive in the motoring mode and will tend to

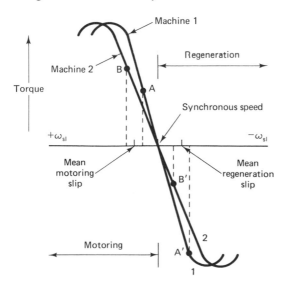

Figure 7.16 Effect of unequal wheel diameters in parallel machine transit car drive.

induce wheel slippage. The increment of speed as a result of slippage will decrease its torque sharing and therefore will tend to have self-correction. In the regenerating mode, on the other hand, if machine 1 exceeds the adhesion limit, its torque sharing will be larger, worsening the condition.

Current-Fed Inverter Control

Some of the control principles of voltage-fed inverters as discussed above are also valid for current-fed inverters. It was mentioned before that a voltage-fed inverter drive with current control has the characteristics of a current-fed inverter drive.

Independent current and slip control. In a current-fed inverter drive, the dc link current and inverter frequency are the two control parameters where the current can be varied by modulating the firing angle of the front-end rectifier. Unfortunately, a current-fed inverter cannot be controlled in an open-loop manner like a voltage-fed inverter. A minimum closed-loop control system of a current-

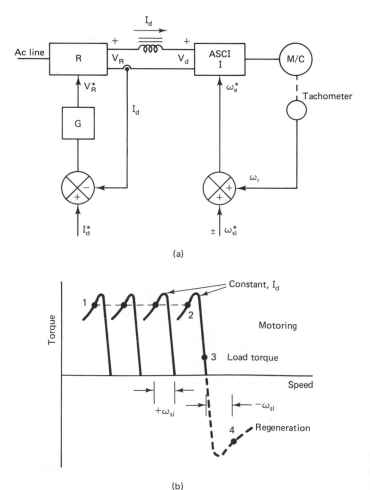

(a)

(b)

Figure 7.17 (a) Independent current and slip control; (b) torque–slip characteristics at constant dc current.

fed inverter where the current and slip are controlled independently is shown in Fig. 7.17(a), and part (b) shows its performance characteristics. The dc link current I_d is controlled by a feedback loop that controls the rectifier output voltage V_R; and the commanded slip is added with the speed signal to generate the frequency command. The machine can be operated in the regeneration mode with a negative slip command when both the V_R and V_d voltage polarities reverse and energy is pumped back to the source. The principal disadvantage of the system is that the machine air gap flux has no control. In the system, the torque can be controlled either by controlling dc link current I_d or slip ω_{sl}. The machine acceleration at constant torque from point 1 to point 2 under constant-current and constant-slip conditions is shown in Fig. 7.17. Intentionally, the operation is in the statically unstable region of the torque–speed curve so that the air gap flux remains below saturation, which was explained in Chapter 2. At steady-state conditions, if the slip is reduced to balance the developed and load torque, the operation will be at point 3, causing flux saturation. On the other hand, if the current I_d is decreased at constant slip to balance the load torque, the machine will operate at weak flux. Because of these problems, the control is rarely used in practice.

Constant Flux Operation with Programmable Current Control

A practical and much improved speed control scheme using a current-fed inverter is shown in Fig. 7.18, where the command current I_d^* is generated as a function of slip ω_{sl}^* so as to maintain the air gap flux constant. It was mentioned before that

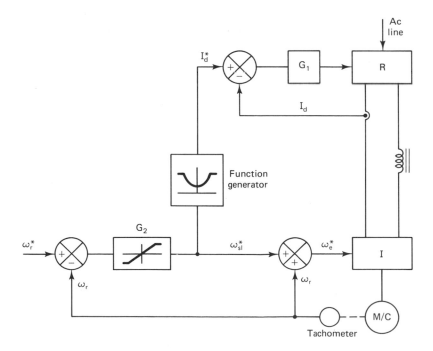

Figure 7.18   Programmable current control for constant-flux operation.

maintaining the rated air gap flux gives improved stability and fast transient response of the drive system.   At zero slip, the developed torque is zero, but the
current I_d has a minimum value that corresponds to the magnetizing current of the
machine.   As the slip is increased, I_d is also increased so as to maintain the
equivalent volts/hertz constant, as was explained in Fig. 2.12.   For the condition
of negative slip, the $I_d^* - \omega_{sl}^*$ relation is symmetrical to that of the first quadrant and
the drive system can be operated satisfactorily in all four quadrants.   The function
generator output can be precomputed accurately for a particular machine.   If the
parameter variations are ignored, the machine behaves as a dc machine with a
constant air gap flux at steady-state conditions.   Instead of controlling the slip
directly from the speed control loop, the speed loop error can control the current
I_d^* and then the slip command ω_{sl}^* can be generated as a function of I_d^*.   For
regenerative operation in this scheme, the slip polarity can be reversed when the
speed loop error becomes negative.   The control performances of both loops are
nearly identical.   These control principles are also applicable to voltage-fed inverters.

Tighter control of the air gap flux may be obtained by an independent flux
control loop as shown in Fig. 7.12, except here the flux loop error controls the
inner current control loop.   It should be noted that in both direct and indirect
control of flux by the current control loop, control is lost when the rectifier saturates
at full voltage.   The system will then continue to operate satisfactorily as a voltage-
fed inverter in the field weakening mode under the slip control.   The flux command
can be reduced in inverse relation to the speed so that the I_d control loop also
remains active in the field weakening mode.   Field weakening control is discussed
further in the next section.

A control scheme known as angle control has been developed (Ref. 8) where
an additional control loop of torque angle (sin θ) is provided over the slip control
loop.   The angle θ is the angle between the air gap flux and the stator current and
is related to torque by the expression

$$T_e = \frac{3}{2} \left(\frac{P}{2} \right) \left| \hat{\psi}_m \right| \left| \hat{I}_s \right| \sin \theta$$

The scheme gives a somewhat improved torque response by rotating the current
vector $\hat{I}_s$ transiently, but the performance is inferior to vector control, which is
described in the following section.

Slip Power Recovery Control

Between the two methods of slip power recovery control (i.e., the static Kramer
method and the static Scherbius method), the former will be described here and
the latter will be treated in the next section using the vector control method.   A
speed control system using the static Kramer method is shown in Fig. 7.19.   As
explained before, this type of drive has characteristics similar to that of a separately
excited dc machine, and therefore the control configuration is analogous to a phase-
controlled rectifier dc drive system.   With a constant air gap flux, the torque is

proportional to dc link current I_d, which is controlled by a feedback loop. Instead of sensing I_d directly, 60-Hz line current is sensed by current transformers and rectified to get the I_d signal. The speed control loop is added over the current control loop, as shown. The control is considerably simple but a disadvantage is that the drive system can be controlled in one quadrant only, as discussed in Chapter 6. If the command speed is increased by a step, the motor accelerates at constant developed torque corresponding to the I_d^* limit set by the speed control loop. As the actual speed approaches the command speed, the dc link current is reduced so as to balance with the load torque. If the speed command is decreased by a step, I_d approaches zero and the machine slows down with load torque. Then as the speed error tends to be zero in the steady state, I_d is restored to balance with the load torque. The air gap flux remains approximately constant during the whole operation, as dictated by the stator voltage and frequency. With a finite load torque, the motor speed is always less than the synchronous speed, as shown in Fig. 7.9(b). For motoring operation at true synchronous speed (i.e., the machine behaves as a synchronous motor), the rotor requires dc excitation current. The dc power can be supplied by operating the inverter as a rectifier, but the inverse polarity of the dc link voltage will short circuit the diode rectifier, bypassing the rotor winding. Synchronous motor operation is possible with the Scherbius drive system and is discussed later.

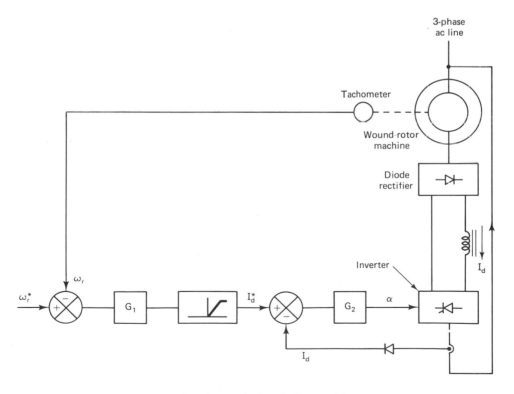

Figure 7.19   Speed control of static Kramer drive system.

7.5 VECTOR CONTROL METHODS

In the scalar control methods of voltage-fed and current-fed inverter drives discussed so far, the voltage or current and the frequency are the basic control variables of the induction motor. In a voltage-fed drive, for example, both the torque and air gap flux are functions of voltage and frequency. This coupling effect is responsible for the sluggish response of the induction motor. If, for example, the torque is increased by incrementing the frequency (i.e., the slip), the flux tends to decrease. But it is compensated by the sluggish flux control loop feeding in additional voltage. This transient dipping of flux reduces the torque sensitivity with slip and therefore lengthens the response time. The explanation is equally valid for a current-fed drive system.

The foregoing limitation can be overcome by applying vector or field-oriented control methods. This control method is applicable to both induction and synchronous machines, and the latter is described in Chapter 8. In the vector control method, an ac machine is controlled like a separately excited dc machine. This analogy is explained in Fig. 7.20. In a dc machine, neglecting the armature demagnetization effect and field saturation, the torque is given by

$$T_e = K_t' I_a I_f \tag{7.50}$$

where I_a is the armature or torque component of current and I_f is the field or flux component of current. In a dc machine, the control variables I_a and I_f can be considered as orthogonal or decoupled "vectors." In normal operation, the field current I_f is set to maintain the rated field flux and torque is changed by changing

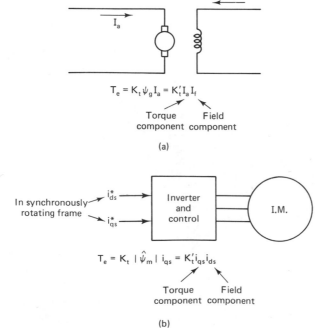

Figure 7.20 Induction motor and dc machine analogy in vector control.

the armature current. Since the current I_f or the corresponding field flux is decoupled from the armature current I_a, the torque sensitivity remains maximum in both transient- and steady-state operations. This mode of control can be extended to an induction motor also if the machine operation is considered in a synchronously rotating reference frame where the sinusoidal variables appear as dc quantities. In Fig. 7.20 the induction motor with inverter and control is shown with two control inputs, i_{ds}^* and i_{qs}^*. The currents i_{ds} and i_{qs} are the direct-axis component and quadrature-axis component, respectively, of the stator current, where both are in a synchronously rotating reference frame. In vector control, i_{ds} is analogous to the field current I_f and i_{qs} is analogous to the armature current I_a of a dc machine. Therefore, the torque can be expressed as

$$T_e = K_t |\hat{\psi}_m| i_{qs} = K_t' i_{qs} i_{ds} \tag{7.51}$$

The basic concept of how i_{ds} and i_{qs} can be established as control vectors in the vector control method is explained in Fig. 7.21 with the help of phasor diagrams in a synchronously rotating d^e–q^e reference frame. For simplicity, the rotor leakage inductance is neglected. The phasor diagram is drawn with the air gap voltage $\hat{V}_g$ aligned with the q^e axis. The stator current $\hat{I}_s$ lags the voltage $\hat{V}_g$ by $(90 - \theta)°$, i.e., $i_{qs} = \hat{I}_s \sin \theta$ is in phase with $\hat{V}_g$ and $i_{ds} = \hat{I}_s \cos \theta$ is in quadrature with $\hat{V}_g$. The current i_{qs} is the active or torque component of the stator current and the corresponding active power across the air gap is $\hat{V}_g i_{qs}$. The current i_{ds} is the reactive or field component of the stator current and is responsible for establishing the air gap flux $\hat{\psi}_m$. The corresponding reactive power across the air gap is $\hat{V}_g i_{ds}$. From the phasor diagram, the developed torque across the air gap is given by $T_e = K_t |\hat{\psi}_m| i_{qs} = K_t' i_{qs} i_{ds}$, where i_{qs} and i_{ds} are shown in Fig. 7.21. The torque equation is therefore identical to that of a dc machine. The variables i_{qs} and i_{ds} are mutually decoupled and can be independently varied without affecting the orthogonal component. For normal operation, as in a dc machine, the current i_{ds} remains constant and the torque is varied by varying the i_{qs} component. Correspondingly, the polar position of $\hat{I}_s$ shifts to $\hat{I}_s'$ as shown. Vector control can be implemented in either Cartesian or polar form, and this is discussed later.

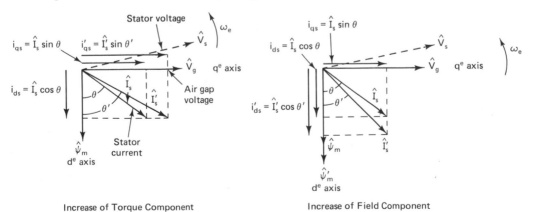

Figure 7.21 Phasor diagrams in direct vector control (in terms of peak values).

The fundamentals of vector control implementation with the machine model can be explained with the help of Fig. 7.22. The inverter is omitted in the figure and it is assumed to generate the ideal phase current waves i_a, i_b, and i_c as dictated by the corresponding reference waves generated by the controller. The machine model is shown on the right. The phase currents i_a, i_b, and i_c are converted to i_{qs}^s and i_{ds}^s components by three-phase/two-phase transformation. These are then converted to a synchronously rotating reference frame by the unit vectors $\cos \omega_e t$ and $\sin \omega_e t$ as shown, before being applied to the machine model. The controller makes two stages of inverse transformation so that the control parameters i_{ds}^* and i_{qs}^* correspond to machine variables i_{ds} and i_{qs}, respectively. The unit vectors assure the alignment of i_{ds} with the $\hat{\psi}_m$ phasor and i_{qs} with the $\hat{V}_g$ phasor. Note that the transformation and inverse transformation do not incorporate any dynamics, and therefore the response to i_{ds} and i_{qs} is instantaneous.

It should be mentioned here that in addition to fast transient response due to decoupling control, the conventional stability problem of an induction motor, that is by crossing the breakdown torque point, does not exist here. The control can easily be designed to have four-quadrant operation. Therefore, the vector-controlled induction motor drives can be used for high-performance applications (i.e., servo drives, steel mill controls, etc.), where traditionally, dc machines have been used.

There are essentially two general methods of vector control. One, called the direct method, was developed by F. Blaschke (Ref. 11), and the other, known as the indirect method, was developed by K. Hasse (Ref. 12). The methods relate to how the unit vector signals $\cos \omega_e t$ and $\sin \omega_e t$ are generated.

Direct Method of Vector Control

The basic block diagram of the direct vector control method for a PWM current-controlled inverter is shown in Fig. 7.23. The principal control parameters i_{ds}^* and i_{qs}^*, which are dc quantities, are converted to a stationary reference frame with the

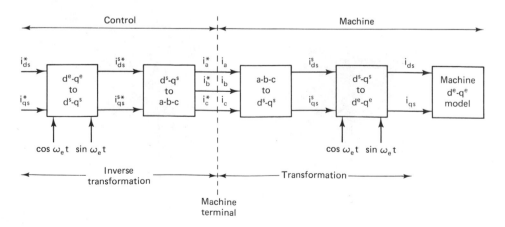

Figure 7.22 Vector control implementation with machine model.

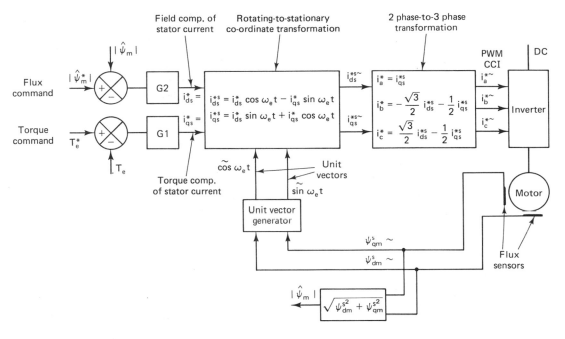

Figure 7.23 Direct method of vector control of a voltage-fed inverter (rotor leakage inductance neglected).

help of cos $\omega_e t$ and sin $\omega_e t$ signals generated from flux signals as shown. The resulting stationary frame signals are then converted to phase current commands for the inverter. The transformation equations, which are discussed in Chapter 2, are shown in the boxes. A flux control loop has been added for precision flux control. The current i_{qs}^* is generated from a torque control loop which may have an outer speed control loop if desired. The current i_{qs}^* becomes negative for negative torque, and correspondingly the phase position of i_{qs} becomes reversed in Fig. 7.21.

The direct vector control method depends on the generation of unit vector signals from the air gap fluxes. The air gap fluxes ψ_{dm}^s and ψ_{qm}^s can be measured directly or estimated from the stator voltage and current signals, as explained in Fig. 7.13. The currents i_{ds} and i_{qs} are to be aligned with the rotating frame d^e and q^e axes, respectively, using the unit vectors. We can write the following relations (see Fig. 7.24):

$$|\hat{\psi}_m| = \sqrt{\psi_{dm}^2 + \psi_{qm}^2} = \sqrt{(\psi_{dm}^s)^2 + (\psi_{qm}^s)^2} \qquad (7.52)$$

$$\psi_{dm}^s = |\hat{\psi}_m| \cos \omega_e t \qquad (7.53)$$

$$\psi_{qm}^s = |\hat{\psi}_m| \sin \omega_e t \qquad (7.54)$$

Equations (7.53) and (7.54) indicate that cos $\omega_e t$ and sin $\omega_e t$ are cophasal to ψ_{dm}^s and ψ_{qm}^s, respectively. The synthesis of unit vectors from ψ_{dm}^s and ψ_{qm}^s by the feedback control principle is shown in Fig. 7.24.

So far we have considered the vector control method neglecting the rotor

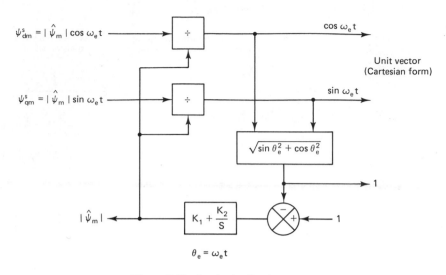

$$\psi_{dm}^s = | \hat{\psi}_m | \cos \omega_e t$$

$$\cos \omega_e t$$

Unit vector
(Cartesian form)

$$\psi_{qm}^s = | \hat{\psi}_m | \sin \omega_e t$$

$$\sin \omega_e t$$

$$\sqrt{\sin \theta_e^2 + \cos \theta_e^2}$$

$$1$$

$$| \hat{\psi}_m |$$

$$K_1 + \frac{K_2}{S}$$

$$1$$

$$\theta_e = \omega_e t$$

Figure 7.24 Synthesis of unit vectors.

leakage inductance.   It can be shown that the rotor leakage flux exerts a considerable amount of influence and therefore cannot be neglected.   In fact, the rotor flux should be considered in both the vector and scalar control methods rather than the air gap flux.   Blaschke has shown that the vector control on the basis of air gap flux may result in an undesirable stability problem.   The air gap flux can be compensated for the rotor leakage as follows:

$$\psi_{qr}^s = L_m i_{qs}^s + L_r i_{qr}^s \tag{7.55}$$

$$\psi_{qm}^s = L_m i_{qs}^s + L_m i_{qr}^s \tag{7.56}$$

Eliminating i_{qr}^s from equation (7.55) yields

$$\psi_{qr}^s = \frac{L_r}{L_m} \psi_{qm}^s - L_{lr} i_{qs}^s \tag{7.57}$$

Similarly, from the d^e-axis equivalent circuit,

$$\psi_{dr}^s = \frac{L_r}{L_m} \psi_{dm}^s - L_{lr} i_{ds}^s \tag{7.58}$$

The synthesis of rotor fluxes from equations (7.57) and (7.58) are shown in Fig. 7.25.   The rotor flux $| \hat{\psi}_r |$ and unit vectors for Fig. 7.23 can then be given as

$$| \hat{\psi}_r | = \sqrt{\psi_{dr}^2 + \psi_{qr}^2} = \sqrt{(\psi_{dr}^s)^2 + (\psi_{qr}^s)^2} \tag{7.59}$$

$$\cos \omega_e t = \frac{\psi_{dr}^s}{| \hat{\psi}_r |} \tag{7.60}$$

$$\sin \omega_e t = \frac{\psi_{qr}^s}{| \hat{\psi}_r |} \tag{7.61}$$

The direct vector control method has so far been discussed using instantaneous current-controlled PWM inverter.   The current control is logical since the developed torque is related to currents rather than voltages.   Although current control

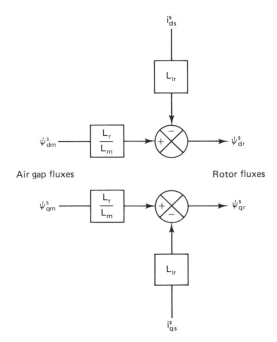

Figure 7.25 Synthesis of rotor fluxes.

is desirable, the inverter can have voltage control within the current loop.   The instantaneous phase current command can be compared with the actual phase current and the error through a PI compensator can generate the phase voltage command, which can then generate the PWM voltage wave through a triangular carrier wave.   Alternatively, i_{ds}^* and i_{qs}^* commands can be compared with the respective actual i_{ds} and i_{qs} currents and the errors can generate the respective voltage commands v_{ds}^* and v_{qs}^* through PI compensators.   These can then be converted to phase voltages to control the inverter, as explained later.

The direct method of vector control described so far can be applied typically above 10% of the base speed because of the difficulty in accurate flux signal synthesis at low speed.   In fact, the flux signals obtained by direct integration of phase voltages can be used only in a higher speed range.   The resulting coupling effect, although small at higher speed, becomes worse as the speed is reduced.   In applications such as servo drives, the drive system must operate at truly zero speed with the best possible transient response.   The accurate stator drop compensation near zero speed is very difficult.   In the low-speed region, the rotor flux can be synthesized more accurately from speed and stator current signals.   The rotor equation of the q^s-axis stationary frame equivalent circuit can be given as [see Fig. 2.18(a)]

$$\frac{d\psi_{qr}^s}{dt} + i_{qr}^s R_r - \omega_r \psi_{dr}^s = 0 \tag{7.62}$$

Adding the term $(L_m R_r / L_r) i_{qs}^s$ on both sides of the equation, we have

$$\frac{d\psi_{qr}^s}{dt} + \frac{R_r}{L_r}(L_m i_{qs}^s + L_r i_{qr}^s) - \omega_r \psi_{dr}^s = \frac{L_m R_r}{L_r} i_{qs}^s \tag{7.63}$$

Substituting equation (7.55) and simplifying yield

$$\frac{d\psi_{qr}^s}{dt} = \frac{L_m}{T_R} i_{qs}^s + \omega_r \psi_{dr}^s - \frac{1}{T_R} \psi_{qr}^s \tag{7.64}$$

Similarly, the equation from the d^s-axis equivalent circuit can be derived as

$$\frac{d\psi_{dr}^s}{dt} = \frac{L_m}{T_R} i_{ds}^s + \omega_r \psi_{qr}^s - \frac{1}{T_R} \psi_{dr}^s \tag{7.65}$$

where $T_R = L_r/R_r$ is the rotor circuit time constant. Equations (7.64) and (7.65) give rotor fluxes as functions of stator current and speed, and the simulation diagram for estimation is given in Fig. 7.26. In the front end, i_{qs}^s and i_{ds}^s signals can be generated from the phase current signals by three-phase/two-phase transformation. The generation of rotor flux $|\hat{\psi}_r|$ and unit vector signals are shown on the right. This method of flux estimation remains valid from zero to the maximum speed range. However, note that the estimation is dependent on machine parameters; the rotor resistance variation, especially, becomes dominant due to the temperature variation and skin effect. In the direct method of unit vector synthesis shown in Fig. 7.24, the parameter variation effect is not significant as long as the machine operates at reasonably high speed. The harmonic distortion of signals causes some problem in the direct vector control method. Installation of a filter gives a frequency-sensitive phase shift and therefore worsens the coupling effect.

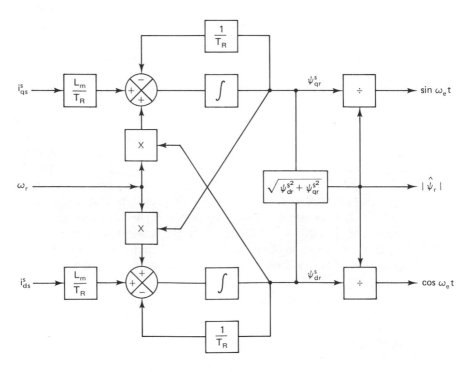

Figure 7.26 Rotor flux estimation from speed and stator currents.

Indirect Method of Vector Control

In the direct method of vector control discussed so far, the synthesis of unit vectors is dependent on the machine terminal conditions. In the indirect method of vector control to be discussed here, this dependence does not arise and therefore the distortion problem does not exist.

Figure 7.27 explains the indirect vector control principle with the help of a phasor diagram. The d^s–q^s axes are fixed on the stator while the d^e–q^e axes rotate at synchronous angular velocity ω_e as shown. At any instant, the q^e electrical axis is at angular position θ_e with respect to the q^s axis. The angle θ_e is given by the sum of rotor angular position θ_r and slip angular position θ_{sl}, where $\theta_e = \omega_e t$, $\theta_r = \omega_r t$, and $\theta_{sl} = \omega_{sl} t$. The rotor flux $\hat{\psi}_r$, consisting of the air gap flux and the rotor leakage flux, is aligned to the d^e axis as shown. Therefore, for decoupling control, the stator flux component of current i_{ds} and the torque component of current i_{qs} are to be aligned to the d^e and q^e axes, respectively.

We can write the following equations from rotating frame d^e–q^e equivalent circuits:

$$\frac{d\psi_{qr}}{dt} + R_r i_{qr} + (\omega_e - \omega_r)\psi_{dr} = 0 \tag{7.66}$$

$$\frac{d\psi_{dr}}{dt} + R_r i_{dr} - (\omega_e - \omega_r)\psi_{qr} = 0 \tag{7.67}$$

Again,

$$\psi_{qr} = L_r i_{qr} + L_m i_{qs} \tag{7.68}$$

$$\psi_{dr} = L_r i_{dr} + L_m i_{ds} \tag{7.69}$$

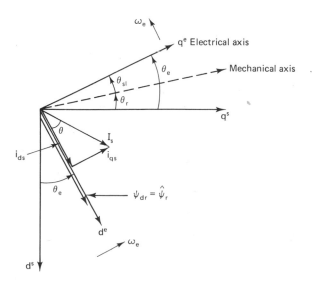

Figure 7.27 Phasor diagram for indirect vector control.

From equations (7.68) and (7.69),

$$i_{qr} = \frac{1}{L_r} \psi_{qr} - \frac{L_m}{L_r} i_{qs} \tag{7.70}$$

$$i_{dr} = \frac{1}{L_r} \psi_{dr} - \frac{L_m}{L_r} i_{ds} \tag{7.71}$$

The rotor currents from equations (7.66) and (7.67) can be eliminated by substituting equations (7.70) and (7.71) as

$$\frac{d\psi_{dr}}{dt} + \frac{R_r}{L_r} \psi_{qr} - \frac{L_m}{L_r} R_r i_{qs} + \omega_{sl} \psi_{dr} = 0 \tag{7.72}$$

$$\frac{d\psi_{dr}}{dt} + \frac{R_r}{L_r} \psi_{dr} - \frac{L_m}{L_r} R_r i_{ds} - \omega_{sl} \psi_{qr} = 0 \tag{7.73}$$

where $\omega_{sl} = \omega_e - \omega_r$.

For decoupling control it is desirable that

$$\psi_{qr} = \frac{d\psi_{qr}}{dt} = 0$$

$$\psi_{dr} = \hat{\psi}_r = \text{constant}$$

$$\frac{d\psi_{dr}}{dt} = 0$$

Substituting the first two conditions, equations (7.72) and (7.73) can be simplified as

$$\omega_{sl} = \frac{L_m}{\hat{\psi}_r} \left(\frac{R_r}{L_r}\right) i_{qs} \tag{7.74}$$

$$\frac{L_r}{R_r} \frac{d\hat{\psi}_r}{dt} + \hat{\psi}_r = L_m i_{ds} \tag{7.75}$$

Again, the torque as a function of rotor flux and stator current can be derived as follows: The stator flux linkage relations can be written from Fig. 2.17 as

$$\psi_{qs} = L_m i_{qr} + L_s i_{qs} \tag{7.76}$$

$$\psi_{ds} = L_m i_{dr} + L_s i_{ds} \tag{7.77}$$

Substituting equations (7.76) and (7.77) in equations (7.68) and (7.69), we get

$$\psi_{qs} = \left(L_s - \frac{L_m^2}{L_r}\right) i_{qs} + \frac{L_m}{L_r} \psi_{qr} \tag{7.78}$$

$$\psi_{ds} = \left(L_s - \frac{L_m^2}{L_r}\right) i_{ds} + \frac{L_m}{L_r} \psi_{dr} \tag{7.79}$$

The torque equation as a function of stator currents and stator fluxes is

$$T_e = \frac{3}{2}\left(\frac{P}{2}\right)(i_{qs}\psi_{ds} - i_{ds}\psi_{qs}) \tag{7.80}$$

Equations (7.78) and (7.79) can be substituted in equation (7.80) to eliminate the stator fluxes. Therefore,

$$T_e = \frac{3}{2}\left(\frac{P}{2}\right)\frac{L_m}{L_r}(i_{qs}\psi_{dr} - i_{ds}\psi_{qr}) \tag{7.81}$$

Substituting $\psi_{qr} = 0$ and $\psi_{dr} = \hat{\psi}_r$, the torque expression is

$$T_e = \frac{3}{2}\left(\frac{P}{2}\right)\frac{L_m}{L_r}i_{qs}\hat{\psi}_r \tag{7.82}$$

The equations above, together with the mechanical equation

$$\left(\frac{2}{P}\right)J\frac{d\omega_r}{dt} = T_e - T_L \tag{2.78}$$

describe the machine model in decoupling control as shown in Fig. 7.28. The inverter is assumed to be current controlled, and the delay between the command and response currents are neglected. The developed torque T_e responds instantaneously with current i_{qs}, but has delayed response due to i_{ds}. The analogy of the model with a separately excited dc machine is obvious.

For implementation of indirect vector control, it is necessary to take equations (7.74) and (7.75) into consideration. Figure 7.29 shows a position servo system using the indirect method of vector control. The flux component of current i_{ds}^* for the desired rotor flux $\hat{\psi}_r$ is determined from equation (7.75) and is maintained constant here. The torque component of current i_{qs}^* is derived from the speed control loop as usual. The set value of slip ω_{sl}^* is related to current i_{qs}^* by equation (7.74). The slip-angle vectors $\sin\theta_{sl}^*$ and $\cos\theta_{sl}^*$, which determine the desired electrical axis with respect to the rotor-mechanical axis, are generated in a feed-forward manner from the ω_{sl}^* signal through a VCO, a counter, and a ROM-based SIN/COS generator. The rotor-position vectors $\cos\theta_r$ and $\sin\theta_r$ are obtained from angle encoder and are added with the slip vectors to obtain the $\cos\theta_e$ and $\sin\theta_e$

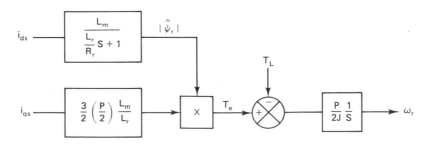

Figure 7.28   Block diagram of machine model with decoupling control.

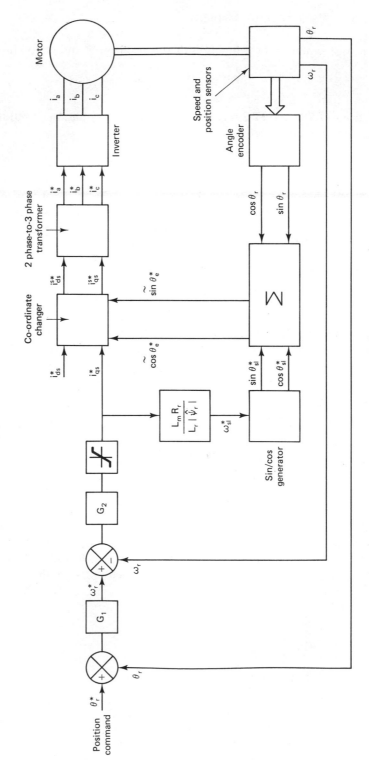

Figure 7.29 Position servo system with indirect vector control.

signals as follows:

$$\cos \theta_e^* = \cos (\theta_r + \theta_{sl}^*) = \cos \theta_r \cos \theta_{sl}^* - \sin \theta_r \sin \theta_{sl}^* \qquad (7.83)$$

$$\sin \theta_e^* = \sin (\theta_r + \theta_{sl}^*) = \sin \theta_r \cos \theta_{sl}^* + \cos \theta_r \sin \theta_{sl}^* \qquad (7.84)$$

The computations for the coordinate changer and two-phase/three-phase transformer are the same as in Fig. 7.23. Instead of processing the slip-angle and rotor-position vectors independently in Cartesian form as shown, the slip and rotor speed signals can be added directly and then the $\cos \theta_e$ and $\sin \theta_e$ signals can be synthesized by a VCO, a counter, and a SIN/COS waveform generator. In this case the motor will physically locate the field at any position and θ_r need not be absolute. Again, in the polar method of vector control, which is discussed later, coordinate changing can be done in polar form.

Figure 7.29 can be modified to incorporate control in the field weakening region. A controller block diagram to extend the operation in the field weakening region is shown in Fig. 7.30. Below the base speed, the machine operates at constant $|\hat{\psi}_r|$ and therefore operation is identical to that shown in Fig. 7.29. Above the base speed, $|\hat{\psi}_r|$ is weakened to be inversely proportional to speed so that the drive system remains under the vector control mode. Note that the flux is being controlled in an open-loop manner by solving equation (7.75).

With indirect vector control, the drive system can be operated in all the four quadrants as in the direct method, and speed can be controlled from truly zero to the full value. However, a rotor position signal becomes mandatory in this method. Again, the controller is dependent on machine parameters, and for ideal decoupling the controller parameters should track the machine parameters, which is extremely difficult to achieve. The dominant parameter to be considered is the rotor resistance, which has been estimated on-line by various methods, giving only limited success in ideal decoupling control.

In both the direct and indirect vector control methods discussed so far, the instantaneous phase current control has been used in the inverter. In the low-

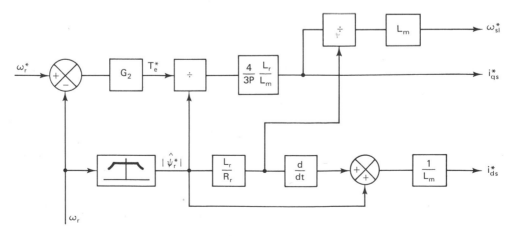

Figure 7.30 Control block diagram to extend operation in field weakening region.

speed region where machine counter emf is low, the current controller will track well. But at a higher speed, the current controller will tend to saturate in part of the cycle because of high counter emf. In this condition, the fundamental current magnitude will be less and its phase will deviate from that of the commanded current. The amplitude and phase error problems can be solved by the block diagram shown in Fig. 7.31. The machine phase currents are converted to the synchronously rotating frame to generate the v_{ds}^* and v_{qs}^* commands through the PI compensators as shown. These voltage commands are then converted into stationary frame instantaneous phase voltages. The rotating frame feedback loops with integral control assure amplitude and phase tracking of currents in the PWM voltage transition region and the supply voltage rises smoothly to the square wave. Obviously, addition of a loop with the integrating compensator adversely affects the system response.

If it is necessary to retain the instantaneous phase current control feature, Fig. 7.31 can also be used where the voltage commands are replaced by the corresponding current commands. Such an overlay current control scheme will remain satisfactory [Ref. 22] in both the unsaturated and saturated ranges of instantaneous current controller. In the low-speed region, the compensator output will track the loop command, but in partial saturation of instantaneous current controller, these outputs will be higher so that the loop error is forced to be zero by integrator action.

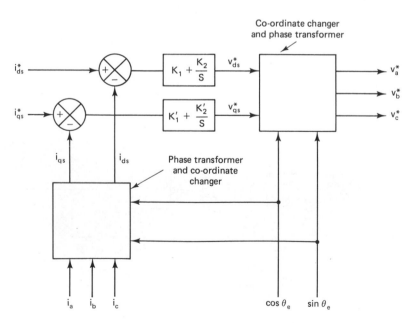

Figure 7.31 Control block diagram to overcome current controller saturation effect.

Control of Current-Fed Inverter

The direct and indirect vector control methods have so far been illustrated with current-controlled voltage-fed inverters. The principles can easily be extended to drives with other types of converters. It should be mentioned here that the vector control is not limited only to ac drives. The principle can be applied in general to three-phase ac systems to regulate the active and reactive powers independently.

Figure 7.32 shows direct method of vector control for a current-fed inverter drive using polar form. The control is valid in both the constant-torque and field weakening regions. The speed is being controlled in the outer loop and the loop error generates the torque command T_e^*, which is being divided by the flux $|\hat{\psi}_r|$ to generate the command current i_{qs}^*. The flux command $|\hat{\psi}_r^*|$ is constant in the constant-torque region but is weakened inversely proportional to speed in the constant-power region. The flux is controlled by a closed loop and the loop error generates the flux component of current i_{ds}^*. The currents i_{qs}^* and i_{ds}^* are converted to polar form as shown. The machine stator current $|\hat{I}_s^*|$ is related to the rectifier current and therefore it is closed-loop controlled by controlling the rectifier firing angle. The inverter frequency is controlled in the phase-locked-loop (PLL) mode so that the stator current is positioned at the desired torque angle θ with respect to the rotor flux. The scheme is sometimes defined as "angle control" of the machine. The rotor flux $|\hat{\psi}_r|$ and torque angle θ can be generated from Fig. 7.26 using the following additional relations:

$$i_{qs} = i_{qs}^s \cos \omega_e t - i_{ds}^s \sin \omega_e t \tag{7.85}$$

$$i_{ds} = i_{qs}^s \sin \omega_e t + i_{ds}^s \cos \omega_e t \tag{7.86}$$

$$\angle \theta = \tan^{-1} \frac{i_{qs}}{i_{ds}} \tag{7.87}$$

The control scheme remains valid from zero to the full speed and in all four quadrants.

A vector control scheme for the super/subsynchronous speed range using a static converter cascade on the rotor side of a wound-rotor machine is shown in Fig. 7.33. The line-side converter (C_1) is commutated by an ac line, whereas the machine-side converter (C_2) is force commutated. In the subsynchronous motoring and supersynchronous regeneration modes, the converter C_2 operates as a rectifier and C_1 as an inverter so that the slip energy can be recovered to the line. In subsynchronous regeneration and supersynchronous motoring modes, the converters reverse their roles and slip energy is fed to the rotor. This type of drive was discussed in detail in Chapter 6. The error from the speed control loop generates the torque command, which correspondingly controls the dc link current I_d by controlling the converter C_1. The converter C_2 is slaved and angle controlled by the phase-locked-loop scheme discussed in Fig. 7.32. Since the converter C_2 is force commutated and is required to control the active slip power only, the

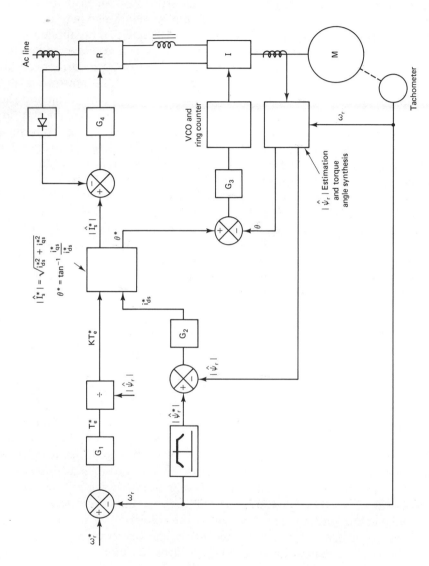

Figure 7.32 Vector control of current-fed inverter drive.

278

torque angle θ can be set to 90°.   For slip power recovery, the rotor current is in phase with the rotor voltage (i.e., $\theta = +90°$), and for feeding slip power the current is out of phase with voltage (i.e., $\theta = -90°$).   Therefore, the polarity of the θ^* angle in different modes of operation can be summarized as follows:

Sign of T_e^*	Super/sub direction	Sign of θ^*
+	SUB (+)	+
−	SUB (+)	−
+	SUPER (−)	−
−	SUPER (−)	+

The generation of the actual θ angle is explained in Fig. 7.33.   The unit vector θ_e in polar form is synthesized from the stator side and can be considered as aligned

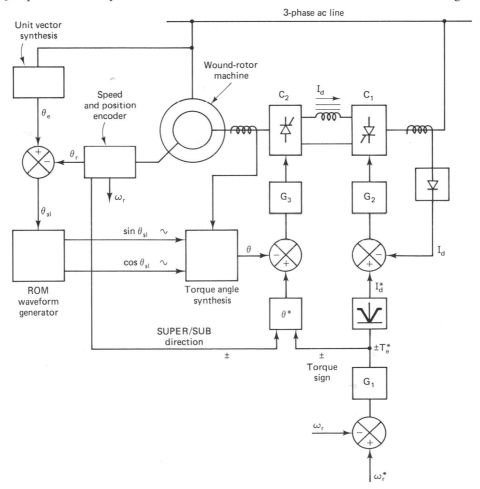

Figure 7.33   Super/subsynchronous speed control of induction motor by converter cascade.

approximately with the air gap flux. The angle θ_e is then subtracted by the rotor angle θ_r from the position encoder to generate the slip angle θ_{sl}, which is then converted to $\sin \theta_{sl}$ and $\cos \theta_{sl}$ signals by a ROM waveform generator. These signals are then combined with the rotor currents using the structure of equations (2.51), (2.52), and (7.85) to (7.87) and the torque angle θ is synthesized.

Control of Cycloconverter

The vector control methods discussed so far for voltage-fed and current-fed inverters are also applicable to cycloconverters supplying the stator of machines. In the usual case where the cycloconverter phase currents are individually controlled, the command phase currents can either be generated by direct method of Fig. 7.23 or by indirect method of Fig. 7.29. There is, however, some advantage in using the polar form of control shown in Fig. 7.32 when a microcomputer is used in the controller. In this method, as discussed before, the three phases of current are handled as a single vector and control is exercised by amplitude $|\hat{I}_s^*|$ and phase angle θ^*, the constituents of a polar current vector. These quantities are then controlled independently by feedback loops. The feedback $|\hat{I}_s|$ and θ parameters can be generated by the scheme of Fig. 7.26 and equations (7.85) to (7.87).

Figure 7.34 shows the vector control of the Scherbius drive system using the direct method. Here, as in Fig. 7.33, the cycloconverter is required to recover the slip energy in subsynchronous motoring and supersynchronous regeneration and to feed slip energy to the rotor in subsynchronous regeneration and supersynchronous motoring. The signals are processed here in Cartesian form instead of polar form shown in Fig. 7.33. The currents I_P and I_Q are the in-phase and quadrature components of rotor current, respectively, with respect to slip voltage. The error from the speed-control loop controls I_P, whereas I_Q can be set to an arbitrary value. The currents I_P and I_Q correspond to i_{qs} and i_{ds}, respectively, in vector control nomenclature. The polarity of the I_P signal is reversed if the machine operates in the supersynchronous region. The method of generating unit vectors $\sin \theta_{sl}$ and $\cos \theta_{sl}$ is the same as in Fig. 7.33. The cycloconverter phase current commands at slip frequency are then obtained by a standard rotating-to-stationary coordinate change and two-phase/three-phase transformation as shown. In extreme cases, if the rotor is considered stationary (i.e., $\theta_{sl} = \theta_e$), the machine acts as a transformer and the cycloconverter operating at line frequency controls active and reactive power circulation by vector control. Obviously, if the motor transitions from the subsynchronous to the supersynchronous range, the phase sequence of slip-angle vectors reverse to correspond to the actual slip-voltage phase sequence. At true synchronous speed, the machine can operate as a synchronous motor or generator, and in this condition the cycloconverter acts as a rectifier and supplies dc excitation current to the machine. At this condition, the slip-angle vectors freeze to dc values, and correspondingly, dc phase current commands are generated for the cycloconverter.

If the Scherbius drive system is used for VSCF generation as discussed in Chapter 6, the control strategy essentially remains the same except that the active and reactive currents I_P and I_Q of the cycloconverter are controlled to control the

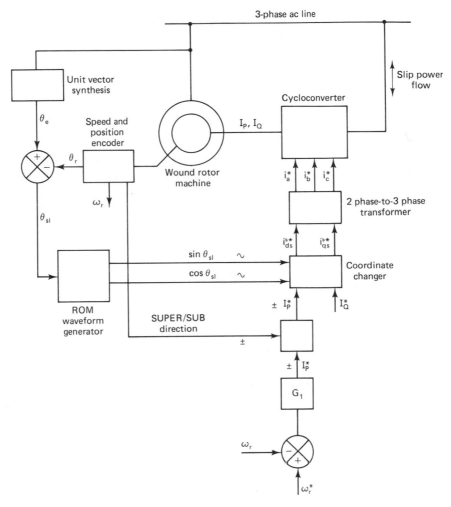

Figure 7.34   Vector control of Scherbius drive.

real and reactive powers, respectively, at the stator terminals by the feedback method: The operation corresponds to the continuous regenerative mode of operation of the Scherbius drive system.   If the stator bus is isolated and the cycloconverter is fed from the machine-shaft-mounted exciter, the unit vector θ_e is to be set independently.   The generation and processing of other signals remain the same as in Fig. 7.34.

7.6 ADAPTIVE CONTROL PRINCIPLES

So far we have discussed control systems with fixed controller parameters.   In a practical drive system, the plant parameters may vary, and as a result the system performance may deteriorate, giving instability in the extreme case.   The problem can be solved by adaptive control techniques. In adaptive control, the controller is forced to adapt to the plant operating condition based on on-line extraction of plant information. A powerful microcomputer is indeed an effective tool for implementation of this control method.

Self-Tuning Regulator

In this method, as the name indicates, the controller parameters are tuned to adapt to the plant parameter variations. An example is that the speed loop gain of a dc machine drive system may be varied proportionately with the shaft moment of inertia (J), provided that such information is available. In a more sophisticated self-tuning control (Ref. 23), the controller parameters track the plant parameters so that the system closed-loop poles, zeros, and gain remain unique under all operating conditions. Figure 7.35 explains the principle of self-tuning regulation. The plant parameter estimation algorithm solves the plant model in discrete-time form and updates the plant parameters on the basis of a recursive least-squares formulation. A tuning algorithm then adjusts the regulator parameters based on the estimate of plant parameters. The regulator parameters may be updated at a slower rate than the main control loop sampling rate assuming that the plant parameters vary slowly. This assumption is not necessarily valid in an ac drive system. For successful operation of the system, the global stability of the system is to be assured.

Model Referencing Control

In a model referencing adaptive control system, the plant response is forced to track the response of a reference model irrespective of plant parameter variation. The reference model with fixed parameters is stored in microcomputer memory and therefore the response of the plant becomes insensitive to parameter variation.

Consider an indirect vector-controlled induction motor servo drive system (Fig. 7.29) and assume that the control parameters have been matched with the

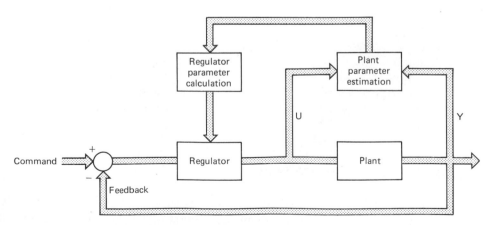

Figure 7.35 Block diagram of self-tuning regulation.

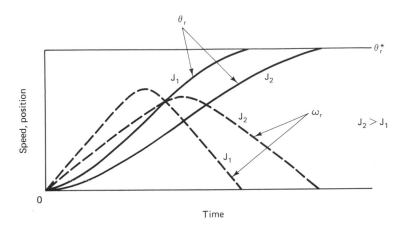

Figure 7.36 Response variation of servo drive with variation of moment of inertia (J).

machine parameters to achieve perfect decoupling. Ideally, the model of such a system is identical to that of a dc machine. The position loop response of the drive system can be given by a second-order transfer function in the constant-torque mode. A typical response for such a drive system with a variable-inertia ($J_2 > J_1$) load is shown in Fig. 7.36. The response variation with variation of inertia may not be desirable in a servo system. Such a problem can be solved by the model referencing adaptive control system shown in Fig. 7.37.

The speed command ω_r^* generated by the position control loop is applied in parallel to the reference model and plant controller as shown. The reference model output speed ω_{rm} is compared with the measured plant speed ω_r, and the resulting error signal e actuates the adaptation algorithm. The feedforward and feedback gains K_F and K_B, respectively, of the plant controller are iterated by the adaptation algorithm so as to dynamically reduce the error e to zero. The algorithm contains a PI-type control law so that the desired K_F and K_B values are locked in when the error vanishes to zero. The plant will be capable of tracking the reference model without saturation provided that the J parameter in the reference model is defined on a worst-case (maximum) basis. Therefore, the desired robustness of the control system is obtained at the sacrifice of optimum response speed.

The adaptation algorithm can be defined as (Ref. 15)

$$K_F = K_{FO} + FV\omega_r^* + \int_0^t GV\omega_r^* \, d\tau \tag{7.88}$$

$$K_B = K_{BO} + LV\omega_r + \int_0^t MV\omega_r \, d\tau \tag{7.89}$$

$$V = De \tag{7.90}$$

where K_{FO} and K_{BO} are initial gain values and F, G, L, M, and D are adaptation law constants. In general, the structure of the reference model and plant should be the same and the parameters should be compatible for satisfactory adaptation.

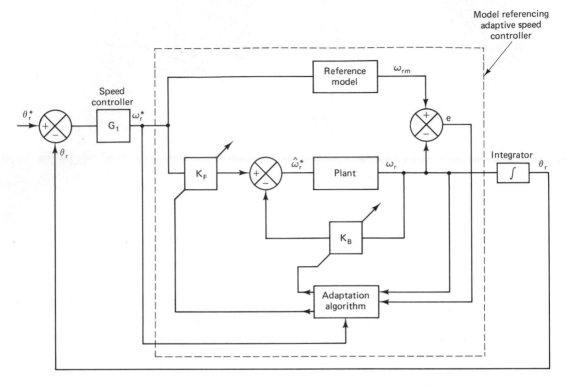

Figure 7.37   Model referencing adaptive servo control system.

In Fig. 7.37, the state space equations are

$$\text{Reference model:}\quad \dot{\omega}_{rm} = A_m\omega_{rm} + B_M\omega_r^* \tag{7.91}$$

$$\text{Plant:}\quad \dot{\omega}_r = A_p\omega_r + B_p\hat{\omega}_r^* \tag{7.92}$$

The other system equations are

$$\omega_r^* = G_1(\theta_r^* - \theta_r) \tag{7.93}$$

$$\hat{\omega}_r^* = K_F\omega_r^* + K_B\omega_r \tag{7.94}$$

$$e = \omega_{rm} - \omega_r \tag{7.95}$$

The parameters K_F and K_B vary with time.   The speed control system within the dashed line in Fig. 7.37 can be represented by an equivalent feedforward time-invariant linear system with a feedback nonlinear time-varying block.   The global stability of such a system can be analyzed by Popov's hyperstability theorem, and correspondingly the parameters F, G, L, M, and D can be determined.

Model Referencing Control with Search Strategy

Another method of model referencing adaptive control which is based on search strategy (Ref. 16) is illustrated in Fig. 7.38.   Here the plant under consideration may be an induction motor drive system using the indirect vector control method.

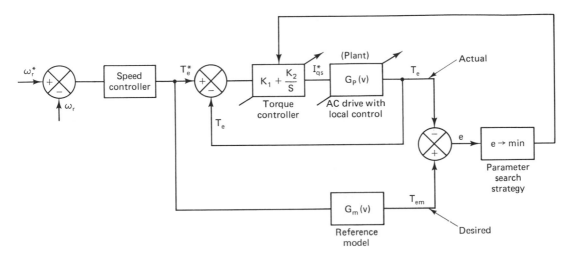

Figure 7.38 Block diagram of self-optimizing controller with search strategy. From "Self Optimizing Controller Employing Microprocessor," by G. Fromme in Conference Records on Microelectrics in Power Electronics and Electrical Drives. Oct. 1982.

The rotor resistance of the motor varies primarily with temperature and the mismatch with the control circuit parameter causes coupling between the direct and quadrature axes. As a result, the torque sensitivity deteriorates and the machine response is dictated by a higher-order state space equation. Providing a high-gain torque control loop can minimize the drift in transient torque response due to resistance variation, but the gain has to be limited to assure stability under worst-case operating conditions.

The parameters of the torque controller can be adapted to compensate the plant parameter variation so that the system tracks with the reference model. In Fig. 7.38, the plant is characterized by the time-domain transfer characteristic $G_p(v)$, which is variable, as indicated. The command torque T_e^* of the drive system is also applied to the reference model and the resulting response T_{em} is compared with the actual torque of the drive system. The error signal iterates the PI compensator parameters K_1 and K_2 dynamically through the parameter search strategy so that the error e is zero or remains bounded within a hysteresis band. The procedure of parameter iteration is analogous to the manual parameter change using potentiometers in the laboratory breadboard by trial and error, except that here it is done dynamically by a microcomputer.

The reference model of the drive system is similar to that in torque control loop and is stored in the microcomputer memory. The reference model is determined on the basis of worst-case parameters of the plant so that the torque loop can physically track the reference mode. The mathematical model of the reference model may be difficult to store and solve in real time. One convenient way is to store impulse response $G_m(v)$ of the model and then solve the reference model in the time domain by the general convolution sum principle as follows:

$$Y(v) = \sum_{\mu=0}^{v} X(\mu)G_m(v - \mu) \tag{7.96}$$

where $Y(v)$ is the response, $X(v)$ the excitation, and $G_m(v)$ the model impulse response.

Sliding Mode Control

A variable structure control technique known as sliding mode control can be applied to ac drive systems (Ref. 19). It is basically an adaptive model reference control (MRAC) but is easier to implement than the conventional MRAC system. The advantage of sliding mode control is that the response of the drive system is insensitive to parameter variation and load disturbance effects, and therefore it is ideally suitable for servo applications, such as robot and machine tool drives. The control method can be extended to drives for transit cars, elevators, rolling mills, and multimachine drives where close speed or position tracking is desirable.

In sliding mode control, the "reference model" or a predefined trajectory in a phase plane is stored in the microcomputer and the drive system is forced to follow or "slide" along the trajectory by a switching control algorithm, irrespective of plant parameter variation and load torque disturbance. The microcomputer detects the deviation of the actual trajectory from the reference trajectory and correspondingly changes the switching topology to restore tracking.

Figure 7.39 illustrates the sliding mode control method of an induction motor drive system. Sliding mode control is implemented directly within the primary position control loop. The controller receives the following inputs:

- The position loop error, defined as X_1
- The position error velocity $X_2 = dX_1/dt$, which is derived from the speed signal by the relation $X_2 = -K\omega_r$
- A constant of amplitude A

The output U of the controller is multiplied by a gain constant and is fed as the command active current i_{qs}^*, which is proportional to the developed torque. In the constant-torque mode, the field current i_{ds}^* which is proportional to rotor flux is maintained constant. With vector control, the motor with the inverter and local control can ideally be represented as second-order system.

The definition of sliding trajectory in the phase plane X_1–X_2 is explained in Fig. 7.40. The curves are described in the fourth and second quadrants, which correspond to forward and reverse rotation of the shaft, respectively. The outer curve (shown by shading) is determined by the limiting values of acceleration, speed, and deceleration of the drive system. Normally, the variation of parameters of the plant will cause drift within a band as shown by the dashed lines. For example, if the inertia J is increased, the maximum acceleration and deceleration limit will shrink. The sliding trajectory or the reference contour in the phase plane which the drive system will be forced to follow has to be described beyond the drift band so that the system becomes controllable and the response is not affected by the drift. The sliding trajectory defined here consists of three straight-line

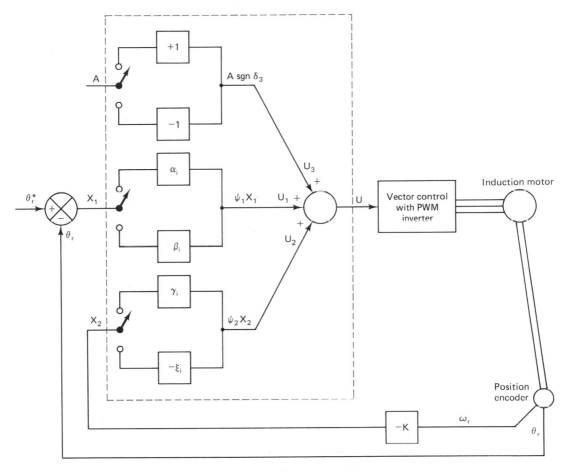

Figure 7.39 Block diagram with sliding mode control

segments, and their equations in the fourth quadrant are as follows:

1. Acceleration segment

$$\delta_1 = -B(X_1 - X_{10}) + X_2 \tag{7.97}$$

where δ_1 is a variable and its sign is given by the conditions:

$\delta_1 = 0$ for point on the segment

$\delta_1 > 0$ for point above the segment

$\delta_1 < 0$ for point below the segment

B = slope which is limited by the acceleration limit curve

X_{10} = initial position error

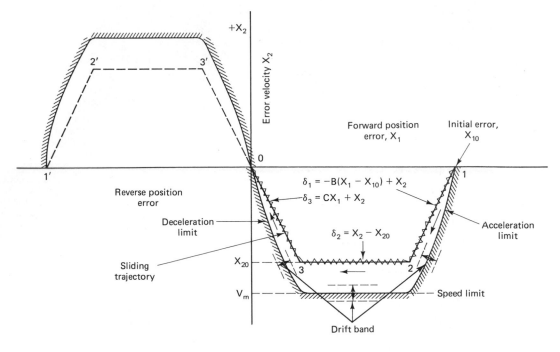

Figure 7.40   Sliding trajectory in phase plane.

2. Constant-velocity segment

$$\delta_2 = X_2 - X_{20} \qquad\qquad (7.98)$$

where δ_2 is a variable as defined by equation (7.97) and $X_{20} = (X_{20} < |V_m|)$ maximum velocity.

3. Deceleration segment

$$\delta_3 = CX_1 + X_2 \qquad\qquad (7.99)$$

where δ_3 is a variable as defined by equation (7.97) and C is the slope, which is limited by the deceleration limit curve.

The actual sliding curve that follows the defined trajectory is given by a zig-zag line in the direction of the arrow, as shown. At steady state, the operating point oscillates at the origin of the phase plane. The trajectory should be defined as close as possible to the limit envelope but beyond the drift band so as to get suboptimal transient response. It can be shown (Ref. 17) that the second-order system theoretically requires only the error signal X_1 and its first derivative X_2 as control inputs. The constant A is injected to eliminate steady-state error due to coulomb friction and loading effects. In the sliding mode controller, all the input signals are transmitted through single-pole double-throw (SPDT) switches and the respective gains to constitute the resultant signal U as

$$U = U_1 + U_2 + U_3 \qquad\qquad (7.100)$$

The sliding mode control law can be defined mathematically as

$$U = A \, \text{sgn} \, \delta_3 + \psi_1 X_1 + \psi_2 X_2 \qquad (7.101)$$

$$\delta = \delta_i \qquad i = 1, 2, 3 \qquad (7.102)$$

where

$$\text{sgn} \, \delta_3 = \begin{cases} +1 & \text{if } \delta_3 \geq 0 \\ -1 & \text{if } \delta_3 < 0 \end{cases}$$

$$\psi_1 = \begin{cases} \alpha_i & \text{if } \delta_i X_1 \geq 0 \\ -\beta_i & \text{if } \delta_i X_1 < 0 \end{cases} \qquad (7.103)$$

$$\psi_2 = \begin{cases} \gamma_i & \text{if } \delta_i X_2 \geq 0 \\ -\xi_i & \text{if } \delta_i X_2 < 0 \end{cases}$$

The gain constants α_i, β_i, γ_i, and ξ_i may vary from segment to segment of the trajectory. These constants for each segment can be determined from the state space model of the system using the following existence equation (Ref. 17):

$$\lim_{\delta_i \to +0} \frac{d\delta_i}{dt} < 0 \quad \text{and} \quad \lim_{\delta_i \to -0} \frac{d\delta_i}{dt} > 0 \qquad (7.104)$$

or

$$\lim_{\delta_i \to 0} \delta_i \frac{d\delta_i}{dt} > 0 \qquad (7.105)$$

In other words, the validity of the existence equation guarantees that the response will cross the trajectory in each switch position, and this is essential for the system to be controllable by sliding mode. In practice, the controller parameters are fine tuned experimentally for best performance of the system. A flowchart of the controller can be drawn which solves δ_i from the appropriate segment equation

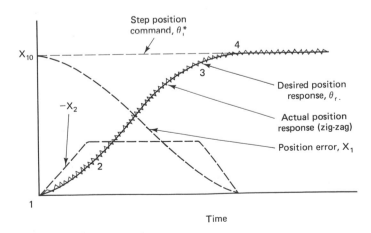

Figure 7.41 Time-domain response in sliding mode control (in fourth quadrant).

and determines the position of the SPDT switches according to the criteria given in equation (7.103).

Figure 7.41 shows the time-domain response in sliding mode control for the step position command, which has been translated from the phase plane trajectory. The jitter in the response is to be closely regulated by good resolution of the signals, the small sampling time of computation, and the large switching frequency of the inverter.

REFERENCES

1. K. Ogata, *State Space Analysis of Control Systems*, Prentice-Hall, Englewood Cliffs, N.J., 1967.

2. W. L. Brogan, *Modern Control Theory*, Quantum, New York, 1974.

3. M. Athans, *Modern Control Theory*, Part 1 MIT Press, Cambridge, Mass., 1974.

4. R. Saucedo and E. Schiring, *Introduction to Continuous and Digital Control Systems*, Macmillan, New York, 1968.

5. G. A. Kaufman and A. B. Plunkett, "A High-Performance Torque Controller Using a Voltage Source Inverter and Induction Machine," *Conf. Rec. IEEE/IAS Annu. Meet.*, pp. 863–872, Oct. 1981.

6. D. W. Novotny and J. H. Wouterse, "Induction Machine Transfer Functions and Dynamic Response by Means of Complex Time Variables," *IEEE Trans. Power Appar. Syst.*, Vol. PAS-95, pp. 1325–1334, July–Aug. 1976.

7. A. B. Plunkett, "A Current Controlled PWM Transistor Inverter Drive," *Conf. Rec. IEEE/IAS Annu. Meet.*, pp. 785–792, Oct. 1979.

8. A. B. Plunkett and D. L. Plette, "Inverter–Induction Motor Drive for Transit Cars," *IEEE Trans. Ind. Appl.*, Vol. IA-18, pp. 26–37, 1977.

9. B. K. Bose, "Adjustable Speed AC Drives–A Technology Status Review," *Proc. IEEE*, Vol. 70, pp. 116–135, Feb. 1982.

10. E. P. Cornell and T. A. Lipo, "Modeling and Design of Controlled Current Induction Motor Drive System," *IEEE Trans. Ind. Appl.*, Vol. IA-13, pp. 321–330, July–Aug. 1977.

11. F. Blaschke, "The Principle of Field Orientation as Applied to the New TRANSVEC-TOR Closed Loop Control System for Rotating Field Machines," *Siemens Review*, Vol. 34, pp. 217–220, May 1972.

12. K. Hasse, "Zur Dynamik drehzahlgeregelter Antriebe mit Stromrichtergespeisten Asynchron-Kurzschluβlaufermaschinen," *Darmstadt, Techn. Hochsch., Diss.*, 1969.

13. R. Gabriel, W. Leonhard, and C. Nordby, "Field Oriented Control of a Standard AC Motor Using Microprocessors," *Conf. Rec. IEEE/IAS Annu. Meet.*, pp. 910–916, 1979.

14. G. Kaufman, L. Garces, and G. Gallagher, "High Performance Servo Drives for Machine Tool Applications Using AC Motors," *Conf. Rec. IEEE/IAS Annu. Meet.*, pp. 604–609, Oct. 1982.

15. Y. D. Landau, *Adaptive Control*, Dekker, New York, 1979.

16. G. Fromme, "Self-Optimizing Controller Employing Microprocessor," *Proc. Microelectron. Power Electron. Electr. Drives*, Darmstadt, pp. 117–125, 1982.

17. U. Itkis, *Control Systems of Variable Structures*, Wiley, New York, 1976.

18. W. Schumacher, "Microprocessor Controlled AC Servo Drive," *Proc. Microelectron. Power Electron. Electr. Drives*, Darmstadt, pp. 311–320, 1982.

19. B. K. Bose, "Sliding Mode Control of Induction Motor," *Conf. Rec. IEEE/Annu. Meet.*, pp. 479–486, 1985.

20. L. J. Garces, "Parameter Adaption for the Speed Controlled Static AC Drive with Squirrel Cage Induction Motor," *Conf. Rec. IEEE/IAS Annu. Meet.*, pp. 843–850, 1979.

21. A. Nabae, K. Otsuka, H. Uchins, and R. Kurosawa, "An Approach to Flux Control of Induction Motors with Variable Frequency Power Supply," *Conf. Rec. IEEE/IAS Annu. Meet.*, pp. 890–896, 1978.

22. K. Hasse, "Control of Cycloconverters for Feeding of Asynchronous Machines," *Proc. IFAC Symp. on Control in Power Elec. and Electrical Drives*, pp. 537–545, 1977.

23. A. Brickwedde, "Microprocessor-Based Adaptive Speed- and Position-Control for Electrical Drives," *Conf. Rec. IEEE/IAS Annu. Meet.*, pp. 411–417, 1984.

8

CONTROL OF SYNCHRONOUS MACHINES

8.0 INTRODUCTION

In Chapter 7 we discussed very comprehensively essentially all the important aspects of induction motor control. The majority of the discussion can also be extrapolated for control of the synchronous machine. Therefore, in this chapter, the discussion is focused on the salient features of synchronous machine control.

The principal differences between an induction machine and a synchronous machine can be summarized as follows:

- The stator supply frequency and speed of a synchronous machine are uniquely related (i.e., the slip ω_{sl} is zero).

- A synchronous machine is normally supplied by dc field excitation on the rotor side, which is unlike an induction machine, where the field is supplied from the stator side. As a result, an induction motor must run at lagging power factor, whereas a synchronous motor can be operated at any arbitrary power factor (i.e., leading, lagging, or unity). At lagging power factor, the stator-supplied field assists the rotor field, whereas at leading power factor the rotor field is bucked. In a permanent-magnet synchronous machine, the rotor-supplied field can be considered as constant. In a synchronous reluctance machine, the rotor does not have any field, and therefore it has to be supplied entirely from the stator side like an induction machine requiring lagging power factor operation.

- A synchronous machine may or may not have an amortisseur winding. With the presence of amortisseur winding, the rotor leakage flux will affect the total rotor flux during transient conditions but will vanish as the electrical transients disappear.

- A synchronous machine may have salient or nonsalient poles, unlike an induction machine which usually has nonsalient poles. The torque in a salient-pole machine is contributed by the reluctance component of the torque and the field component of the torque. For control purposes we will neglect the saliency effect.

In this chapter we discuss the control of synchronous machines, which may be supplied by voltage-fed inverters, current-fed inverters, or cycloconverters. The control will be categorized into two groups: scalar control and vector control.

8.1 SCALAR CONTROL

A synchronous machine drive system may have essentially two different modes of operation. One is the open-loop true synchronous motor mode, where the machine speed is controlled by an independent oscillator. The other is known as a self-control mode, where the variable-frequency inverter (or cycloconverter) control pulses are derived from the rotor position sensor. In the later mode, the supply frequency is no longer independent but is linked with the rotor speed, which might vary with the variation of load torque.

Open Loop Volts/Hertz Control

An example of independent frequency control is the open loop volts/hertz control shown in Fig. 8.1. This method of speed control is very popular in multiple reluctance or permanent-magnet machine drives, where close speed tracking is essential in such applications as fiber-spinning mill. Here a number of machines are connected in parallel and are supplied by the same inverter, and the machine speed is uniquely related to the command frequency ω_e^*. Maintaining a constant volts/hertz ratio makes the air gap flux constant, permitting maximum torque availability of the machines. At a certain frequency and voltage condition, if the load torque increases, the developed torque increases to match it, until the stability limit is reached. The speed can be varied from zero to the full value by gradually varying the frequency. Beyond the base speed, the dc link voltage saturates and the machine enters into the field weakening constant-power region, where the torque decreases with an increase in the frequency. The air gap flux can be independently controlled by a closed loop, if desired.

Self-Control Mode

A synchronous machine operating in the self-control mode can be defined as an electronically commutated motor (ECM), brushless dc motor, or commutatorless-brushless motor. The reason for this definition is that in this mode of operation, the synchronous machine is analogous to dc machine in the following way: Inter-

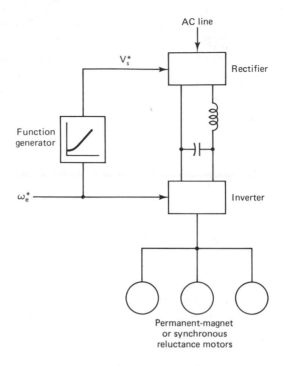

Figure 8.1 Volts/hertz control of multiple synchronous machines.

nally, a dc machine can be viewed as an ac synchronous machine in which the field is stationary but the armature with multiphase ac winding is rotating. The ac supply to the armature is derived from the input dc supply through commutators and brushes, which can be viewed as a mechanical shaft-position-sensitive inverter. A synchronous machine in the self-control mode is somewhat analogous, but here the field is rotating and the armature is stationary, and it is supplied by a shaft-position-controlled electronic inverter. The advantages of replacement of mechanical commutators and brushes in a dc machine by electronic commutation are obvious. Self-control modes of a synchronous machine may be valid whether fed by a voltage-fed inverter, a current-fed inverter, or a cycloconverter.

The self-control principle is illustrated in Fig. 8.2 by a current-fed inverter. The machine rotor has a shaft position sensor and the sensor signal is processed through a delay angle control circuit to generate the inverter firing pulses as shown. This self-control mode relates the inverter frequency uniquely with the machine speed. Figure 8.3 shows the fundamental-frequency phasor diagram for load commutation of the inverter where the stator current I_s leads the stator voltage V_s by the angle ϕ. For simplicity, the saliency and stator resistance are neglected. The field flux ψ_f which is related to the rotor position is established independently by the field current. The magnitude of the stator current controlled by the rectifier determines the magnitude of the armature reaction flux $\psi_a = I_s L_s$, and the phasor can be positioned at a desirable angle by the firing delay angle α_d. In the figure, the phasor ψ_a leads ψ_f by the angle ϕ' so that the resultant stator flux ψ_s (including leakage) induces the stator voltage V_s, which lags angle ϕ with respect to the stator

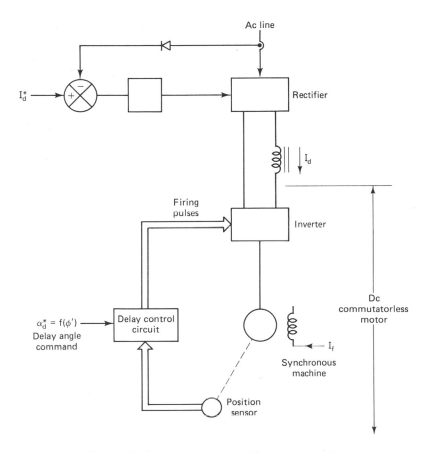

Figure 8.2   Synchronous motor self-control principle.

current I_s.   The angle ϕ' can be given as

$$\phi' = 180° - \alpha \tag{8.1}$$
$$= v + 90° + \phi$$

where v is the angle of the ψ_s phasor.   The magnitudes of I_s, ϕ', and I_f can be controlled with speed to establish the optimum commutation condition, which will be described later.   It is worth mentioning here that in a dc machine the fluxes ψ_a and ψ_f are orthogonal (i.e., the angle $\phi' = 90°$).

Figure 8.2 illustrates a current-controlled brushless dc machine with separate field excitation.   A shunt- or series-type machine is also possible by connecting

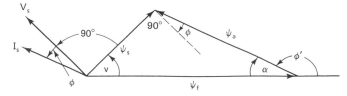

Figure 8.3   Phasor diagram of synchronous machine at leading power factor (motoring).

the field winding in parallel or in series with the dc link, respectively. The self-control with voltage-fed inverters and cycloconverters will be discussed later.

The self-controlled synchronous machine has the advantages that it cannot fall out of step by the steady-state stability limit and rarely shows any transient stability problem. The machine shows a dc machine-like dynamic behavior. This type of control is almost universally used for synchronous machines.

Figure 8.4 shows an optical-type incremental rotor position encoder and the corresponding encoder waves for a four-pole machine. The disk has a large number of slots in the outer perimeter but two $90°$ slots at an inner radius. There are four stationary optical sensors ($S_1 - S_4$), of which S_4 is mounted at the outer perimeter and S_1, S_2, and S_3 are at the inner radius with $120°$ electrical spacing as shown. A sensor consists of a light-emitting diode and a phototransistor pair, and a logic "1" is generated when the sensor is within the slot. The sensors S_1 to S_3 generate

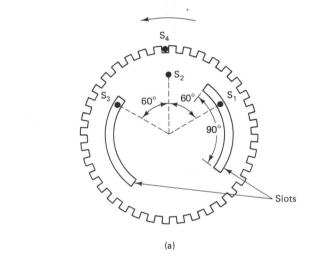

(a)

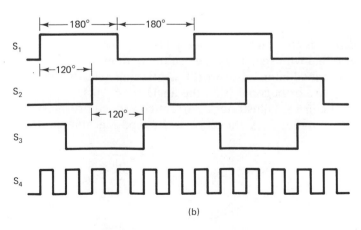

(b)

Figure 8.4   Rotor position encoding (four-pole machine): (a) optical type encoder; (b) encoder waves. From *IEEE Press*, B. K. Bose, © 1979 IEEE.

square waves at 120° phase shift and S_4 generates a high-frequency square wave which is used for inverter firing angle control and speed sensing.

Figure 8.5 shows the block diagram of a delay control circuit and Fig. 8.6 shows the corresponding waves for phase *a* only in both the motoring and generating modes. The resulting idealized voltage and current waves for phase *a* of the motor are shown in Fig. 8.7. Each channel of the delay control circuit consists of a programmable up/down-counter followed by a *D*-flip-flop. The outputs of the flip-flops are combined in a pulse distributor to generate the thyristor firing pulses. The counters are operated in the down mode in motoring but in the up mode in regeneration. At the leading edge of the reference signal, the delay angle α_d is loaded in the counter and is decremented (or incremented) with the sensor S_4 pulse train. At the terminal count, a clock pulse is applied to the *D* flip-flop. The outputs of *D* flip-flops, are square waves which are like the reference signals but phase delayed by the α_d angle, are then logically combined and ANDed with a clock to generate 120°-wide thyristor firing pulses as shown. It should be noted that the reference signal wave shown in Fig. 8.7 does not coincide with the flux ψ_f wave but has a fixed phase relation ($\alpha_d = k\phi'$) with it. This should be evident from Fig. 8.3, where I_s leads the ψ_f phasor by angle ϕ'.

Load commutation with constant margin angle. A complete speed control system for a synchronous machine using the self-control principle is shown in Fig. 8.8. Here the inverter is load commutated with constant margin angle γ (see Fig. 5.9) and the stator flux linkage ψ_s is maintained constant. The control in the upper part is identical to that of a dc machine fed by a phase-controlled rectifier;

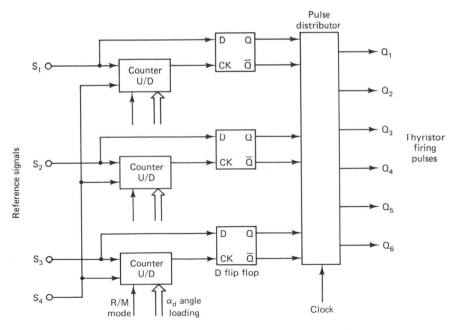

Figure 8.5 Delay control block diagram. From *IEEE Press*, B. K. Bose, © 1979 IEEE.

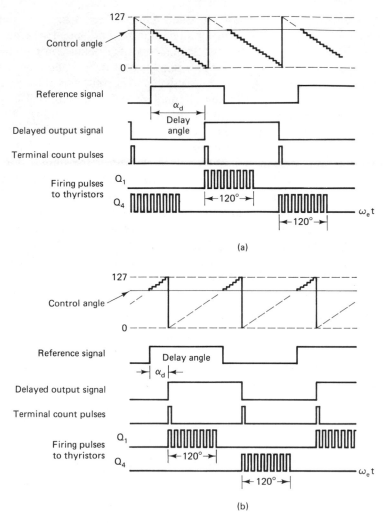

Figure 8.6 Waveforms of delay control circuit: (a) motoring mode; (b) regeneration mode. From *IEEE Press*, B. K. Bose, © 1979 IEEE.

that is, the speed is controlled in the outer loop, and dc link current I_d, which is related to torque, is controlled in the inner loop. The additional complexity arises with synchronous machines because the field current I_f and phase angle ϕ' are to be controlled so that the phasor diagram shown in Fig. 8.3 is satisfied at all load and speed conditions. The field current is controlled by the closed loop and the command value I_f^* can be generated by the following relation:

$$\psi_f^* = \sqrt{\psi_s^{*2} + \psi_a^{*2} + 2\psi_s^* \, \psi_a^* \sin (90 + \phi^*)} \tag{8.2}$$

where

$$\psi_a^* = K_a I_d^* \tag{8.3}$$

$$\phi^* \simeq \beta^* = \gamma^* + \mu \tag{8.4}$$

$$I_f^* = K_f \psi_f^* \tag{8.5}$$

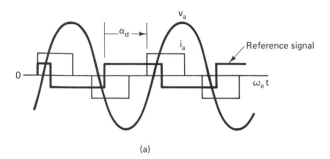

(a)

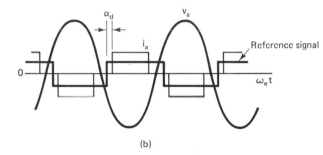

(b)

Figure 8.7 Idealized motor voltage and current waves: (a) motoring mode; (b) regeneration mode. From *IEEE Press*, B.K. Bose, © 1979 IEEE.

In the equations above, ψ_s^* and γ^* are constant and ψ_a^* is generated from the command dc link current I_d^*, which is uniquely related to the stator fundamental current for the ideal six-step waveform. The overlap angle μ can be measured by the scheme shown in Fig. 8.9. The voltage across each inverter thyristor is sensed with proper isolation and then the respective turn-off angle γ is determined through a comparator. The γ angles are logically ORed and fed to an *RS* flip-flop as shown. The flip-flop is set by the gate firing pulse, but is reset by the leading edge of the γ angle, giving a μ angle at the output. Alternatively, the overlap angle μ can be estimated mathematically by the principle discussed in Chapter 3. Instead of controlling the stator flux by an open loop principle as above, a closed-loop method can also be used. The control angle $\phi^{*\prime}$ can be generated from equation (8.1). The additional input angle v can be determined by measuring the angle $90° + v$ between the V_s and ψ_f phasors. The phase position of the V_s wave can be measured by a zero-crossing detector and the ψ_f position is known by the reference wave of the position sensor.

Instead of mathematically deriving the I_f^* and $\phi^{*\prime}$ control parameters on-line, these can be precomputed approximately as a function of I_d^* and then implemented by function generators (Ref. 3). The function generators are easy to implement but the accuracy of control is lost due to variation in the parameters, and unless the design is made conservatively, commutation failure may occur.

It should be mentioned here that instead of constant γ angle control, it is possible to implement constant margin (turn-off) time t_{off} control, where $t_{\text{off}} = \gamma/\omega_e$. The constant t_{off} control has the advantage that the machine power factor remains optimum in the full speed range while assuring satisfactory commutation of inverter thyristors.

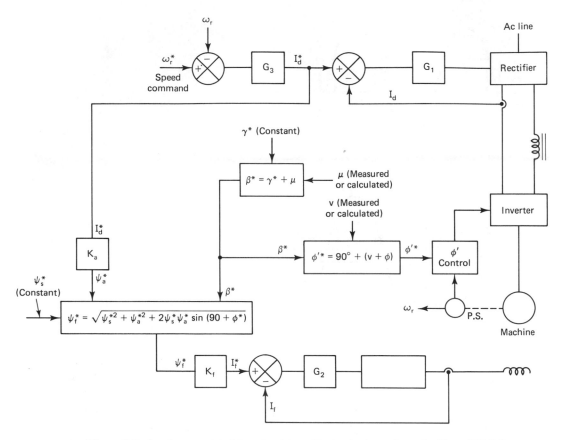

Figure 8.8 Load-commutated inverter drive with constant margin angle. From IFAC Symposium on "Control in Power Electronics and Electrical Drives," by J. Leimgruber (1977) by Pergamon Press, London.

The current-fed inverter can be controlled so that the machine terminal power factor is approximately unity by making $\beta^* = 0$ in Fig. 8.8. Unity power factor operation requires a force-commutated inverter, but as a result, operation down to zero speed is possible.

Self-control with terminal voltage sensing. The shaft position sensor can be used for self-control of the machine down to zero speed, although load commutation as described above is valid only above a certain minimum speed. Further discussion on self-control from zero speed will be given later. The position sensor can be eliminated, and instead, machine terminal voltage sensing can be used for self-control if the speed does not fall below a minimum value. The method of generating inverter firing signals by terminal voltage sensing is analogous to that of rectifier control by line voltage sensing, discussed in Chapter 3. The machine terminal voltage waves contain large commutation spikes and are therefore difficult to process in analog form. Attempting to filter these transients causes frequency-sensitive phase shift which is difficult to compensate. A zero-crossing method of

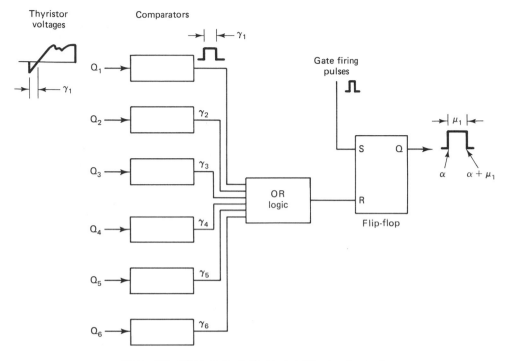

Figure 8.9   Measurement of commutation overlap angle.

terminal voltage sensing is shown in Fig. 8.10. The machine line voltages are processed through zero-crossing detectors to generate the three square waves with 120° phase shift as shown. The details of the zero-crossing detector with optical isolation scheme is shown in Fig. 8.10(b). The opto-transistor flip-flop is insensitive to commutation spikes and gives a square-wave output which is in phase with the input line voltage. Any phase error due to threshold voltage of the opto-coupler can be compensated as a function of the machine speed. The voltage-sensing circuit also provides input to a frequency multiplier clock whose operating principle is explained in Fig. 8.11. As shown in the block diagram, the machine fundamental frequency is multiplied by a factor of 360. A conventional PLL chip with a limited-frequency locking range may not be satisfactory for this application. The output square waves of the voltage sensing circuit are passed through edge detectors to generate a pulse train at $6f_e$ frequency, which is then used to load an 8-bit latch from a counter and then reset the counter with a delay as shown. The counter is clocked by the frequency $f_c/360$ obtained from a crystal clock through a down-counter. The latch loads the digital word $(1/6f_e) \times (f_c/360)$ to a programmable down-counter which is clocked by the frequency $f_c/6$. This results in an output pulse train at frequency $f_0 = 360f_e$. The inverter firing pulses can be generated with the help of the delay control circuit shown in Fig. 8.5. The explanatory waveforms of one channel are shown in Fig. 8.12. Here the reference signals S_1 to S_3 are the outputs of the voltage-sensing circuit, S_4 is the input from the frequency multiplier, and the delay angle α_d is as shown in Fig. 8.12. This means that the incoming thyristors are fired at advance angle β with reference to the respective

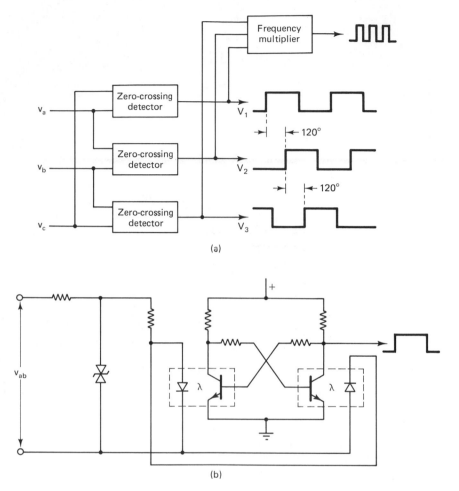

Figure 8.10 Terminal voltage-sensing principle: (a) sensing circuit block diagram; (b) zero-crossing detector with optical isolation. From *IEEE Press*, B. K. Bose, © 1980 IEEE.

line voltage. The α_d^* angle command can be generated from the relation

$$\alpha_d^* = 180° - \beta = 180° - (\mu + \gamma^*) \qquad (8.6)$$

where γ is a constant turn-off angle and μ is the overlap angle, which can be measured or a calculated value can be used. The α_d angle can be controlled easily for the drive system to operate in both the motoring and regeneration modes.

Phase-locked turn-off time control (Ref. 8). The phase-locked-loop method discussed in Chapter 3 can easily be extended to the load-commutated inverter drive of synchronous machines. For example, the angle γ can be maintained constant by PLL control. Instead of the turn-off angle, the turn-off time can be maintained constant by the PLL method, as illustrated in Fig. 8.13. The turn-off

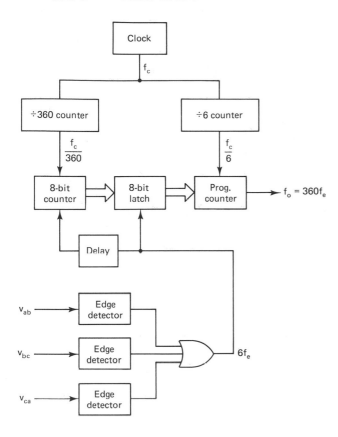

Figure 8.11 Frequency multiplier principle. From *IEEE Press*, B.K. Bose, © 1980 IEEE.

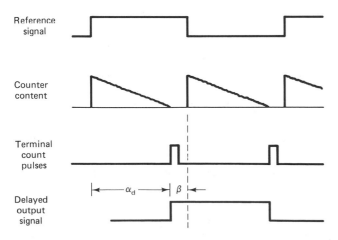

Figure 8.12 Waveforms of delay control circuit.

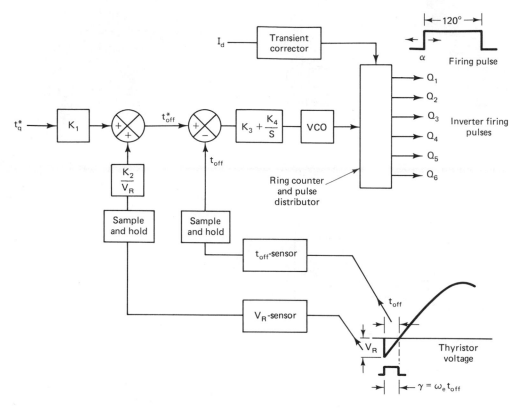

Figure 8.13   Phase-locked-loop turn-off time control of load-commutated inverter.

time t_{off} $(= \gamma/\omega_e)$ can be formulated as

$$t_{\text{off}} = K_1 t_q + \frac{K_2}{V_R} \tag{8.7}$$

where t_q is the specification sheet turn-off time of the inverter thyristor, V_R is the thyristor inverse voltage immediately after commutation, and K_1 and K_2 are weighting constants.   In Fig. 8.13, the commanded t_q^* is multiplied by K_1 and added to K_2/V_R, where the V_R value is measured across the thyristor.   The commanded t_{off}^* is compared with the measured t_{off} and the error through a PI compensator and voltage-controlled oscillator, and a ring-counting circuit generates the inverter firing pulses as shown.   The measured t_{off} signal is available at 60° intervals, which is satisfactory at steady state and slowly varying transients.   Otherwise, any large transient in dc current i_d will tend to cause commutation failure.   The effect of transient i_d change within a 60° interval can be corrected by a transient corrector as indicated.   This input acts as an overriding signal to correct the leading edge of the firing pulse of the incoming thyristor.

The phase-locked-loop control method can also be extended to the regeneration mode and field weakening region.   When the speed loop error is negative, indicating regeneration demand, the control angle γ (or the equivalent time) can be advanced to be near 180°.   In the field weakening mode, when the dc link rectifier voltage reaches full value, the flux ψ_s is reduced by control of the field current I_f with an inverse relation to speed so that the angle γ remains nearly constant. The reduction of flux obviously causes a reduction of torque in the constant-power field weakening region.

Closed-loop power factor control. A square-wave voltage-fed inverter drive where the machine power factor is controlled by a closed loop is shown in Fig. 8.14. The command power factor (preferably unity) is compared with the machine terminal power factor, which is computed from the stator voltage and current signals. The error delays the reference signal from the position sensor until the unity power factor condition is achieved. This principle can also be extended to other converter types. Of course, the power factor computation may be difficult due to distortion in the voltage and current waves. In Fig. 8.14, the speed is controlled by controlling the dc voltage and the air gap flux is controlled by field current I_f, as shown.

Cycloconverter control. A cycloconverter-fed synchronous machine can be self-controlled so that the machine always operates at unity power factor condition. This class of drives is sometimes called an ac brushless motor. Figure 8.15 shows a control block diagram with cycloconverter control, and Fig. 8.16

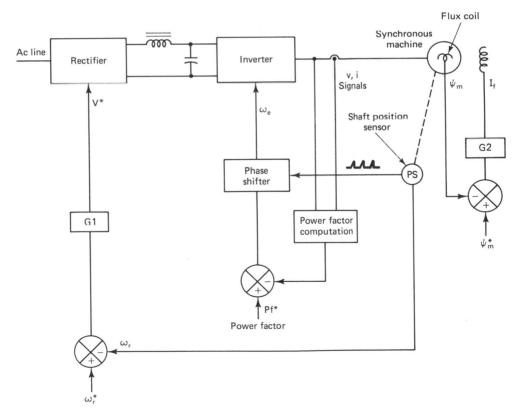

Figure 8.14 Square-wave voltage-fed inverter control at constant power factor.

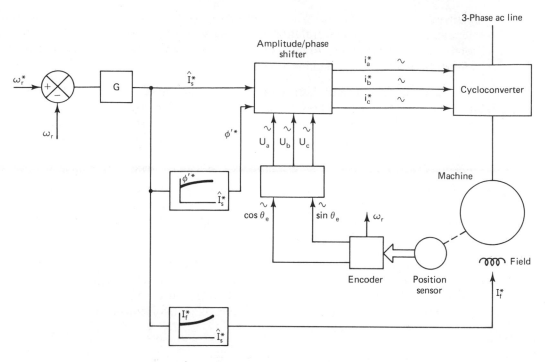

Figure 8.15   Cycloconverter drive at unity power factor.

shows the corresponding phasor diagram at unity-power-factor condition.   The
figure is identical to Fig. 8.3 except that the phase angle ϕ between the voltage
and current phasors is reduced to zero.   The flux phasor diagram remains essentially
invariant with speed but varies with stator current I_s so that the phasor ψ_s remains
orthogonal to the ψ_a phasor and the locus of point A describes a circle.   The angle
ϕ' and field current I_f can then be described graphically as a function of stator
current $\hat{I}_s$ as shown in Fig. 8.16(b).   These relations can be implemented by
function generators as shown in Fig. 8.15.   The error in the speed loop generates

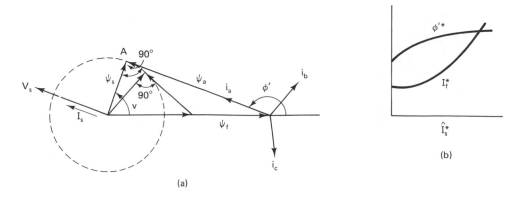

Figure 8.16   (a) Phasor diagram for motoring at unity power factor (ψ_s = constant);
(b) phase angle and field current as function of stator current.

the stator current command $\hat{I}_s^*$, which is the principal control variable in the system. The parameter $\hat{I}_s^*$ is a dc quantity and is proportional to the torque demand of the machine.   The other control parameters $\phi'*$ and I_f^* are generated from $\hat{I}_s^*$ through function generators.   The position sensor and encoder provide the sinusoidally varying unit signals $\cos \theta_e$ and $\sin \theta_e$ and the speed signal ω_r, where $\theta_e = \omega_e t$.   The $\cos \theta_e$ wave is aligned to be in phase with the ψ_f phasor.   The two-phase quadrature waves are then converted to three-phase unit signals by the relations

$$U_a = \cos \omega_e t \tag{8.8}$$

$$U_b = \cos (\omega_e t - 120°) = \frac{\sqrt{3}}{2} \sin \omega_e t - \frac{1}{2} \cos \omega_e t \tag{8.9}$$

$$U_c = \cos (\omega_e t + 120°) = -\frac{\sqrt{3}}{2} \sin \omega_e t - \frac{1}{2} \cos \omega_e t \tag{8.10}$$

Each of the three-phase unit signals is then multiplied by the magnitude $\hat{I}_s^*$ and phase shifted by the angle $\phi'*$ to generate the corresponding phase current command signals as follows:

$$i_a^* = \hat{I}_s^* \ U_a \underline{/\phi'*} \tag{8.11}$$

$$i_b^* = \hat{I}_s^* \ U_b \underline{/\phi'*} \tag{8.12}$$

$$i_c^* = \hat{I}_s^* \ U_c \underline{/\phi'*} \tag{8.13}$$

The cycloconverter phase currents can then be controlled by the hysteresis-band bang-bang method.   The phasor diagram in the regeneration mode of the system is shown in Fig. 8.17, where V_s and I_s are opposite in phase.   The control remains identical with motoring except that the angle ϕ' is negative.   The polarity of ϕ' can be controlled by the polarity of the speed loop error.

As described above, cycloconverter control can easily be extended to a PWM voltage-fed inverter operating in the current control mode.   This control principle is analogous to vector control, which is described in the following section.

8.2 VECTOR CONTROL

The concept of vector control can easily be extended to synchronous machines. The self-controlled synchronous machines discussed in the preceding section are analogous to dc machines, but the field current and stator torque component of

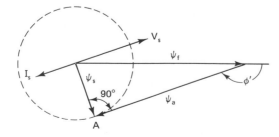

Figure 8.17 Phasor diagram for regeneration at unity power factor.

current are not decoupled as in a dc machine. In the constant-torque region, if the developed torque must be increased at the constant stator flux, the field current I_f is required to be boosted to satisfy the phasor diagram of Fig. 8.16. The response of the field current is sluggish because of the large time constant, and as a result, the response of a self-controlled synchronous machine is slow. The response can be improved considerably by using vector control, where the transient magnetizing current demand to maintain the rated flux can temporarily be supplied from the stator side. In fact, the flux can be intentionally boosted by vector control during start-up conditions to speed up the response time. It should be mentioned here that in induction motor vector control, the rotor flux which is supplied from the stator side can be made constant by maintaining the flux component of current i_{ds} constant, and torque can be increased almost instantly by increasing the torque component of current i_{qs}. The synchronous machine torque response with vector control will therefore be somewhat sluggish because of the flux correction by the stator-fed magnetizing current.

Cycloconverter Control

The vector control method can be applied to voltage-fed inverters, cycloconverters, and force-commutated current-fed inverters. We will illustrate here the application of vector control with a cycloconverter drive, which is shown in Fig. 8.18, and

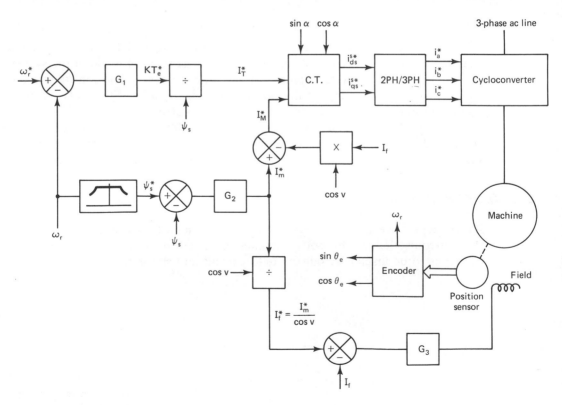

Figure 8.18 Vector control of cycloconverter drive.

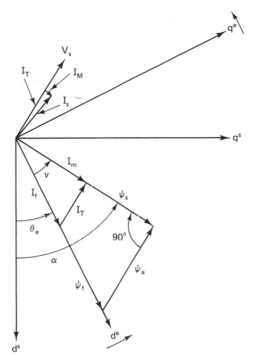

Figure 8.19 Phasor diagram of synchronous machine illustrating transient boosting of magnetizing current (I_M).

Fig. 8.19 shows the corresponding phasor diagram. The error in the speed control loop generates the torque command T_e^*, which is divided by the magnitude of the stator flux to generate the stator torque component of current I_T^*. The flux ψ_s^* which is generated as a function of speed is controlled by a closed loop. The flux loop error generates the magnetizing current command I_m^* required to establish the desired ψ_s. At steady state, the phasor diagram in Fig. 8.19 indicates that the phasors ψ_s and ψ_a' are at quadrature. This means that at steady state $I_s = I_T$ and is cophasal with the stator voltage V_s, which is at quadrature lead with repect to the ψ_s phasor. The current phasors I_m, I_f, and I_T form a triangle that is proportional to the corresponding mmf triangle. At steady state, the field current I_f can be related to I_m by

$$I_f = \frac{I_m}{\cos v} \tag{8.14}$$

where v is the angle between the I_m and I_f phasors. The control equation (8.14) is implemented in Fig. 8.18, where I_f is controlled by a feedback loop. The flux component of stator current I_M^* is given by the relation

$$I_M^* = I_m^* - I_f \cos v \tag{8.15}$$

In steady-state operation, $I_M^* = 0$, so that equation (8.14) is satisfied. The currents I_T^* and I_M^* (if any) are converted to stationary reference frame d^s–q^s and then to the a–b–c frame, keeping I_T and I_M aligned with the ψ_a' and ψ_s phasors, respectively. The cycloconverter can then be controlled by the hysteresis-band bang-bang method

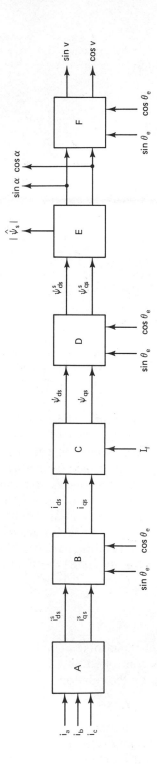

A: $i_{ds}^s = -\dfrac{1}{\sqrt{3}}(i_b - i_c)$

$i_{qs}^s = \dfrac{2}{3}i_a - \dfrac{1}{3}i_b - \dfrac{1}{3}i_c$

B: $i_{ds} = i_{ds}^s \cos\theta_e + i_{qs}^s \sin\theta_e$

$i_{qs} = -i_{ds}^s \sin\theta_e + i_{qs}^s \cos\theta_e$

C: See equations 8.17 and 8.18

D: $\psi_{ds}^s = \psi_{ds} \cos\theta_e - \psi_{qs} \sin\theta_e$

$\psi_{qs}^s = \psi_{ds} \sin\theta_e + \psi_{qs} \cos\theta_e$

E: $|\hat{\psi}_s| = \sqrt{\psi_{ds}^{s2} + \psi_{qs}^{s2}}$

$\sin\alpha = \dfrac{\psi_{qs}^s}{|\hat{\psi}_s|}$

$\cos\alpha = \dfrac{\psi_{ds}^s}{|\hat{\psi}_s|}$

F: $\sin v = \sin(\alpha - \theta_e)$

$\quad = \sin\alpha \cos\theta_e - \cos\alpha \sin\theta_e$

$\cos v = \cos(\alpha - \theta_e)$

$\quad = \cos\alpha \cos\theta_e + \sin\alpha \sin\theta_e$

Figure 8.20 Conversion block diagram and summary of mathematical relations. "Field Oriented Close-Loop Control of a Synchronous Machine with the New Transvektor Control System," by K.H. Bayer in Siemans Review xxxix (1972) No. 5

so that the actual phase currents follow the command currents. Assume that in the constant-torque mode there is a sudden increase in torque demand. For the constant flux ψ_s, the current I_m^* will remain constant but the angle v will increase and will therefore require a higher field current I_f^*. But the response delay of I_f will cause a finite I_M^* by equation (8.15), which will assist to maintain constant ψ_s. As the current I_f builds up gradually, I_M^* will be diminished until it vanishes completely at steady state when equation (8.14) is satisfied. During transient conditions, the stator current $\hat{I}_s$ is given by

$$\hat{I}_s = \sqrt{I_T^2 + I_M^2} \tag{8.16}$$

and the machine power factor deviates from unity as shown.

The conversion block diagram for the control system that generates the unit vectors sin α and cos α, the stator flux $|\hat{\psi}_s|$, and cos v is shown in Fig. 8.20, which also summarizes the mathematical relations of each block. The phase currents of the machine are sensed and first converted to the d^s–q^s frame and then to the d^e–q^e frame by the unit vectors sin θ_e and cos θ_e available from the position encoder. The rotating frame currents i_{ds} and i_{qs} are combined with I_f to solve the following transient relations of ψ_{ds} and ψ_{qs} from the equivalent circuit of Fig. 2.29:

$$\psi_{ds} = (i_{ds} + I_f)L_{md} \frac{R_{dr} + SL_{ldr}}{R_{dr} + S(L_{ldr} + L_{md})} + i_{ds}L_{ls} \tag{8.17}$$

$$\psi_{qs} = i_{qs}L_{mq} \frac{R_{qr} + SL_{lqr}}{R_{qr} + S(L_{lqr} + L_{mq})} + i_{qs}L_{ls} \tag{8.18}$$

where S is the Laplace operator.

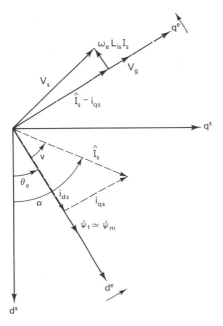

Figure 8.21 Phasor diagram of permanent-magnet synchronous motor (surface magnet).

The transient equations above can be solved by either analog or digital computation. The parameters ψ_{ds} and ψ_{qs} are converted to the stationary reference frame, then to flux magnitude $|\hat{\psi}_s|$ and unit vectors $\sin \alpha$ and $\cos \alpha$, and finally to $\sin v$ and $\cos v$ parameters by block F. The complex computations required for vector control of a synchronous machine are to be carefully justified against the advantage of improvement in the response time. Again it should be noted that since the machine parameters enter into computation of the conversion block diagram, the phasor diagram of Fig. 8.19 can no longer be accurately satisfied. Of course, compensation for parameter variation can be provided if a powerful microcomputer is used.

The stator current-based flux synthesis discussed above is necessary if the machine speed is controlled down to low value. If the machine speed exceeds typically 10 percent of the base speed, then ψ_{ds}^s and ψ_{qs}^s signals can be derived from stator voltages by integrating the 3-phase/2-phase converted voltages.

Permanent-Magnet Machine Control

In a permanent-magnet synchronous machine, the rotor field flux ψ_f and the corresponding equivalent field current I_f can be considered as constant. For a PM machine where the magnets are mounted on the rotor surface, the large effective airgap makes the saliency and armature reaction effect negligible. Therefore, $\psi_f \simeq \psi_m$, and for maximum torque sensitivity with stator current $i_{ds}^* = 0$ and $\hat{I}_s^* = i_{qs}$, as shown by the solid lines in Fig. 8.21. The stator current slightly lags the stator voltage V_s due to the stator leakage inductance drop, and the resistance drop is neglected. Note that air gap flux is considered reference phasor here instead of stator flux. Figure 8.22 shows the vector control principle for the PM motor. It has been derived from the induction motor vector control diagram shown in Fig. 7.29 with the modifications $\omega_{sl} = 0$, $\theta_r = \theta_e$, and $i_{ds}^* = 0$. The machine can be controlled for motoring and regeneration with constant air gap flux by controlling the magnitude of i_{qs}^* in either polarity. Such a drive gives truly brushless dc motor performance.

Figure 8.23 shows a permanent-magnet motor position servo using the polar form of vector control. The control response is enhanced by using an additional torque control loop. The torque loop drives the current $\hat{I}_s^* = i_{qs}^*$, as indicated in the phasor diagram of Fig. 8.21. The phasor $\hat{I}_s$ is aligned with the q^e axis by the

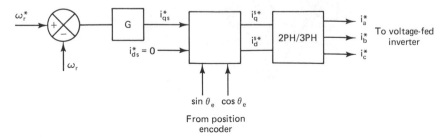

Figure 8.22   Vector control of permanent-magnet motor (surface magnet).

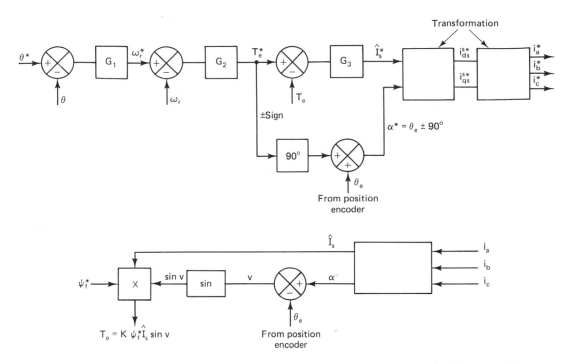

Figure 8.23 Polar form of vector control of permanent-magnet motor (surface magnet).

α^* angle vector so that

$$i_{ds}^s = \hat{I}_s^* \cos \alpha^* \tag{8.19}$$

$$i_{qs}^s = \hat{I}_s^* \sin \alpha^* \tag{8.20}$$

where $\alpha^* = \theta_e + 90°$ as shown. The phase position of $\hat{I}_s$ can be reversed during regeneration by changing the sign of $90°$ by the torque command polarity. The computation of feedback torque T_e is shown by a block diagram in Fig. 8.23. Here the flux ψ_f^* is considered as constant and the torque angle, which may deviate transiently from $90°$, is given by the general relation $v = \alpha - \theta_e$.

REFERENCES

1. H. Le-Huy, R. Perret, and D. Roye, "Microprocessor Control of a Current-Fed Synchronous Motor Drive," *Conf. Rec. IEEE/IAS Annu. Meet.*, pp. 873–880, 1979.

2. H. Le-Huy, A. Jakubowicz, and P. Perret, "A Self-Controlled Synchronous Motor Drive Using Terminal Voltage Sensing," *Conf. Rec. IEEE/IAS Annu. Meet.*, pp. 562–569, 1980.

3. B. K. Bose and T. A. Lipo, "Control and Simulation of a Current-Fed Linear Inductor Machine," *IEEE Trans. Ind. Appl.*, Vol. IA-15, pp. 591–600, Nov.–Dec. 1979.

4. J. Leimgruder, "Stationary and Dynamic Behaviour of a Speed Controlled Synchronous Motor with Cos φ or Commutation Limit Line Control," *Conf. Rec. IFAC Symp. on Control in Power Elec. and Electrical Drives*, pp. 463–473, 1977.

5. H. Stemmler, "Drive System and Electric Control Equipment of the Gearless Tube Mill," *Brown Boveri Rev.*, pp. 120–128, Mar. 1970.

6. K. H. Bayer, H. Waldmann, and M. Weibelzahl, "Field-Oriented Close-Loop Control of a Synchronous Machine with the New Transvector Control System," *Siemens Rev.*, Vol. 39, pp. 220–223, 1972.

7. T. Nakano, H. Ohsawa, and K. Endoh, "A High Performance Cycloconverter-Fed Synchronous Machine Drive System," *Conf. Rec. IEEE/IAS Int. Sem. Power Conv. Conf.*, pp. 334–341, 1982.

8. B. K. Bose, "Microcomputer Based Control Apparatus for a Load Commutated Inverter Synchronous Machine Drive System," U.S. Patent 4,276,505, June 30, 1981.

9. W. Leonhard, "Control of AC Machines with the Help of Microelectronics," *Conf. Rec. IFAC Symp. on Control in Power Elec. and Electrical Drives*, pp. 769–792, 1983.

9

MICROCOMPUTERS

9.0 INTRODUCTION

A power electronic system in general can be controlled either by dedicated hardware or by microcomputer. The microcomputer, or digital control, has some advantages, which can be summarized as follows:

- Careful design of a control system with microcomputer can significantly reduce the controller hardware cost. This advantage will become more and more significant with continuation of the present trends in microcomputer technology. More powerful VLSI microcomputers with higher speed and more functional integration are being made available. Semicustom- or custom-designed VLSI controller chips with software, or software plus dedicated hardware, or dedicated hardware control for specific applications of large-volume production items, can be very cost-effective. Lower volume and weight of the controller and lower power consumption are the added advantages.

- The reliability of the controller can be significantly improved with the microcomputer. The MTBF (mean time between failures) of an LSI or VLSI chip is significantly higher than that of multiple electronic circuit components connected together. Past experience has shown that the reliability of the microcomputer is greatly superior to that of other components of the power electronics system. This, of course, assumes that the microcomputer hardware is designed properly so that none of its specifications, such as temperature rise, goes beyond the safe limit. Military- and industrial-grade components with an extended temperature range can be used when desired.

- Digital circuits do not cause any problems of drift, and parameter variation effects are nonexistent, as in analog circuits. Internal computations are 100% accurate and with appropriate scaling, overflow and underflow problems can be avoided.

- The radiated EMI problem caused by large voltage and current transients in power electronic circuits is avoided by nominal shielding because of large integration. Conducted EMI can be avoided by adequate filtering of power supply and I/O signals.

- Microcomputer control of the local power electronic subsystem permits compatibility with the higher-level host computer in the integrated industrial control environment. The capability of forward and backward information flow is becoming more important with the present trend toward computer automation in the factory.

- Universal hardware can be designed for a class of power electronic systems in which software modules are tailored to suit the particular application. The software control is very flexible and software modules can easily be added, altered, deleted, or upgraded as the physical system changes.

- Sophisticated control functions, which may involve complex computation and decision making, can be incorporated in the microcomputer, thereby enhancing the performance (e.g., efficiency and response) of power electronic systems. Such control functions were previously discarded with dedicated hardware control.

- Microcomputer control permits easy monitoring, warning, data acquisition, and diagnostic functions.

Some limitations of the microcomputer are:

- Although digital computation is 100% accurate, there is a definite error due to quantization and sampling effects when interfacing with the analog world through A/D and D/A conversions. This is evident when sampling an analog sine voltage wave by an A/D converter and then retrieving it through a D/A converter. The errors can be minimized by increasing the bit size and sampling rate of the analog data acquisition system. The signal resolution will suffer if the bit size for signal processing is lower than that of A/D and D/A converters.

- The microcomputer control response is definitely more sluggish than that of dedicated hardware control. With dedicated hardware control, the signals are processed simultaneously in parallel paths with negligible delay. On the other hand, a microcomputer processes signals in a serial manner and thus consumes more time. For multiple tasks, time has to be scheduled in a multiplexed manner, further slowing the process. Sluggishness in processing (i.e., computation delay) may cause a stability problem in feedback control loops and distortion in the computed signals. There are various techniques for enhancing the speed of computation, which will be discussed later.

- Microcomputer implementation of control may be expensive and time consuming, primarily because of the software development effort. Because of its advantages, however, the additional cost may be justified, especially for large-volume production items.

- Software implementation of control does not provide ready access of variables to instruments such as oscilloscopes and multimeters as in a dedicated hardware control system. The parameters (i.e., gain and limit values) cannot easily be monitored or changed under operating conditions. Often a diagnostic or debugging instrument is custom designed for the foregoing purposes.

In this chapter we discuss the basic fundamentals of the microcomputer, and then some state-of-the-art microcomputers with peripheral components are reviewed.* Microcomputer applications are discussed in Chapter 10.

9.1 MICROCOMPUTER FUNDAMENTALS

The basic elements of a microcomputer are shown in Fig. 9.1. It consists of a central processing unit (CPU) set, read-only memory (ROM), random access memory (RAM), analog and digital inputs and outputs, and interrupt controller. The CPU set generally consists of a basic microprocessor, clock generator, bus controller, and bus drivers. The CPU is the heart of microcomputer and is responsible for unified operation of all the other components. It is linked to the other components by an address bus, data bus, and control bus. The width of the data bus usually determines the bit size of a computer. The program of a computer consists of a set of instructions for the CPU and is normally stored in ROM. The instructions are fetched through the data bus, decoded, and then executed. The program

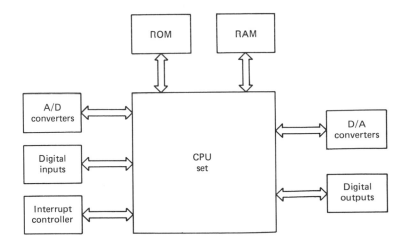

Figure 9.1 Basic elements of a microcomputer.

* The contents of this chapter should be supplemented with manufacturers' user's manuals and specification sheets.

counter in the CPU sets up the sequential addresses of the program, which are sent through the address bus. The data to be manipulated by the program are normally stored in RAM or accessed through analog/digital inputs. The processed data are stored in RAM or sent out through analog/digital outputs. In a "jump" instruction, the program counter can jump to a new memory location determined by the jump address. In a subroutine "call," the program counter stores the next address of the main program in the "stack" and then branches off to service the subroutine. At "return" after completion, main program execution begins after retrieving the address from the "stack pointer." The microcomputer has the capability to attend to an unlimited number of subroutines in a nested manner.

Interrupts

The orderly sequence of program execution can be broken by interrupts. The purpose of interrupts is to draw the computer's attention to more important functions than normal program execution. When an interrupt signal is received, the microcomputer suspends the main program execution, executes the characteristic subroutine, and then returns to the main program. If a function is not time critical, it can be attended by the "poll" method. In this method, the computer periodically interrogates a logic input signal and when a transition is detected, the characteristic subroutine is serviced.

An interrupt can be initiated externally by hardware or internally through hardware or software. With an interrupt request, the CPU completes the instruction under execution, "pushes" the next address in the stack, and vectors into appropriate memory locations to service the interrupt routines. The contents of several registers that may be affected by the routines can be saved in stacks. A number of interrupt signals can be serviced by the microcomputer on a priority basis in a nested manner. The interrupt signals can be globally or selectively enabled or disabled, as desired.

Instruction Set

Each microcomputer has a characteristic set of instructions in assembly language mnemonics which are the building blocks of an assembly language source program. The source program is converted to object or machine code by the macro assembler. The time to execute an instruction (i.e., the instruction cycle time) basically determines the computation speed of a microcomputer. If a function is non-time critical, the source program can be writen in a high-level language such as FORTRAN, PASCAL, PL/M, C or so on, and then translated into object code by the compiler.

The assembly language instruction set can generally be classified into the following groups:

Arithmetic operations: add, subtract, multiply, divide, increment, decrement, and so on.

Logical operations: AND, OR, EXCLUSIVE-OR, compare, rotate, complement, and so on.

Data transfer: move data between registers, or between memory and registers
Program branching: conditional and unconditional jumps, subroutine call and
return, and so on.

The arithmetic-logic unit (ALU) within the CPU performs the arithmetic and logic
operations. The ALU generates the flag bits which specify conditions that arise
as a result of arithmetic and logical manipulations. Flags typically include carry,
zero, sign, and parity. The program jumps can be initiated on the basis of the
status of one or more flags.

Memory and I/O Interface

A CPU has to interface with memory and I/O as shown in the figure. In a single-
chip microcomputer, the memory and I/O may be integrated on the chip monolithically. In a single-chip or multichip microcomputer, extra custom-designed
hardware can always be integrated, depending on the application.

ROM interface. A ROM stores a program which may include frozen data
such as constant parameters and look-up tables. The contents of ROM can only
be read by the CPU. A ROM is nonvolatile memory (i.e., the information is
retained if the power supply fails). A ROM may be mask programmable or
electrically programmable and erasable (EPROM). The EPROM can be programmed by the EPROM programmer and erased by exposing it to ultraviolet
light. In initial product development, an EPROM is used because the program
can easily be altered. When the program is frozen, mask-programmed ROM is
generated, which becomes economical in high-volume product applications.

The memory is accessed by the CPU with the help of READ and CHIP
SELECT control lines. The CHIP SELECT control signal is generated from
ADDRESS lines through a decoder. A memory or I/O device should have an
access time compatible with that of the CPU. If a device has a slower access time,
the CPU can be held in the WAIT mode by making its READY line low until the
information transfer is completed.

RAM interface. A RAM, unlike a ROM, is a read/write or scratch-pad
memory and stores data to be manipulated by the program. In the initial program
development stage, the program may be located in RAM until the EPROM is
burned in when program development is complete. A RAM memory is volatile
and therefore its content can be saved by a backup battery; or else the content can
be transferred to a nonvolatile memory, such as floppy disk, before removing the
power supply. The RAM interface to the CPU is similar to that of ROM except
that it has READ/WRITE control lines.

A RAM chip may be static or dynamic. A dynamic RAM requires periodic
refreshing to save the storage content, but it is transparent to the user. The
individual cell size of a dynamic RAM is smaller and consumes less power, and
therefore dynamic RAM chip is available in larger capacity.

Digital I/O.   The CPU communicates with the outside world through I/O devices.   The information from or to I/O devices can be accessed by READ/ WRITE control lines as in memory devices (memory mapped).   The digital inputs may be individual bit logic signals or parallel byte/word signals coupled to external digital components.   Similarly, the digital outputs may be in the form of individual logic control signals or parallel byte/word signals to other digital components.   The digital data transfer may take place either serially through a universal synchronous/ asynchronous receiver/transmitter or in parallel through the parallel I/O ports. Devices such as a graphics terminal with keyboard, floppy disks, displays, and so on, are interfaced digitally with the microcomputer to input information and display or to store results during program development.

Analog I/O.   In process control applications, a microcomputer interfaces the analog signals through analog-to-digital and digital-to-analog converters.   The input analog signals may be unipolar or bipolar and may be converted to digital words through n-bit converters.   The bit size determines the accuracy of the digital signal and the conversion time determines the maximum throughput to the microcomputer.   For multiple-input channels, either independent A/D converters or a single converter with an analog multiplexer can be used.   The D/A converter converts a digital word to a unipolar or bipolar analog signal almost instantaneously and its bit size determines the resolution of the analog signal.   The bit size for microcomputer signal processing should be compatible with A/D and D/A converter bit size.

9.2 DESCRIPTION OF MICROCOMPUTERS

In this section we review briefly some state-of-the-art Intel microcomputers.*   Although microcomputers are available from a number of vendors, Intel microcomputers are being emphasized because by far the majority of applications in power electronic systems use this type of microcomputer.

Figure 9.2 shows an overview of the Intel microcomputer evolution where the horizontal axis indicates the year of development and the vertical axis shows the general performance.   The performance can be considered as a weighted average of bit size, computation speed, integration of functional components, improved hardware and software features, and so on.   The first available microcomputer was the 4-bit type 4004, which was followed by the 8-bit 8008, but these have become nearly obsolete.   The 8080 is, in fact, the first-generation microcomputer widely used in power electronic systems.   For a period it was the industry's most commonly used microcomputer.   Again, the development of larger LSI and VLSI techniques, faster computation speed, a single +5-V power supply, and many improved performance features of recent microcomputers have pushed the 8080 toward obsolescence.   The microcomputer evolution shown in Fig. 9.2 indicates two general directions: the multichip and single-chip families.   The single-chip

---

* The reader needs to upgrade the contents of this chapter as new devices are introduced.

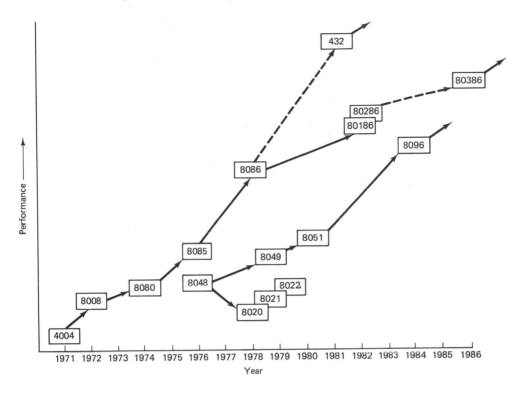

Figure 9.2   Intel microcomputer evolution.

family, which started with the 8048 in 1976, is shown on the right and the multichip family is shown on the left.

8085 Microcomputer

The 8085 microprocessor is an 8-bit CPU and has an architecture similar to the 8080 but with higher integration (i.e., the clock generator and system controller are integrated by the CPU chip). The salient features of the 8085 processor are shown in Table 9.1. The earlier versions, the 8085A (with a 1.3-μs instruction cycle time) and the 8085A-2 (with a 0.8-μs instruction cycle time), which used NMOS technology, have been upgraded in speed and power consumption with HMOS technology. The processor needs a single +5-V power supply and its instruction set is identical to that of 8080, except for two additional instructions (RIM and SIM).

The standard 8085 minimum system with peripheral chips 8155 and 8355/8755 is shown in Fig. 9.3. The 8155 chip contains 256 bytes of static RAM, two programmable 8-bit I/O ports, one programmable 6-bit I/O port, and a programmable 14-bit binary down-counter/timer. The 8355 chip contains 2048 bytes of ROM and two general-purpose 8-bit I/O ports, where each line of I/O port can be individually programmed for input or output. The 8755 chip is identical to the 8355, except that the ROM is replaced by an EPROM. The access time of the 8155

TABLE 9.1 SALIENT FEATURES OF THE 8085 MICROPROCESSOR Courtesy of Intel Corporation

- **Single +5V Power Supply with 10% Voltage Margins**
- **3 MHz, 5 MHz and 6 MHz Selections Available**
- **20% Lower Power Consumption than 8085A for 3 MHz and 5 MHz**
- **1.3 μs Instruction Cycle (8085AH); 0.8 μs (8085AH-2); 0.67 μs (8085AH-1)**
- **100% Compatible with 8085A**
- **100% Software Compatible with 8080A**
- **On-Chip Clock Generator (with External Crystal, LC or RC Network)**

- **On-Chip System Controller; Advanced Cycle Status Information Available for Large System Control**
- **Four Vectored Interrupt Inputs (One is Non-Maskable) Plus an 8080A-Compatible Interrupt**
- **Serial In/Serial Out Port**
- **Decimal, Binary and Double Precision Arithmetic**
- **Direct Addressing Capability to 64K Bytes of Memory**

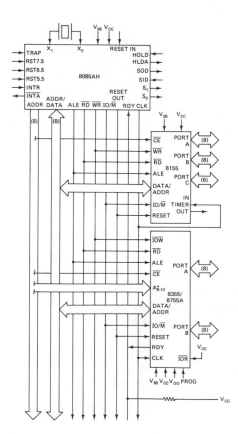

Figure 9.3 Standard 8085 minimum system with 8155 and 8355/8755 chips. Courtesy of Intel Corporation.

and 8355/8755 is comparable to that of the 8085. The microcomputer system is available in the form of a system design kit (SDK-85).

The data bus of 8085 is multiplexed with the lower byte of the address bus and can be isolated in the peripheral chips by the ALE strobe. The 8085 has five interrupt inputs: TRAP, RST 7.5, RST 6.5, RST 5.5, and INTR, and their priority levels are in the order indicated. The TRAP is a nonmaskable external hardware-initiated interrupt. The restart (RST) interrupts are internally invoked by software with the help of an RST instruction. The external hardware interrupts can be expanded with the help of an interrupt controller (such as the 8259). All the interrupts except TRAP can be selectively enabled or disabled by software. The serial input data line (SID) can be loaded into the accumulator by a RIM instruction, and the serial output data line (SOD) can be set or reset by a SIM instruction. The READY (RDY) line should be controlled if the CPU interfaces a slow-access-time device. If the READY line is held low, the CPU will go into the WAIT mode and permit data transmission to be completed. The HOLD line should be held high for the bus to be controlled by the DMA controller or accessed by a parallel processor.

The software of the 8085 is 100% compatible with the 8080. For time-critical real-time applications, an assembly language program supported by the 8080/8085 macro assembler (ASM 80/85) is used. For non-time-critical applications, a high-level language supported by one of the following software programs can be used:

- FORTRAN-80
- BASIC-80
- PASCAL-80
- PL/M-80

The packages operate under the ISIS-II operating system in the Intel development system.

8051 Microcomputer

The 8051 is a stand-alone, 8-bit single-chip microcomputer and is commonly used for real-time control applications. The salient features of the 8031/8051/8751 microcomputer are summarized in Table 9.2, and Fig. 9.4 shows the architecture of the microcomputer. The microcomputer is available in three versions. The generic 8051 contains $4K \times 8$ ROM, 128×8 RAM, four programmable 8-bit ports, two 16-bit programmable timer/counters, two exernal interrupt channels, serial I/O lines, and an on-chip oscillator and clock circuits. The 8751 is identical to the 8051, except that it has user-programmable/erasable ROM (EPROM). The 8031 version does not contain any ROM; otherwise, it is identical to the 8051/8751 chip. The microcomputer requires a single $+5$-V power supply, and its memory is expandable up to 64K bytes of program memory and/or up to 64K bytes of data storage. The 8051 chip evolved from the predecessor single-chip microcomputers

TABLE 9.2   SALIENT FEATURES OF THE 8031/8051/8751 MICROCOMPUTER Courtesy of Intel Corporation

- 8031 - Control Oriented CPU With RAM and I/O
- 8051 - An 8031 With Factory Mask-Programmable ROM
- 8751 - An 8031 With User Programmable/Erasable EPROM

■ 4K x 8 ROM/EPROM	■ Boolean Processor
■ 128 x 8 RAM	■ MCS-48® Architecture Enhanced with:
■ Four 8-Bit Ports, 32 I/O Lines	• Non-Paged Jumps
■ Two 16-Bit Timer/Event Counters	• Direct Addressing
■ Full-Duplex Serial Channel	• Four 8-Register Banks
■ External Memory Expandable to 128K	• Stack Depth Up to 128-Bytes
■ Compatible with MCS-80®/MCS-85®	• Multiply, Divide, Subtract, Compare
Peripherals	■ Most Instructions Execute in 1μs
	■ 4μs Multiply and Divide (unsigned)

8048, 8049, 8020, 8021, and 8022, and therefore its software is upwardly compatible with these devices. With a 12 MHz crystal, the typical instruction cycle time of the 8051 is 1 μs. It supports 8-bit unsigned multiplication and division instructions and the typical execution time is 4 μs. The microcomputer has Boolean processing capability, which is described later.

CPU architecture. The CPU can manipulate four memory spaces: 64K-byte program memory, 64K-byte external data memory, internal data memory, and 16-bit program counter spaces. The internal data memory is further subdivided into 128-byte internal data RAM and 128-byte special function registers (SFRs). The internal data RAM contains four register banks (each with eight registers) and 128 addressable bit locations, as shown in Fig. 9.5(a). The stack area of the microcomputer can be defined within the internal data RAM. The special function registers include arithmetic registers (ACC, B, PSW), data and stack pointers (DPH, DPL, SP), and registers (P0 to P4, IP, IE, SBUF, SCON, TH, TL, TMOD, TCON) that provide interface between the CPU and on-chip peripheral functions. There are also 128 addressable bits within special-functions registers, as shown in Fig. 9.5(b).

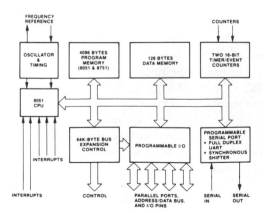

Figure 9.4 Block diagram of 8051 microcomputer. Courtesy of Intel Corporation.

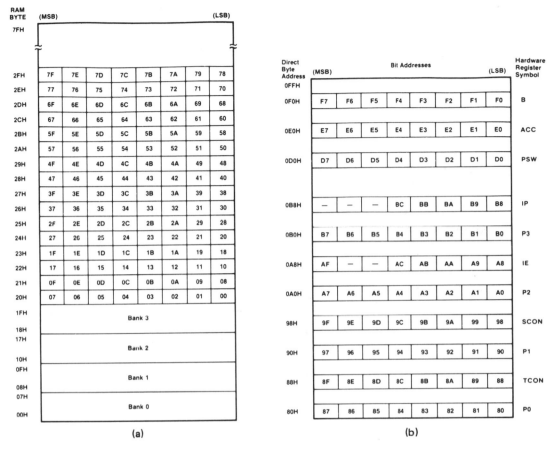

Figure 9.5 Bit address map of 8051 microcomputer: (a) data RAM bit address space; (b) special-function register bit address space. Courtesy of Intel Corporation.

Boolean processor. The Boolean function processing capability is an integral part of the 8051 microcomputer. It is an addressable bit processor with its own instruction set, its own accumulator (the carry flag in the PSW register), a bit-addressable internal data RAM, and special function registers as shown in Fig. 9.5. The special function registers, which have addresses divisible by 8, contain directly addressable bits. On any addressable bit, the Boolean processor can perform the bit operations of set, clear, complement, jump-if-set, jump-if-not-set, jump-if-set-then-clear, and move to/from carry. Between any addressable bit (or its complement) and the carry flag, it can perform the bit operation of logical AND/OR with the result returned to the carry flag.

Interrupt system. The 8051 has five hardware-activated interrupts, of which two are external and three are internal. A summary of interrupt sources and the starting address of the interrupt service program for each interrupt source follows.

Source	Starting Address
External request 0 ($\overline{\text{INT0}}$)	03H
Internal timer/counter 0 (TO)	0BH
External request 1 ($\overline{\text{INT1}}$)	13H
Internal timer/counter 1 (T1)	1BH
Internal serial port	23H

The priority level of the interrupts is determined by the interrupt priority (IP) register. The interrupts can be enabled/disabled selectively or globally by the interrupt enable (IE) register.

I/O Ports. The 8051 has 32 I/O lines, which can be configured as four 8-bit parallel ports (Fig. 9.6): P0, P1, P2, and P3. Each line can be individually and independently programmed as input or output and each can be configured dynamically (i.e., on-the-fly) under software control.

In various operation modes, some of these I/O lines can be used for special input or output functions. The instructions that access external memory use P0 as the multiplexed low-order address/data bus and P2 as the high-order address bus. Ports P1 and P3 are then used as standard I/O. The eight lines of P3 have special functions of external interrupts ($\overline{\text{INT0}}$, $\overline{\text{INT1}}$), counter inputs (T0, T1), two serial data lines (RXD, TXD), and two read/write strobes ($\overline{\text{RD}}$, $\overline{\text{WR}}$).

Within a single port, I/O functions may be combined in many ways: input and output may be performed using different pins at the same time or the same pins at different times; in parallel in some cases and in serial in others; as test pins or (for P3) as additional special functions.

Timer/counters. The microcomputer has two independent 16-bit timer/up-counters which can be used for measuring time intervals, pulse widths, and counting events, and for causing periodic interrupts. The timer/counter 0 (T0) can be programmed into the following four modes:

Mode 0: 8-bit timer with prescaler/8-bit counter with prescaler

Mode 1: 16-bit timer/counter

Mode 2: 8-bit auto-reload timer/counter

Mode 3: 8-bit timer/counter

Timer/counter 1 (T1) can also be programmed in four modes, of which the first three modes are identical to those of T0. The operating modes of the timer/counters are controlled by TMOD and TCON registers under a special function register group.

Instruction set. The 8051 microcomputer instruction set is given in Table 9.3. The set includes 51 fundamental operations under the following five functional groupings shown at the foot of page 330.

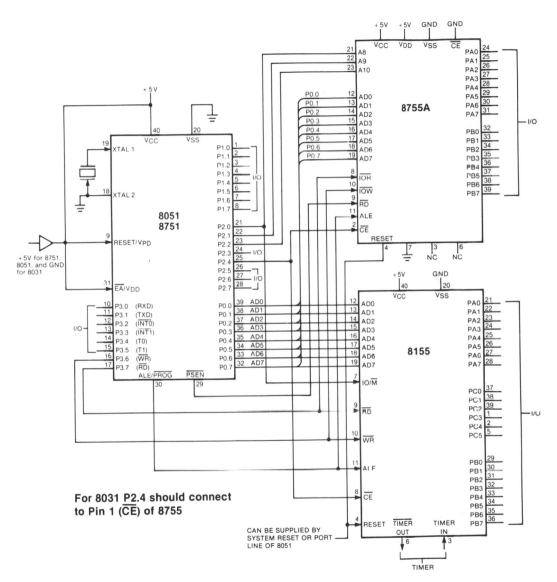

Figure 9.6   8051 three-chip microcomputer system. Courtesy of Intel Corporation.

TABLE 9.3 INSTRUCTION SET FOR THE 8051 MICROCOMPUTER Courtesy of Intel Corporation

ARITHMETIC OPERATIONS

Mnemonic		Description	Byte	Cyc
ADD	A,Rn	Add register to Accumulator	1	1
ADD	A,direct	Add direct byte to Accumulator	2	1
ADD	A,@Ri	Add indirect RAM to Accumulator	1	1
ADD	A,#data	Add immediate data to Accumulator	2	1
ADDC	A,Rn	Add register to Accumulator with Carry	1	1
ADDC	A,direct	Add direct byte to A with Carry flag	2	1
ADDC	A,@Ri	Add indirect RAM to A with Carry flag	1	1
ADDC	A,#data	Add immediate data to A with Carry flag	2	1
SUBB	A,Rn	Subtract register from A with Borrow	1	1
SUBB	A,direct	Subtract direct byte from A with Borrow	2	1
SUBB	A,@Ri	Subtract indirect RAM from A w/Borrow	1	1
SUBB	A,#data	Subtract immed. data from A w/Borrow	2	1
INC	A	Increment Accumulator	1	1
INC	Rn	Increment register	1	1
INC	direct	Increment direct byte	2	1
INC	@Ri	Increment indirect RAM	1	1
DEC	A	Decrement Accumulator	1	1
DEC	Rn	Decrement register	1	1
DEC	direct	Decrement direct byte	2	1
DEC	@Ri	Decrement indirect RAM	1	1
INC	DPTR	Increment Data Pointer	1	2
MUL	AB	Multiply A & B	1	4
DIV	AB	Divide A by B	1	4
DA	A	Decimal Adjust Accumulator	1	1

LOGICAL OPERATIONS

Mnemonic		Destination	Byte	Cyc
ANL	A,Rn	AND register to Accumulator	1	1
ANL	A,direct	AND direct byte to Accumulator	2	1
ANL	A,@Ri	AND indirect RAM to Accumulator	1	1
ANL	A,#data	AND immediate data to Accumulator	2	1
ANL	direct,A	AND Accumulator to direct byte	2	1
ANL	direct,#data	AND immediate data to direct byte	3	2
ORL	A,Rn	OR register to Accumulator	1	1
ORL	A,direct	OR direct byte to Accumulator	2	1
ORL	A,@Ri	OR indirect RAM to Accumulator	1	1
ORL	A,#data	OR immediate data to Accumulator	2	1
ORL	direct,A	OR Accumulator to direct byte	2	1
ORL	direct,#data	OR immediate data to direct byte	3	2
XRL	A,Rn	Exclusive-OR register to Accumulator	1	1
XRL	A,direct	Exclusive-OR direct byte to Accumulator	2	1
XRL	A,@Ri	Exclusive-OR indirect RAM to A	1	1
XRL	A,#data	Exclusive-OR immediate data to A	2	1
XRL	direct,A	Exclusive-OR Accumulator to direct byte	2	1
XRL	direct,#data	Exclusive-OR immediate data to direct	3	2
CLR	A	Clear Accumulator	1	1
CPL	A	Complement Accumulator	1	1
RL	A	Rotate Accumulator Left	1	1

TABLE 9.3 *Continued*

RLC	A	Rotate A Left through the Carry flag	1	1
RR	A	Rotate Accumulator Right	1	1
RRC	A	Rotate A Right through Carry flag	1	1
SWAP	A	Swap nibbles within the Accumulator	1	1

DATA TRANSFER

Mnemonic		Description	Byte	Cyc
MOV	A,Rn	Move register to Accumulator	1	1
MOV	A,direct	Move direct byte to Accumulator	2	1
MOV	A,@Ri	Move indirect RAM to Accumulator	1	1
MOV	A,#data	Move immediate data to Accumulator	2	1
MOV	Rn,A	Move Accumulator to register	1	1
MOV	Rn,direct	Move direct byte to register	2	2
MOV	Rn,#data	Move immediate data to register	2	1
MOV	direct,A	Move Accumulator to direct byte	2	1
MOV	direct, Rn	Move register to direct byte	2	2
MOV	direct,direct	Move direct byte to direct	3	2
MOV	direct,@Ri	Move indirect RAM to direct byte	2	2
MOV	direct,#data	Move immediate data to direct byte	3	2
MOV	@Ri,A	Move Accumulator to indirect RAM	1	1
MOV	@Ri,direct	Move direct byte to indirect RAM	2	2
MOV	@Ri,#data	Move immediate data to indirect RAM	2	1
MOV	DPTR,#data16	Load Data Pointer with a 16-bit constant	3	2
MOVC	A,@A+DPTR	Move Code byte relative to DPTR to A	1	2
MOVC	A@A+PC	Move Code byte relative to PC to A	1	2
MOVX	A,@Ri	Move External RAM (8-bit addr) to A	1	2
MOVX	A,@DPTR	Move External RAM (16-bit addr) to A	1	2
MOVX	@Ri,A	Move A to External RAM (8-bit addr)	1	2
MOVX	@DPTR,A	Move A to External RAM (16-bit addr)	1	2
PUSH	direct	Push direct byte onto stack	2	2
POP	direct	Pop direct byte from stack	2	2
XCH	A,Rn	Exchange register with Accumulator	1	1
XCH	A,direct	Exchange direct byte with Accumulator	2	1
XCH	A,@Ri	Exchange indirect RAM with A	1	1
XCHD	A,@Ri	Exchange low-order Digit ind. RAM w/A	1	1

BOOLEAN VARIABLE MANIPULATION

Mnemonic		Description	Byte	Cyc
CLR	C	Clear Carry flag	1	1
CLR	bit	Clear direct bit	2	1
SETB	C	Set Carry flag	1	1
SETB	bit	Set direct bit	2	1
CPL	C	Complement Carry flag	1	1
CPL	bit	Complement direct bit	2	1
ANL	C,bit	AND direct bit to Carry flag	2	2
ANL	C,/bit	AND complement of direct bit to Carry	2	2
ORL	C,bit	OR direct bit to Carry flag	2	2
ORL	C,/bit	OR complement of direct bit to Carry	2	2
MOV	C,bit	Move direct bit to Carry flag	2	1
MOV	bit,C	Move Carry flag to direct bit	2	2

TABLE 9.3 *Continued*

PROGRAM AND MACHINE CONTROL

Mnemonic		Description	Byte	Cyc
ACALL	addr11	Absolute Subroutine Call	2	2
LCALL	addr16	Long Subroutine Call	3	2
RET		Return from subroutine	1	2
RETI		Return from interrupt	1	2
AJMP	addr11	Absolute Jump	2	2
LJMP	addr16	Long Jump	3	2
SJMP	rel	Short Jump (relative addr)	2	2
JMP	@A + DPTR	Jump indirect relative to the DPTR	1	2
JZ	rel	Jump if Accumulator is Zero	2	2
JNZ	rel	Jump if Accumulator is Not Zero	2	2
JC	rel	Jump if Carry flag is set	2	2
JNC	rel	Jump if No Carry flag	2	2
JB	bit,rel	Jump if direct Bit set	3	2
JNB	bit,rel	Jump if direct Bit Not set	3	2
JBC	bit,rel	Jump if direct Bit is set & Clear bit	3	2
CJNE	A,direct,rel	Compare direct to A & Jump if Not Equal	3	2
CJNE	A,#data,rel	Comp. immed. to A & Jump if Not Equal	3	2
CJNE	Rn.#data,rel	Comp. immed. to reg. & Jump if Not Equal	3	2
CJNE	@Ri,#data,rel	Comp. immed. to ind. & Jump if Not Equal	3	2
DJNZ	Rn,rel	Decrement register & Jump if Not Zero	2	2
DJNZ	direct,rel	Decrement direct & Jump if Not Zero	3	2
NOP		No operation	1	1

Notes on data addressing modes:

Rn   —Working register R0–R7

direct   —128 internal RAM locations, any I/O port, control or status register

@Ri   —Indirect internal RAM location addressed by register R0 or R1

#data   —8-bit constant included in instruction

#data16 —16-bit constant included as bytes 2 & 3 of instruction

bit   —128 software flags, any I/O pin, control or status bit

Notes on program addressing modes:

addr16   —Destination address for LCALL & LJMP may be anywhere within the 64-
             Kilobyte program memory address space.

addr11   —Destination address for ACALL & AJMP will be within the same 2-Kilobyte
             page of program memory as the first byte of the following instruction.

rel   —SJMP and all conditional jumps include an 8-bit offset byte. Range is + 127–
             128 bytes relative to first byte of the following instruction.

All mnemonics copyrighted © Intel Corporation 1979.

1. Arithmetic operations

2. Logical operations

3. Data transfer

4. Boolean variable manipulation

5. Program and machine control

Combining with various addressing modes for Boolean (1-bit), nibble (4-bit), byte (8-bit), and address (16-bit) data types produces a total of 111 instructions.

Each assembly language instruction consists of an operation mnemonic and one to four operands. The mnemonic abbreviates the basic operation to be performed and the operands, separated by commas, mention which variables are involved, what data to use, or what instruction to execute next. Instructions that need two data operands specify the destination first, followed by the source variable, except for the move operation between two directly addressable bytes, where the source operand is first and the destination operand is second. The number of bytes for each instruction and the corresponding number of cycles are indicated on the right. A cycle consists of 12 oscillator periods, and with a 12 MHz crystal a one-cycle instruction takes 1 μs. The byte size and cycle length help in estimating the program memory size and execution time, respectively.

8051 system. Although the 8051 is a stand-alone microcomputer, it can be expanded with various I/O chips. Figure 9.6 illustrates a three-chip microcomputer system using 8755 and 8155 chips. The external ROM, RAM, timer, and additional I/O considerably expand the system capacity. The multiplexed address/data bus of port P0 has been demultiplexed by using the ALE strobe. The same ALE line can be used to feed the program pulse for EPROM programming. The chip enable ($\overline{\text{CE}}$) of both chips is generated by address line P 2.4. The $\overline{\text{PSEN}}$ output is a control signal that enables the external program memory to the bus during normal fetch operations. The $\overline{\text{EA}}/V_{DD}$ pin is held high if the 8051 executes instructions from the internal ROM. When the line is held low, all instructions are fetched from external program memory. The pin can also receive the 21V EPROM programming supply voltage. A low-to-high transition on the RESET/V_{PD} pin resets the 8051. If V_{PD} is held high, while V_{CC} drops low, V_{PD} will provide standby power to the RAM. When V_{PD} is held low, the RAM's current is drawn from V_{CC}.

Software support. The 8051 is supported by the following software packages:

- ASM-51: macro assembler
- CONV-51: 8048 assembly language source code to 8051 assembly language source code conversion program
- RL-51: linker and relocator program
- PL/M-51

8086 microcomputer

The 8086 microcomputer is based on the 8086 microprocessor (iAPX 86/10) and its salient features are summarized in Table 9.4. Figure 9.7 shows the CPU functional diagram. It is a 16-bit CPU that is available in three clock rates: 5, 8, and 10 MHz, and the corresponding instruction cycle times typically are 1.2, 0.75, and

TABLE 9.4 SALIENT FEATURES OF 8086 MICROPROCESSOR
Courtesy of Intel Corporation

• Direct addressing capability to 1 Mbyte of memory	• 8- and 16-bit signed and un-signed arithmetic in binary or decimal, including multiply and divide
• Architecture designed for as-sembly language and high level languages	
• 14 Word, by 16-bit register set with symmetrical operations	• Range of clock rates: 5 MHz for 8086 8 MHz for 8086-2 10 MHz for 8086-1
• 24 operand addressing modes	
• Bit, byte, word, and block op-erations	• MULTIBUS system-compatible interface

0.6 μs, respectively. The processor has a 20-bit address bus and therefore can address up to 1 megabyte of memory. The lower 16 address lines are multiplexed with a data bus and the upper four address lines are multiplexed with status information lines which can be separated by latches with the help of the ALE strobe (see Fig. 9.9).

The internal functions of the CPU can be divided into two major units: the execution/control unit (EU) and the bus interface unit (BIU). The EU performs the basic processing functions, as it contains the data, pointer, and index registers and the arithmetic-logic unit (ALU). It accepts prefetched instructions from the BIU and returns unrelocated operand addresses to it. It then receives memory operands via the BIU, processes them, and passes the results to the BIU for storage. The BIU prefetches instructions before they are required by the EU. It buffers them in a queue that may contain up to six bytes of instruction stream, awaiting their decoding and instruction. This mechanism enhances the execution speed of the 8086. The BIU also provides the functions related to operand fetch and store, address relocation, and bus control, all in parallel with EU operation.

The memory addresses are logically subdivided into segments of 64K bytes each, which can be allocated to code, data, and stack, with each segment falling on 16-byte boundaries. Figure 9.8 shows the memory organization of the 8086. All memory references are made relative to base addresses contained in the code segment (CS), stack segment (SS), data segment (DS), and extra data segment (ES) registers. The bytes or words within a segment are addressed using 16-bit offset addresses, or effective addresses within the 64K-byte segment. A 20-bit physical address is constructed by adding the 16-bit offset address to the 16-bit segment address with four low-order zero bits appended.

The word (16-bit) operands can be located on even or odd address boundaries. Physically, the memory is organized as a high bank ($D_{15}-D_8$) and a low bank (D_7-D_0) of 512K 8-bit bytes addressed in parallel by the processor's address lines, $A_{19}-A_0$. Byte data with even addresses are transferred on the D_7-D_0 bus lines, while odd-addressed byte data ($A_0 = 1$) are transferred on the $D_{15}-D_8$ bus lines. The CPU provides two enable signals, $\overline{BHE}$ and A_0, to selectively allow reading from or writing into either an odd-byte location, even-byte location, or both, as

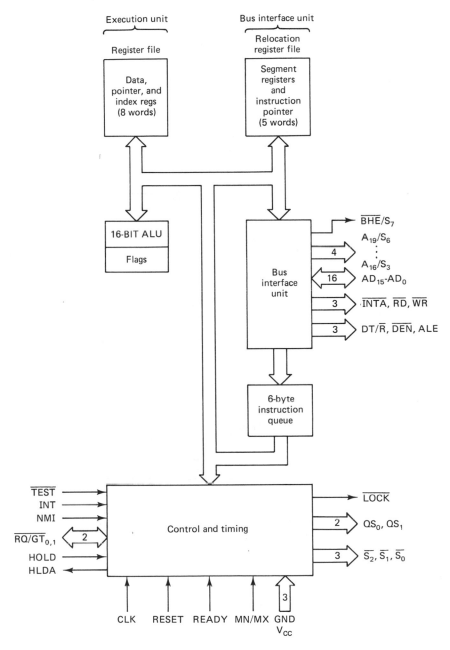

Figure 9.7 8086 CPU functional diagram. Courtesy of Intel Corporation.

summarized below:

$\overline{BHE}$	A_0	Function
0	0	16-bit word from/to addressed locations
0	1	Upper 8 bits from/to odd-addressed location
1	0	Lower 8 bits from/to even-addressed location
1	1	No selection of location

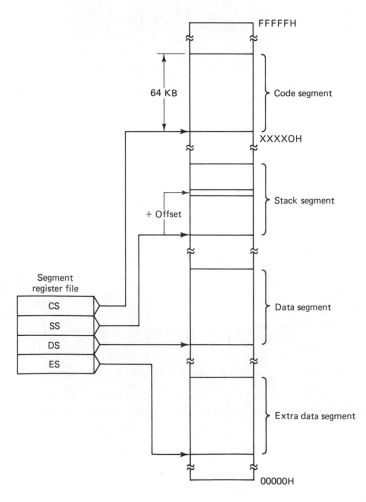

Figure 9.8 Memory organization of 8086. Courtesy of Intel Corporation.

The 8086 microprocessor can be used either in minimum-mode or maximum-mode configuration by connecting the strap pin MN/$\overline{\text{MX}}$ to V_{CC} or ground, respectively. Figure 9.9 shows a typical system configuration in minimum mode. The basic CPU group consists of the 8086 processor, the 8284 clock generator, the 8282 octal latches, and the 8286 octal bus transceivers as shown. A wait-state generator can be coupled to the clock generator to lengthen the bus cycle time if the CPU interfaces a slow-access-time device. The power-on-reset of the CPU is generated by the *RC* network. In a minimum system, the control signals $\overline{\text{WR}}$, M/$\overline{\text{IO}}$, DT/$\overline{\text{R}}$, $\overline{\text{DEN}}$, ALE, and $\overline{\text{INTA}}$ are generated by the CPU directly. The M/$\overline{\text{IO}}$ line distinguishes memory-referenced I/O from directly accessed I/O. The 8286 transceivers are used as buffers which permit higher drive capability of data bus. The data enable ($\overline{\text{DEN}}$) and data transmit/receive (DT/$\overline{\text{R}}$) lines are coupled to this chip. The address and data buses are coupled to 2K bytes of static RAM (four 2142s), 4K bytes of EPROM (two 2716s), and an 8080 microcomputer system. The upper byte, lower byte, or both bytes of RAM can be selectively addressed, but for the EPROM both the bytes are always accessed.

A typical system configuration in the maximum mode of the 8086 is shown in Fig. 9.10. The configuration is somewhat similar to that of the minimum mode except that now the bus controller 8288 generates all the timing and control signals from the three status bits $\overline{S}_0$, $\overline{S}_1$, and $\overline{S}_2$, which are compatible with multibus operation. The minimum-mode control signal pins can now take on additional functions, such as extra direct memory access (DMA) control and bus locking capabilities for use in multiprocessor applications.

The 8086 has an instruction set which supports arithmetic, logic, data transfer,

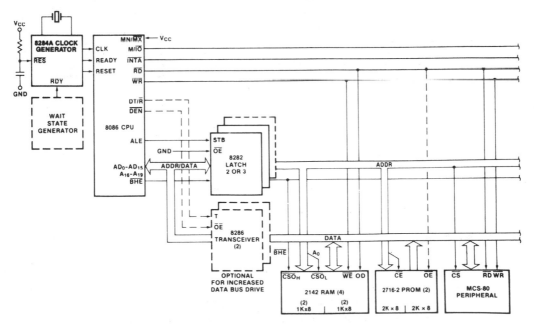

Figure 9.9 Typical system configuration with 8086 in minimum mode. Courtesy of Intel Corporation.

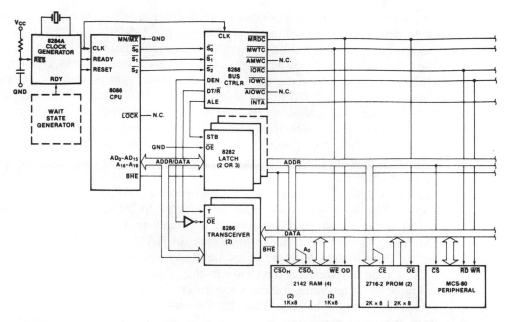

Figure 9.10   Typical system configuration with 8086 in maximum mode. Courtesy of Intel Corporation.

string manipulation, control transfer, and processor control instructions. The arithmetic instructions include 8- and 16-bit signed and unsigned multiplications and divisions. The 8086-based microcomputer is supported by the following software development programs:

- ASM-86/88: macro assembler
- PL/M-86/88
- PASCAL-86/88
- FORTRAN-86/88
- LINK-86/88 and LOC-86/88: linkage and relocation utilities
- CONV-86/88: conversion of 8080/8085 assembly language source code to 8086/8088 assembly language source code
- OH-86/88: object-to-hexadecimal converter
- LIB-86/88: library manager

8088 Microcomputer

The 8088 microcomputer is based on the 8088 microprocessor (iAPX 88/10), which is essentially identical to the 8086 microprocessor except it has an 8-bit data bus. For systems where the 8086 features are desirable but 8-bit resolution is adequate, the 8088-based microcomputer is used. Table 9.5 gives the salient features of the 8088 CPU, indicating its availability in two clock rates, 5 MHz and 8 MHz. Most internal functions of the 8088 are identical to the equivalent 8086 functions. The

TABLE 9.5   SALIENT FEATURES OF THE 8088 MICROPROCESSOR Courtesy of Intel Corporation

- 8-Bit Data Bus Interface
- 16-Bit Internal Architecture
- Direct Addressing Capability to 1 Mbyte of Memory
- Direct Software Compatibility with iAPX 86/10 (8086 CPU)
- 14-Word by 16-Bit Register Set with Symmetrical Operations
- 24 Operand Addressing Modes
- Byte, Word, and Block Operations

- 8-Bit and 16-Bit Signed and Unsigned Arithmetic in Binary or Decimal, Including Multiply and Divide
- Compatible with 8155-2, 8755A-2 and 8185-2 Multiplexed Peripherals
- Two Clock Rates: 5 MHz for 8088 8 MHz for 8088-2

8088 handles the external bus in the same way that the 8086 does, the difference being that it handles only 8 bits at a time. The 16-bit operands are fetched in two consecutive bus cycles. Both processors appear identical to the software engineer with the identical instruction set, but the execution time may be different. The software packages that support 8086 also support 8088. Internally, there are only three differences between the 8088 and the 8086, as summarized below.

- The instruction queue length of the 8088 is four bytes, whereas the 8086 has a length of six bytes.
- The instruction prefetching algorithm of 8088 is slightly different. The 8088 BIU will fetch a new instruction to load into the queue each time there is a one-byte space in the queue. The 8086 waits until a two-byte space is available.
- The internal execution time of the instruction set is affected by the 8-bit interface. All 16-bit fetches and writes from/to memory take an extra four clock cycles. The CPU is also limited by the speed of instruction fetches.

Figure 9.11 shows a typical system configuration in the minimum mode. In this mode, the 8088 can be used with either a multiplexed or a demultiplexed bus. The multiplexed bus configuration, as shown in Fig. 9.11, is compatible with the peripheral chips 8155 (RAM–I/O ports–timer), 8355/8755 (ROM–I/O ports), and 8185 (additional RAM).

80186 Microcomputer

The 80186 microcomputer is based on the 80186 microprocessor (iAPX 186) and is essentially a high-integration version of the 8086 processor. The salient features of the microprocessor are summarized in Table 9.6, and Fig. 9.12 shows its block diagram. Its architecture is similar to that of the enhanced 8086–2 CPU with the integration of clock generator, two DMA channels, three 16-bit timers, an interrupt controller, a wait-state generator, a memory and peripheral chip select logic, and a local bus controller. The processor does not have the MN/$\overline{\text{MX}}$ mode of operation as in 8086. All the processors in the 8086/8088/80186/80286 family have the same

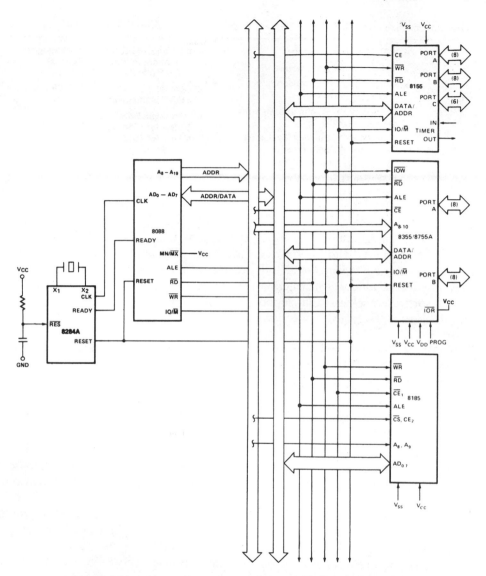

Figure 9.11 Typical system configuration with 8088 in minimum mode. Courtesy of Intel Corporation.

basic registers and instruction set. The 80186 processor has 10 more instructions than are in the 8086/8088 instruction set.

The 80186 can address up to 1M byte of memory, which is organized in sets of 64K-byte segments. The memory is addressed using a 16-bit base segment and a 16-bit offset, as in the 8086.

The 80186 provides six memory chip select outputs for three address areas: upper memory, lower memory, and midrange memory. One each is provided for upper memory (UCS) and lower memory (LCS), and four are provided for mid-range memory (MCSO-3). The range for each chip selection is user programmable,

TABLE 9.6   SALIENT FEATURES OF THE 80186 MICROPROCESSOR Courtesy of Intel Corporation

- **Integrated Feature Set**
  - **Enhanced 8086-2 CPU**
  - **Clock Generator**
  - **2 Independent, High-Speed DMA Channels**
  - **Programmable Interrupt Controller**
  - **3 Programmable 16-bit Timers**
  - **Programmable Memory and Peripheral Chip-Select Logic**
  - **Programmable Wait State Generator**
  - **Local Bus Controller**
- **High-Performance 8 MHz Processor**
  - **2 Times the Performance of the Standard iAPX 86**
  - **4 MByte/Sec Bus Bandwidth Interface**
- **Direct Addressing Capability to 1 MByte of Memory**

- **Completely Object Code Compatible with All Existing iAPX 86, 88 Software**
  - **10 New Instruction Types**
- **Compatible with 8282/83/86/87, 8288, 8289 Bus Support Components**
- **Complete System Development Support**
  - **Development Software: Assembler, PL/M, Pascal, Fortran, and System Utilities**
  - **In-Circuit-Emulator (I²ICE™-186)**
  - **iRMX™ 86, 88 Compatible (80130 OSF)**
- **Optional Numeric Processor Extension**
  - **iAPX 186/20 High-Performance 80-bit Numeric Data Processor**

but one chip selection remains active at a time. In addition, the processor can generate chip selection for up to seven peripheral devices. These chip selects are active for seven contiguous blocks of 128 bytes above a programmable base address.

Altogether, the processor can service 16 interrupts invoked by hardware and software. The software interrupts are generated by specific instructions or the

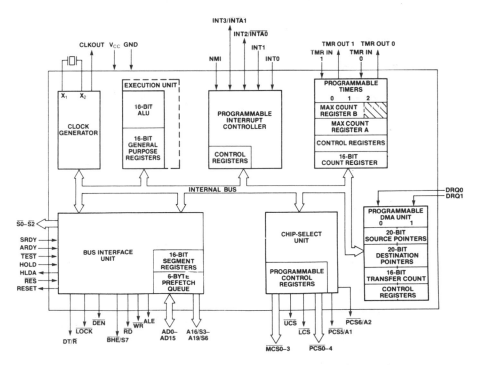

Figure 9.12   Block diagram of 80186. Courtesy of Intel Corporation.

results of conditions specified by instructions. There are five hardware interrupts, of which one is nonmaskable (NMI). The interrupts can be programmed into several modes by the interrupt controller.

The 80186 has three 16-bit programmable timers. Two of these are connected externally to count external events, generate nonrepetitive waveforms, and so on. The third timer is used for real-time coding and time-delay applications. It can also be used as a prescaler to the other two, or as a DMA request source.

The DMA controller provides two independent high-speed DMA channels, which permit data transfer between memory and I/O spaces (memory to I/O) or within the same space (memory to memory or I/O to I/O). Data can be transferred either in bytes or in words to or from even or odd addresses. The maximum DMA transfer can occur at the rate of 2M bytes/s and the channels can be programmed such that one channel has priority over the other.

Figure 9.13 shows a typical 80186-based microcomputer. The memory of the microcomputer is mapped into upper memory (RESET ROM), midrange memory (PROGRAM RAM), and lower memory (LOW RAM). The midrange and lower memories provide selective high/low byte transfer capability. The system interfaces a video terminal and a disk drive system. The disk has DMA transfer capability to and from the microcomputer. The serial I/O and disk interface hardware interfacing the terminal and disk, respectively, are addressed by peripheral chip select outputs $\overline{PCS0}$/A1 and $\overline{PCS4}$/A2. The octal latch 8282/8283 demultiplexes the address bus (and provides the buffer), which addresses the memories. The octal bus transceiver 8286/8287 provides a buffer to the data bus, which interfaces the serial I/O.

80286 Microcomputer

The 80286 microcomputer is based on the 80286 CPU (iAPX 286/10), which is an advanced, high-performance microprocessor with specially optimized capabilities for multiuser and multitasking systems. The salient features of the processor are given in Table 9.7. The processor can perform up to six times faster than the standard 5 MHz 8086, depending on the application, and provides upward software compatability with the 8086, 8088, and 80186 microprocessors. The processor operates in two modes: the 8086 real address mode and a protected virtual address mode. Both modes execute a superset of the 8086/8088 instruction set. In the real address mode, programs use real addresses with up to 1 megabyte of address space. Programs use virtual addresses in the protected virtual address mode (called the protected mode). In this mode, the CPU automatically maps 1-gigabyte virtual addresses per task into a 16-megabyte real address space. This mode also provides memory protection to isolate the operating system and ensure the privacy of each task's program and data.

The base architecture of the 80286 CPU is same as that of the 8086 and contains the same set of registers, instructions, and addressing modes. The 24-bit address bus and 16-bit data bus are demultiplexed. This feature, together with pipeline addressing capability, enhances the speed of program execution. The

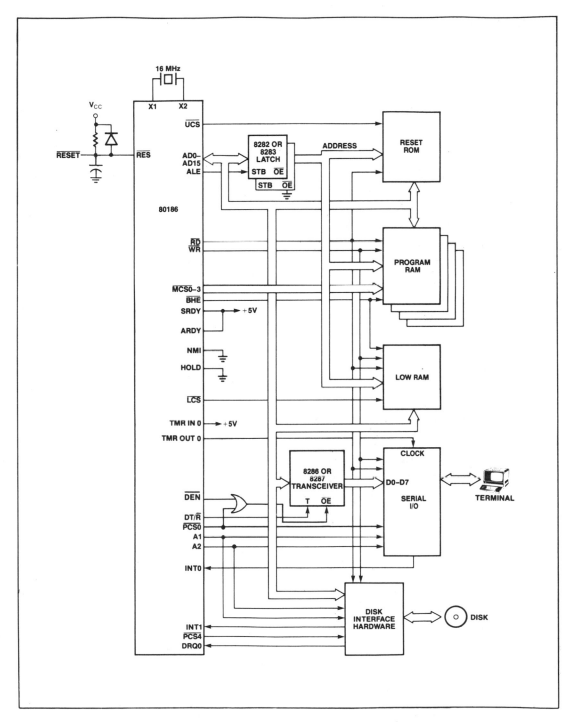

Figure 9.13 Typical 80186-based microcomputer. Courtesy of Intel Corporation.

TABLE 9.7 SALIENT FEATURES OF THE 80286 MICROPROCESSOR Courtesy of Intel Corporation

■ High Performance 8 and 10 MHz Processor (Up to six times iAPX 86)	■ Optional Processor Extension: —iAPX 286/20 High Performance 80-bit Numeric Data Processor
■ Large Address Space: —16 Megabytes Physical —1 Gigabyte Virtual per Task	■ Complete System Development Support: —Development Software: Assembler, PL/M, Pascal, FORTRAN, and System Utilities
■ Integrated Memory Management, Four-Level Memory Protection and Support for Virtual Memory and Operating Systems	—In-Circuit-Emulator (ICE™-286)
■ Two iAPX 86 Upward Compatible Operating Modes: —iAPX 86 Real Address Mode —Protected Virtual Address Mode	■ High Bandwidth Bus Interface (8 or 10 Megabyte/Sec) ■ Available: —Standard Temperature Range

processor has a nonmaskable interrupt (NMI) and a maskable INTR input which can be used with an interrupt controller (8259).

The processor is supported by the following vendor-supplied software programs:

- iAPX-286: evaluation package; permits a programmer to develop an elementary assembly language program and to evaluate the processor
- PL/M-286
- PASCAL-286
- FORTRAN-286
- iAPX-286: software development package, a complete system development package supported by the following programs:
  —iAPX-286: macro assembler (ASM-286)
  —iAPX-286: binder (BND-286)
  —iAPX-286: mapper (MAP-286)
  —iAPX-286: librarian (LIB-286)
  —iAPX-286: system builder (BLD-286)
  —iAPX-286: simulator (SIM-286)

A typical 80286 microcomputer system configuration is shown in Fig. 9.14. The CPU is interfaced with a clock generator (82284), a bus controller (82288), a latch (8282/8283), and a transceiver (8286/8287) to generate the address bus, data bus, and control signals shown. The actual ROM and RAM interface is not shown. The interrupt controller (8259), which is discussed later, permits eight external interrupt signals. The optional processor extension shown by the dashed lines may include numeric data processor 80287 (discussed later). This can perform numerical calculations and data transfer concurrently with CPU program execution. The optional decode logic shown in the figure generates advanced memory and I/O select signals from the overlapped address and data of the 80286 bus cycle. This minimizes delays caused by address and decoding propagation.

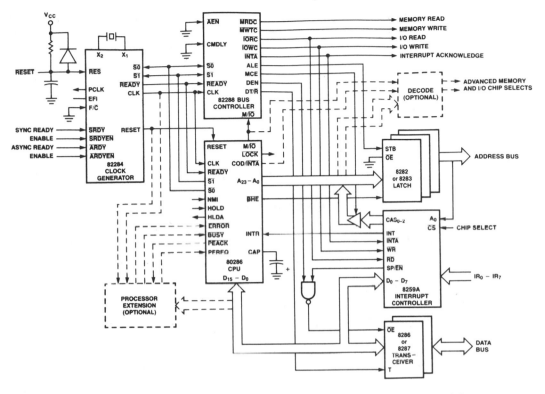

Figure 9.14   Typical 80286 microcomputer system configuration. Courtesy of Intel Corporation.

8096 Microcomputer

The 8096 is a 16-bit single-chip stand-alone microcomputer designed for real-time control applications.   The salient features of the 8096 can be summarized as follows:

- 8K-byte on-chip ROM
- 232-byte register space (RAM)
- 10-bit, eight-channel A/D converter
- Five 8-bit I/O ports
- Full-duplex serial port
- High-speed pulse I/O
- Pulse-width-modulated output
- Eight interrupt sources
- Four 16-bit software timers and two 16-bit hardware timers
- Watchdog timer
- Hardware signed and unsigned multiply/divide

The 8096 is the generic name of the microcomputer and several versions exist where the chip may be with or without ROM and/or with (four or eight channels) or without an A/D converter. With a 12 MHz input frequency, the 8096 can do 16-bit addition in 1.0 μs and a 16 × 16-bit multiply or 32/16-bit divide in 6.5 μs.

The block diagram of the 8096 is shown in Fig. 9.15, and Fig. 9.16 shows its memory map. The several functional components can be identified as the CPU, programmable high-speed I/O unit, A/D converter, serial port, and a PWM output which can be used for D/A conversion. The major components of the CPU are the register/arithmetic-logic unit (RALU) and register file, and communication to the outside world is done either through the special-function register (SFR) or the memory controller. The addressable memory space is 64K bytes, of which OOH through FFH are internal RAM space and 2080H through 3FFFH are internal program storage ROM. The RAM space is divided into a register file and the SFR. All the I/O operations are controlled through the SFR, as indicated in Fig. 9.16. The upper 16 RAM locations (FOH through FFH) can be used as continuous memory with a power supply through the V_{PD} pin.

The I/O ports connect to the internal bus through the bus buffers. Port 0 is an input port that shares its pins with the analog inputs to the A/D converter. Ports 1, 2, 3, and 4 are bidirectional ports. Port 2 shares its pins with serial I/O, external interrupt, PWM output, and the hardware timer reset and input. Ports 3 and 4 share pins with the 16-bit multiplexed address/data bus for accessing off-chip memory. There are four high-speed trigger inputs (HSI) which look for transition of input lines and therefore record the times at which external events

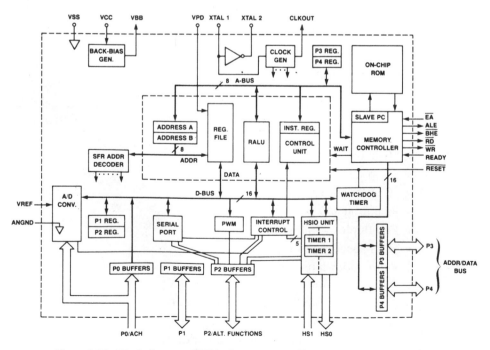

Figure 9.15   Block diagram of 8096 microcomputer. Courtesy of Intel Corporation.

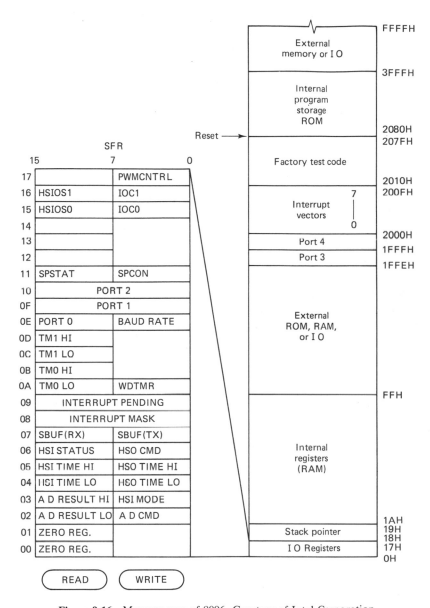

Figure 9.16 Memory map of 8096. Courtesy of Intel Corporation.

occur. The high-speed output (HSO) unit, consisting of six pulse generator outputs, provide a trigger to external events at preset times. The HSO can be programmed to generate interrupts at preset times. Up to four such "software timers" can be in operation at a time.

The A/D converter is a unipolar 10-bit successive approximation type with a fixed conversion time of 42 μs at the 12 MHz clock. The PWM output has a variable duty cycle but repeats every 64 μs. Changes in the duty cycle are made by writing to the PWM register at location 17H. The output can therefore be used for D/A conversion.   The watchdog timer is a 16-bit counter that is incremented every state time (three crystal clock periods).   When it overflows, the RESET is activated.   This feature is provided as a means of graceful recovery from a software upset.   The counter must be cleared by the software before it overflows and a RESET is activated.

A complete instruction set for the 8096 is given in Table 9.8.   The flag bits of the program status word (PSW) shown on the right of the table can be defined as follows:

- Z: zero bit
- N: negative bit
- V: overflow bit
- VT: overflow trap bit
- C: carry bit
- ST: sticky bit

The PSW is located in the higher byte of the interrupt mask register.   The instruction set contains arithmetic and logic operations for 8-bit and 16-bit signed and unsigned data.   Instruction execution times average 1 to 2 μs in typical applications.

The 8096 microcomputer is supported by the following software programs:

- ASM-96: macro assembler
- PL/M-96
- RL-96: linker/relocator
- LIB-96: librarian
- SBE-96: single-board emulation package

9.3 PERIPHERAL COMPONENTS

Several microprocessor peripheral chips were reviewed in the preceding section. A few more selected digital and analog peripheral components are described here. A more detailed description of the available components may be found in component data books.

TMS320 Signal Processor

TMS320 is the generic name of high-speed and numeric-intensive signal processing microcomputers manufactured by Texas Instruments.   The TMS32010 is the original microcomputer that has been updated by the TMS32020 version.   The key features of TMS32010 are summarized in Table 9.9.   The 16-bit microcomputer has 160-ns instruction cycle time (TMS 32010-25) which includes 16×16-bit multiply instruction.   The chip has 144×16-bit on-chip RAM and 4096×16-bit external addressable ROM/RAM program memory.   In a maskable ROM version

TABLE 9.8 INSTRUCTION SET OF 8096 MICROCOMPUTER Courtesy of Intel Corporation

Mnemonic	Oper-ands	Operation (Note 1)	Z	N	C	V	VT	ST	Notes
ADD/ADDB	2	D ← D + A	√	√	√	√	√	—	
ADD/ADDB	3	D ← B + A	√	√	√	√	√	—	
ADDC/ADDCB	2	D ← D + A + C	√	√	√	√	√	—	
SUB/SUBB	2	D ← D − A	√	√	√	√	√	—	
SUB/SUBB	3	D ← B − A	√	√	√	√	√	—	
SUBC/SUBCB	2	D ← D − A + C − 1	√	√	√	√	√	—	
CMP/CMPB	2	D − A	√	√	√	√	√	—	
MUL/MULU	2	D, D + 2 ← D * A	—	—	—	—	—	√	2
MUL/MULU	3	D, D + 2 ← B * A	—	—	—	—	—	√	2
MULB/MULUB	2	D, D + 1 ← D * A	—	—	—	—	—	√	3
MULB/MULUB	3	D, D + 1 ← B * A	—	—	—	—	—	√	3
DIV/DIVU	2	D ← (D, D + 2)/A D + 2 remainder	—	—	—	√	√	—	2
DIVB/DIVUB	3	D ← (D, D + 1)/A D + 1 remainder	—	—	—	√	√	—	3
AND/ANDB	2	D ← D and A	√	√	0	0	—	—	
AND/ANDB	3	D ← B and A	√	√	0	0	—	—	
OR/ORB	2	D ← D or A	√	√	0	0	—	—	
XOR/XORB	2	D ← D (excl. or) A	√	√	0	0	—	—	
LD/LDB	2	D ← A							
ST/STB	2	A ← D							
LDBSE	2	D ← A; D + 1 ← SIGN(A)	—	—	—	—	—	—	3,4
LDBZE	2	D ← A; D + 1 ← 0	—	—	—	—	—	—	3,4
PUSH	1	SP ← SP − 2; (SP) A	—	—	—	—	—	—	
POP	1	A ← (SP); SP ← SP + 2	—	—	—	—	—	—	
PUSHF	0	SP ← SP − 2; (SP) ← PSW; PSW ← 0000H	0	0	0	0	0	0	
POPF	0	PSW ← (SP); SP ← SP + 2	√	√	√	√	√	√	
SJMP	1	PC ← PC + 11-bit offset	—	—	—	—	—	—	5
LJMP	1	PC ← PC + 16-bit offset	—	—	—	—	—	—	5
INDJMP	1	PC ← (A)	—	—	—	—	—	—	
SCALL	1	SP ← SP − 2; (SP) ← PC; PC ← PC + 11-bit offset	—	—	—	—	—	—	5
LCALL	1	SP ← SP − 2; (SP) ← PC; PC ← PC + 16-bit offset	—	—	—	—	—	—	5
RET	0	PC ← (SP); SP ← SP + 2	—	—	—	—	—	—	
J(conditional)	1	PC ← PC + 8-bit offset	—	—	—	—	—	—	5
JC		Jump if C = 1	—	—	—	—	—	—	5
JNC		Jump if C = 0	—	—	—	—	—	—	5

Note

1. If the mnemonic ends in "B", a byte operation is performed, otherwise a word operation is done. Operands D, B, and A must conform to the alignment rules for the required operand type. D and B are locations in the register file; A can be located anywhere in memory.
2. D, D + 2 are consecutive WORDS in memory; D is DOUBLE-WORD aligned.
3. D, D + 1 are consecutive BYTES in memory; D is WORD aligned.
4. Changes a byte to a word.
5. Offset is a 2's complement number.

TABLE 9.8 *Continued*

Mnemonic	Oper-ands	Operation (Note 1)	Z	N	C	V	VT	ST	Notes
JE		Jump if Z = 1	—	—	—	—	—	—	5
JNE		Jump if Z = 0	—	—	—	—	—	—	5
JGE		Jump if N = 1	—	—	—	—	—	—	5
JLT		Jump if N = 0	—	—	—	—	—	—	5
JGT		Jump if N = 0 and Z = 0	—	—	—	—	—	—	5
JLE		Jump if N = 1 or Z = 1	—	—	—	—	—	—	5
JH		Jump if C = 1 and Z = 0	—	—	—	—	—	—	5
JNH		Jump if C = 0 or Z = 1	—	—	—	—	—	—	5
JV		Jump if V = 1	—	—	—	—	—	—	5
JNV		Jump if V = 0	—	—	—	—	—	—	5
JVT		Jump if VT = 1; Clear VT	—	—	—	—	0	—	5
JNVT		Jump if VT = 0; Clear VT	—	—	—	—	0	—	5
JST		Jump if ST = 1	—	—	—	—	—	—	5
JNST		Jump if ST = 0	—	—	—	—	—	—	5
JBS		Jump if ST = 1	—	—	—	—	—	—	5,6
JBC		Jump if Specified Bit = 0	—	—	—	—	—	—	5,6
DJNZ	1	D ← D − 1; if D ≠ then PC ← PC + 8-bit offset	—	—	—	—	—	—	5
DEC/DECB	1	D ← D − 1	√	√	√	√	√	—	
NEG/NEGB	1	D ← 0 − D	√	√	√	√	√	—	
INC/INCB	1	D ← D + 1	√	√	√	√	√	—	
EXT	1	D ← D; D + 2 ← Sign (D)	√	√	0	0	—	—	2
EXTB	1	D ← D; D + 1 ← Sign (D)	√	√	0	0	—	—	3
NOT/NOTB	1	D ← Logical Not (D)	√	√	0	0	—	—	
CLR/CLRB	1	D ← 0	1	0	0	0	—	—	
SHL/SHLB/SHLL	1	C ← msb — — — — — lsb ← 0	√	√	√	√	√	—	7
SHR/SHRB/SHRL	1	0 → msb — — — — — lsb → C	√	√	√	0	—	√	7
SHRA/SHRAB/SHRAL	1	msb → msb — — — — — lsb → C	√	√	√	0	—	√	7
SETC	0	C ← 1	—	—	1	—	—	—	
CLRC	0	C ← 0	—	—	0	—	—	—	
CLRVT	0	VT ← 0	—	—	—	—	0	—	
RST	0	PC ← 2080H	0	0	0	0	0	0	8
DI	0	Disable All Interrupts	—	—	—	—	—	—	
EI	0	Enable All Interrupts	—	—	—	—	—	—	
NOP	0	PC ← PC + 1	—	—	—	—	—	—	
SKIP	0	PC ← PC + 2	—	—	—	—	—	—	
NORML	2	Normalize	√	1	—	—	—	—	7

Note

1. If the mnemonic ends in "B", a byte operation is performed, otherwise a word operation is done. Operands D, B and A must conform to the alignment rules for the required operand type. D and B are locations in the register file; A can be located anywhere in memory.
5. Offset is a 2's complement number.
6. Specified bit is one of the 2048 bits in the register file.
7. The "L" (Long) suffix indicates double-word operation.
8. Initiates a Reset by pulling RESET low. Software should re-initialize all the necessary registers with code starting at 2080H.

TABLE 9.9 SALIENT FEATURES OF TMS 32010

- 160-ns instruction cycle
- 144-word on-chip data RAM
- ROMless version—TMS32010
- 1.5K-word on-chip program ROM-TMS320M10
- External memory expansion to a total of 4K words
    at full speed
- 16-bit instruction/data word
- 32-bit ALU/accumulator
- 16 × 16-bit multiply in 160-ns
- 0 to 15-bit barrel shifter
- Eight input and eight output channels
- 16-bit bidirectional data bus with 50-megabits-per-
    second transfer rate
- Interrupt with full context save
- Signed two's complement fixed-point arithmetic
- NMOS technology
- Single 5-V supply
- Two versions available
    TMS32010-20 . . . 20.5 MHz Clock
    TMS32010-25 . . . 25.0 MHz Clock

(320M10), 1536 words of 4K-word memory are available on the chip. It has an internal clock generator which requires an external crystal, or it has the option of an external frequency source.

The TMS320 family utilizes what is called modified Harvard architecture, which permits overlap of instruction fetch and execution of consecutive instruction, and thus permits faster program execution. Speed enhancement also has been possible with the hardware multiplier, which performs signed or unsigned multiplication in a single 160-ns cycle. There is also a hardware barrel shifter that performs 0-15 bits left shift for shifting data on the way to ALU. In addition, extra hardware in auxiliary registers, which provides indirect data RAM addresses, can be configured in an autoincrement/decrement mode for single cycle manipulation of data tables.

The TMS32010's assembly language instruction set supports both numeric-intensive operations and general-purpose high-speed control operations. The following development systems and software supports are available for TMS32010:

- The TMS32010 Evaluation Module (EVM) is a stand-alone, single board with a software library that permits inexpensive use of a microcomputer for evaluation purposes. It can communicate with the host computer (e.g., VAX) and several peripherals to receive an assembly language code.

- The XDS/320 Simulator program verifies the application program and is currently available in VAX computer.

- The XDS/320 Emulator permits real-time in-circuit emulation.

The TMS32020 microcomputer is a considerable enhancement from its predecessor TMS32010 and its key improvements can be summarized as follows:

- 544-word on-chip data RAM, 256 words of which may be programmed as data or program or data memory
- 128K words of data/program space
- Sixteen input and sixteen output channels
- Larger and improved instruction set—supports floating point operation
- On-chip 16-bit timer
- Three external interrupts
- Wait-states capability for external slow devices
- Multiprocessor and global data memory interface

8087 Numeric Data Processor

The 8087 is a floating-point numeric data processor chip and works in conjunction with the 8086 or 8088 processor. The two-chip system is defined as *i*APX 86/20 or *i*APX 88/20, respectively, and its salient features are summarized in Table 9.10. The 8087 executes instructions as a coprocessor to a maximum mode of 8086 or 8088, and adds 68 numeric processing instructions to the *i*APX 86/10 (or *i*APX 88/10) instruction set. A slightly modified chip (80287) operates with an 80286 microprocessor, and the composite chip is defined as *i*APX 286/20. Typical execution times for selected 86/20 numeric instructions are given below:

Floating-Point Instructions	Time (μs) with 5-MHz Clock
Add/subtract	17
Multiply (single precision)	19
Multiply (double precision)	27
Divide	39
Square root	36
Tangent	90
Exponentiation	100

TABLE 9.10   SALIENT FEATURES OF THE 8087 PROCESSOR Courtesy of Intel Corporation

■ High Performance 2-Chip Numeric Data Processor	■ Support 8 Data Types: 8-, 16-, 32-, 64-Bit Integers, 32-, 64-, 80-Bit Floating Point, and 18-Digit BCD Operands
■ Standard iAPX 86/10, 88/10 Instruction Set Plus Arithmetic, Trigonometric, Exponential, and Logarithmic Instructions For All Data Types	■ 8x80-Bit Individually Addressable Register Stack plus 14 General Purpose Registers
■ All 24 iAPX 86/10, 88/10 Addressing Modes Available	■ 7 Built-in Exception Handling Functions
■ Conforms To Proposed IEEE Floating Point Standard	■ MULTIBUS System Compatible Interface

The programs for the *i*APX 86/20 and 88/20 can be written in ASM-86, FORTRAN-86, or PASCAL-86. The execution time of the instructions appears to be large and should be used judiciously in real-time control systems. Some of the instructions can be implemented faster using the lookup-table method with some sacrifice of accuracy.

8251A Communication Interface

The 8251A is the advanced version of the 8251 universal synchronous/asynchronous receiver/transmitter (USART), designed for data communication with microcomputers, such as the 8080, 8085, 8048, 8051, 8086, and 8088. The features of the 8251A are given in Table 9.11, and Fig. 9.17 shows its block diagram. The chip is programmed by the CPU to operate in several modes in serial data communication. The USART accepts data characters from the CPU in parallel format and converts them into a serial stream for transmission. Simultaneously, it can receive serial data streams and convert them into parallel data characters for the CPU. It will inform the CPU whenever it can accept a new character for transmission or whenever it has received a character for the CPU. The CPU can read the complete status of the USART, including transmission errors and control signals, at any time.

The 8251A is tied to the CPU data bus and data transfer is controlled by read/write control. The control words, command words, and status information are also transferred through the data bus. The control/status or data are determined by the signal $C/\overline{D}$. The control words program baud rate, character length, number of stop bits, synchronous or asynchronous operation, even or odd parity, and so on.

The transmitter buffer receives parallel data from the data bus buffer, converts them into a serial bit stream, and transmits on the TXD line. The TXRDY output signals the CPU when it is ready to receive a data character. It is reset automatically when the CPU writes a character into the 8251A. On the other hand, the 8251A receives serial data from the modem or I/O device. Upon receiving the character,

TABLE 9.11   SALIENT FEATURES OF THE 8251A Courtesy of Intel Corporation

- Synchronous and Asynchronous Operation
- Synchronous 5–8 Bit Characters; Internal or External Character Synchronization; Automatic Sync Insertion
- Asynchronous 5–8 Bit Characters; Clock Rate—1, 16 or 64 Times Baud Rate; Break Character Generation; 1, 1½, or 2 Stop Bits; False Start Bit Detection; Automatic Break Detect and Handling
- Synchronous Baud Rate—DC to 64K Baud

- Asynchronous Baud Rate—DC to 19.2K Baud
- Full-Duplex, Double-Buffered Transmitter and Receiver
- Error Detection—Parity, Overrun and Framing
- Compatible with an Extended Range of Intel Microprocessors
- 28-Pin DIP Package
- All Inputs and Outputs are TTL Compatible
- Available
  —Standard Temperature Range
  —Extended Temperature Range

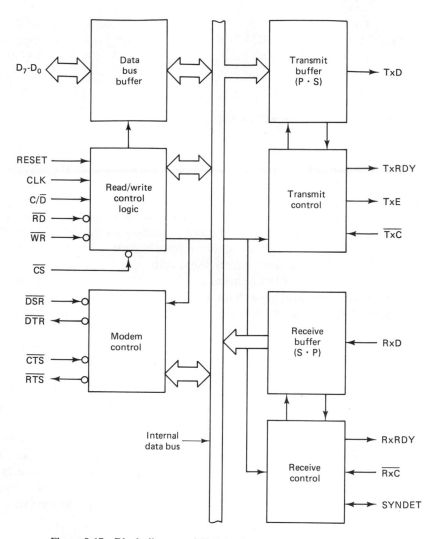

Figure 9.17   Block diagram of 8251A. Courtesy of Intel Corporation.

the RXRDY output is raised high to signal the CPU that it has a character ready for the CPU to fetch. The transmitter and receiver clocks ($\overline{\text{TXC}}$, $\overline{\text{RXC}}$) are tied to the external baud rate generator and determine the rate at which the data are transmitted or received. In the synchronous mode, the frequency is equal to the baud rate, but in the asynchronous mode it is selectable as a multiple ($\times 1$, $\times 16$, $\times 64$) of the baud rate. The 8251A can interface a telephone line through a modem and the modem control inputs and outputs can be used to simplify the interface to almost any modem.

8255A Peripheral Interface

The programmable peripheral interface chip 8255A is a general-purpose programmable I/O device to interface peripheral equipment to the microcomputer system bus. It has 24 I/O lines, which may be individually programmed in two groups of 12 lines, and it has three modes of operation.

Figure 9.18 shows the block diagram of 8255A. The 8-bit data bus is tied to the CPU, and data transfer is controlled by the $\overline{RD}/\overline{WR}$ command. The control words and status information are also transferred through the data bus. The port select signals A_0 and A_1, in conjunction with the $\overline{RD}$ and $\overline{WR}$ inputs, control the selection of one of the three ports A, B, and C. The functional configuration of each port is programmed by the system software. Essentially, a control word determines the chip's functional configuration.

The three modes of 8255A are summarized in Fig. 9.19. Each mode has a number of submodes. The port characteristics can be summarized as follows:

- Port A: one 8-bit data output latch/buffer and one 8-bit data input latch
- Port B: one 8-bit data input/output latch/buffer and one 8-bit data input buffer

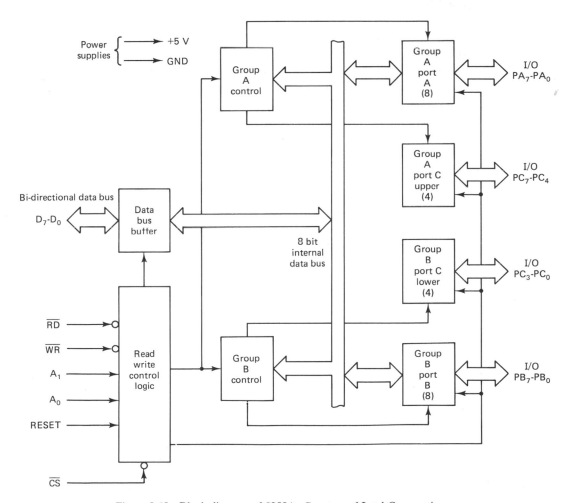

Figure 9.18 Block diagram of 8255A. Courtesy of Intel Corporation.

- Port C: one 8-bit data output latch/buffer and one 8-bit data input buffer (no latch)

8259A Interrupt Controller

The interrupt controller 8259A can receive up to eight external interrupt signals and vector them to appropriate memory locations in the microprocessor to execute the service routines. With eight similar chips, up to 64 interrupt signals can be handled. The interrupt signals can be serviced on a priority basis and the individual inputs can be selectively masked, or the entire interrupt structure can be globally disabled within the CPU.

A block diagram of the 8259A is shown in Fig. 9.20. The interrupt controller acts as an overall manager of the interrupt system. The interrupt request register receives the external interrupt signals (IR0–IR7), and the interrupt mask register stores the bits that mask the appropriate IR lines. The in-service register stores the interrupt levels that are being serviced. The priority resolver determines the priority of the requested signals compared with the signals being serviced, and permits servicing if these are of higher priority. This is defined as the nested mode of operation. The chip is tied to the data bus of the CPU through the data bus buffer and the read/write control logic determines the transfer of status and control words. The cascade buffer/comparator is used for control of multiple 8259A chips.

The 8259A controller can be programmed to operate in the following modes:

Fully nested mode: This is the default mode where the interrupt requests have priority in order from 0 to 7 (i.e., 0 has the highest priority and 7 has the lowest priority).

Rotating priority mode: In this mode, the interrupt requests can have either equal priority or a specific given priority. In the former case, the interrupts

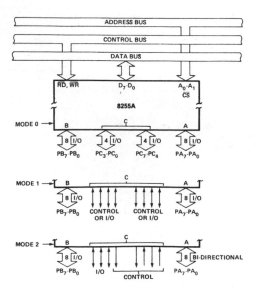

Figure 9.19 Modes of 8255A. Courtesy of Intel Corporation.

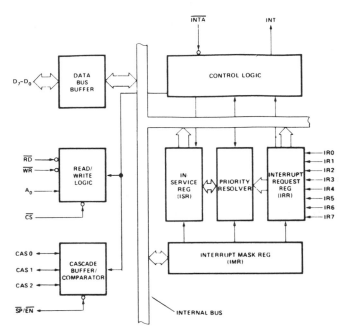

Figure 9.20 Block diagram of 8259 interrupt controller. Courtesy of Intel Corporation.

are serviced by rotation so that an interrupt once serviced automatically gets the lowest priority. In the latter case, the programmer can change priorities by programming the lowest priority.

Special mask mode: In this mode, an interrupt service routine can dynamically alter system priority structure and masking during its execution.

Poll command: With this command, the interrupt is disabled and service to interrupt-requesting devices is achieved by software using a Poll command.

Once the 8259A chip is programmed, its operation sequence when interfacing a typical 8086 microprocessor can be summarized as follows:

- One or several interrupt request lines go high and the corresponding bits in the request register are set.
- The priority is resolved and if of higher priority, an interrupt signal is sent to CPU.
- The CPU acknowledges by sending an $\overline{INTA}$ signal to the 8259A.
- The highest-priority service register bit is set and the corresponding request register bit is reset. The CPU then issues a second $\overline{INTA}$ signal, during which the 8259A releases an 8-bit pointer (address) in the data bus. The pointer vectors to the particular routine, which may be diverted elsewhere in memory by the JUMP instruction.

9513 System Timing Controller

The Advanced Micro Devices AM 9513 system timing controller chip is used for versatile timing and sequencing applications with a host microprocessor. The dedicated multiple timing applications relieve the microprocessor of time-critical operations. The 9513 can be programmed for frequency synthesis, high-resolution duty-cycle waveforms, digital timing functions, time-of-day clocking, coincidence alarms, complex pulse generation, high-resolution baud rate generation, frequency-shift keying, stopwatch timing, waveform analysis, and so on.

A general block diagram of the 9513 is shown in Fig. 9.21(a). The chip contains five general-purpose independent 16-bit counters. The counters can be programmed to count up or down in either binary or BCD. The counters can be concatenated to form an effective length of up to 80 bits. The chip is interfaced

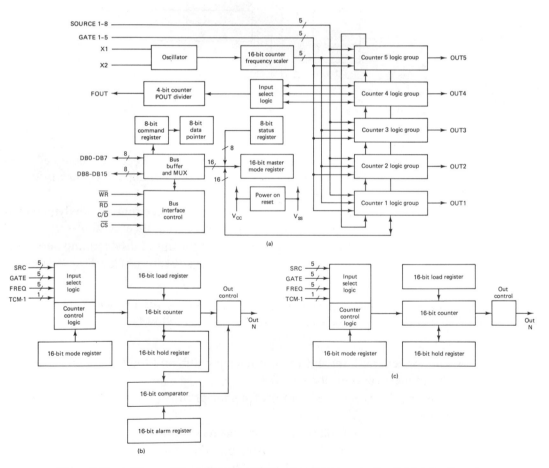

Figure 9.21 (a) General block diagram of 9513 system timing controller; (b) details of counter logic groups 1 and 2; (c) details of counter logic groups 3, 4, and 5.

with the host processor through an 8- or 16-bit data bus and $\overline{WR}$, $\overline{RD}$, C/$\overline{D}$, and $\overline{CS}$ bus interface controls. The internal data bus is 16 bits wide, and in the 8-bit mode the 16-bit information is multiplexed to the low-order data bus lines. An internal oscillator excited by an external crystal provides a general source for the counter inputs. The oscillator output is frequency scaled by BCD or binary scaling to provide several subfrequencies. One of these can be selected and further divided (1 to 16) by the FOUT divider to provide a programmable frequency output FOUT. The details of counter and logic groups are shown in Fig. 9.21(a) and (b). Each 16-bit counter is interfaced with a load register and a hold register. The load register can automatically reload the counter to any predefined value and thus control the effective count period. The hold register can save the count on-the-fly without disturbing the counting. It can also be used as a second load register to generate varieties of output waveforms. Counters 1 and 2 have additional alarm registers, comparators, and the extra logic necessary to operate in a 24-hour time-of-day mode.

Each counter has a dedicated output, and the output pulse polarities are individually programmable. The gate inputs may be used to control the individual counters or the same gate may control up to three counters. The source inputs provide external signals which may be counted by any of the counters. Any source line may be routed to any or all of the counters and the FOUT divider. In fact, the FOUT divider can receive any of the 15 inputs shown. A 16-bit counter mode register controls the gating, counting, output, and source select functions within each counter logic group. The counters can be programmed to operate in various modes, which can be summarized as follows:

Mode A: software-triggered strobe with no hardware gating

Mode B: software-triggered strobe with level gating

Mode C: hardware-triggered strobe

Mode D: rate generator with no hardware gating

Mode E: rate generator with level gating

Mode F: nonretriggerable one-shot

Mode G: software-triggered delayed pulse one-shot

Mode H: software-triggered delayed one-shot with hardware gating

Mode I: hardware-triggered delayed pulse strobe

Mode J: variable duty-cycle rate generator with no hardware gating

Mode K: variable duty-cycle rate generator with level gating

Mode L: hardware-triggered delayed pulse one-shot

Mode N: software-triggered strobe with level gating and hardware retriggering

Mode O: software-triggered strobe with edge gating and hardware retriggering

Mode Q: rate generator with synchronization

Mode R: retriggerable one-shot

Mode V: frequency-shift keying

Analog I/O

The physical system is generally analog in nature and the microcomputer interfaces the analog world through A/D and D/A converters. A typical analog-to-digital data acquisition system (MP22) manufactured by Burr Brown is shown in Fig. 9.22. The system is a hybrid module consisting of a 12-bit CMOS A/D converter, instrumentation amplifier, input multiplexer, address decoder, and control logic. The data acquisition system can accept 16 single-ended or eight differential input channels through the two eight-channel analog multiplexers. A channel is selected by the 4-bit address lines A1–A4. The analog input voltage may be unipolar or bipolar. The unipolar input may be set for any range between +5 V (±5 V for

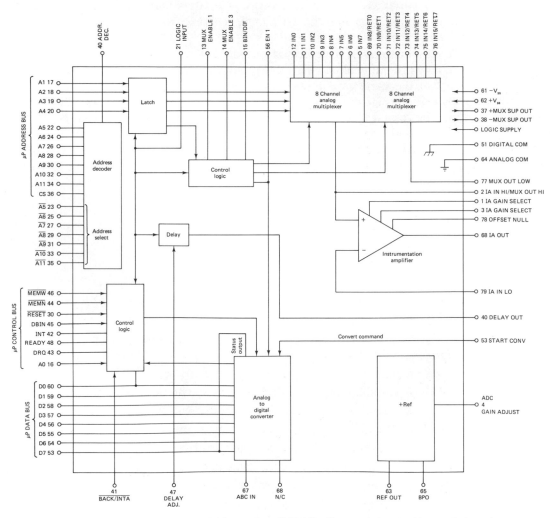

Figure 9.22 Analog data acquisition system (MP 22). Courtesy of and with permission of Burr-Brown.

bipolar) and $+10$ mV (± 10 mV for bipolar) and simple binary output is obtained with 12-bit resolution.   For a bipolar signal, the output is in offset binary form with 11-bit resolution, with the MSB as the sign bit.   The offset binary can be converted to two's-complement form by complementing the MSB.   The instrumentation amplifier provides a programmable gain from 1 to 500 to the low input signal.   The A/D converter is a successive-approximation type with a typical conversion time of 45 μs.   The output is coupled to the microcomputer data bus, and two successive bytes give the 12-bit information.   The data acquisition system is usually memory referenced and the location is selected by the address decoder (the particular $\overline{A}$ output is grounded).   The delay timer provides a time delay between channel selection and start of A/D conversion.   The control logic generates signals to halt or interrupt the CPU while conversion takes place, and to signal the CPU when conversion is complete and data are to be read.   The system can also be used in the direct memory access mode by connecting to the DMA controller.

The module requires a power supply of ± 15 V and $+5$ V.   The system has a linearity within $\pm 0.1\%$ FSR, and gain and offset errors can be adjusted to be zero.

A typical 12-bit D/A converter (DAC), type HS 9338 manufactured by Hybrid Systems, is shown in Fig. 9.23.   The device is microcomputer compatible and has double buffer registers with the input data registers configured as three 4-bit bytes (nibbles).   The DAC is coupled to the 16-bit data bus of the host computer in the higher 12 bits, whereas for an 8-bit computer, only the higher byte is connected. The input register can be selectively loaded in the high byte, middle byte, or low byte by the HBE, MBE, and LBE control signals, respectively.   By selecting these and making the LDAC logic high the DAC register can be loaded directly from the data bus.   The DAC accepts simple binary input for 0 to $+10$ V or 0 to $+5$ V unipolar output.   For bipolar outputs in the ranges ± 10 V and ± 5 V, the digital input should be in offset binary form with 11-bit resolution.   The DAC has a linearity of $\pm 0.01\%$ FSR with a typical settling time of 2.5 μs.

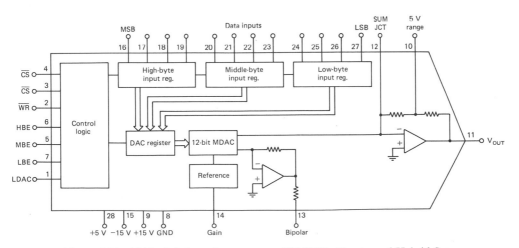

Figure 9.23   12-bit digital–analog converter (HS 9338).   Courtesy of Hybrid Systems.

REFERENCES

1. J. Schreiber, "Present State and Development of Microelectronics," *Proc. Microelectron. Power Electron. Electr. Drives*, Darmstadt, pp. 9–14, Oct. 12–14, 1982.
2. *Intel 8080 Microcomputer Systems User's Manual*, Sept. 1975.
3. *Intel MCS-85 User's Manual*, Jan. 1978.
4. *Intel MCS-51 Family of Single-Chip Microcomputers User's Manual*, Jan. 1981.
5. *Intel MCS-86 User's Manual*, Feb. 1979.
6. *Intel Microprocessor and Peripheral Handbook*, 1983.
7. *Intel Microcontroller Handbook*, 1984.
8. TMS32010 *User's Guide*, Texas Instruments, 1983.

10

MICROCOMPUTER
APPLICATION

10.0 INTRODUCTION

The basic fundamentals of microcomputers and some state-of-the-art microcomputers with peripheral chips were reviewed in Chapter 9. By this time it should be obvious that most of the control and signal-processing functions discussed so far can be implemented digitally by microcomputers.

In this chapter we first review the different application areas of microcomputers in power electronic systems. Then the general methodologies for the design of system, hardware, and software are discussed. Practical design considerations have been reviewed, wherever possible.

10.1 APPLICATION AREAS

The general functions of a microcomputer in a power electronic system can be summarized as follows:

- Gate firing control of phase-controlled converters
- PWM or stepped-wave signal generation
- Feedback control
- Processing of feedback signals
- Sequencing control
- Monitoring and warning

- Diagnostics
- Miscellaneous computation and control

Some microcomputer functions in a power electronic system can be described as follows:

Gate Firing Control

A number of gate firing control schemes for phase-controlled converters were described in Chapter 3 using analog/digital dedicated hardware. Here, two typical schemes which are based on the microcomputer will be described. The gate firing control can be implemented by a dedicated single-chip microcomputer, a custom-designed LSI chip, or may be a function in a main control microcomputer or VLSI chip. Figure 10.1 shows the block diagram of a single-chip microcomputer for gate firing control of a three-phase bridge converter. The microcomputer receives the converter voltage command V_d^* from the feedback control loop and translates it into the firing angle control pulses for the six thyristors Q_1 to Q_6. The microcomputer also receives line voltage synchronizing 60° base-interrupt pulses, the line voltage logic waves ϕ_A, ϕ_B, and ϕ_C, and a clock input supplied by a phase-locked-loop generator, as shown in Fig. 10.1. The generation of 60° base interrupts and line voltage logic signals is shown in Fig. 10.2. An interrupt signal is generated at each crossover point of line voltage and provides a reference point for firing angle delay. The status of line voltage logic waves ϕ_A, ϕ_B, and ϕ_C in a 60° interval identifies the thyristors to be fired. The look-up table for identification of thyristors is given in Table 10.1. Assuming that the conduction is always continuous in both the rectification and inversion modes, the dc voltage command V_d^* can be converted

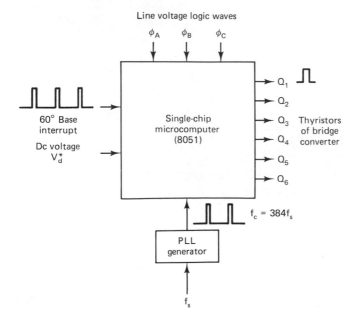

Figure 10.1 Block diagram of 8051 microcomputer used for gate firing control.

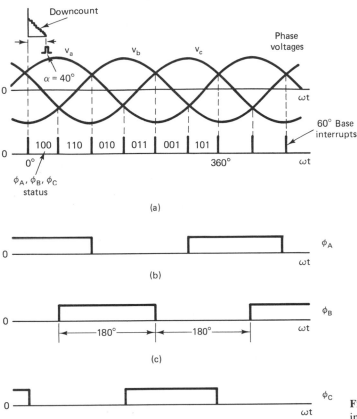

Figure 10.2 Generation of (a) 60° base interrupt signals; and (b) ϕ_A, (c) ϕ_B, and (d) ϕ_C waves.

to firing angle α by a separate $\cos^{-1}$ look-up table which satisfies the relation

$$\alpha = \cos^{-1}\frac{V_d^*}{1.35V_L} \tag{10.1}$$

where V_L is the rms line voltage. The α angle is an 8-bit digital word (D_7-D_0) where the most significant two bits D_7 and D_6 identify the 60° angular segment, as shown in the table. The lower 6 bits (D_5-D_0) give the α angle magnitude for a maximum of 60° [i.e., the resolution is $60/(2^6 - 1) = 0.95°$ per bit]. A larger bit size for the α angle should be chosen if better resolution is desired.

Figure 10.3 shows a flowchart for firing angle control of the converter. A cycle of computation that includes the gate firing control and main program begins every 60° with the base-interrupt signal. The firing angle is determined every 60° interval from the V_d^* command using the $\cos^{-1}$ look-up table. At the instant of base interrupt, the microcomputer loads the α angle to a down-counter* and enters the main program. The down-counter is clocked by the PLL generator (see Fig. 3.41) of clock frequency $f_c = Nf_s$, where f_s is the line frequency and $N = 384$. Any change in the line frequency reflects a corresponding change in f_c, and therefore

---

* Note that the 8051 microcomputer contains up-counters only.

TABLE 10.1 TRUTH TABLE FOR FIRING ANGLE CONTROL

D_1	D_6	ϕ_A	ϕ_B	ϕ_C	Q_1	Q_2	Q_3	Q_4	Q_5	Q_6	Source	Angle Range
0	0	1	0	0	1	0	0	0	0	1	v_{ab}	
0	0	1	1	0	1	1	0	0	0	0	v_{ac}	
0	0	0	1	0	0	1	1	0	0	0	v_{bc}	
0	0	0	1	1	0	0	1	1	0	0	v_{ca}	0–60°
0	0	0	0	1	0	0	0	1	1	0	v_{ca}	
0	0	1	0	1	0	0	0	0	1	1	v_{cb}	
0	1	1	0	0	0	0	0	0	1	1	v_{cb}	
0	1	1	1	0	1	0	0	0	0	1	v_{ab}	
0	1	0	1	0	1	1	0	0	0	0	v_{ac}	
0	1	0	1	1	0	1	1	0	0	0	v_{bc}	60–120°
0	1	0	0	1	0	0	1	1	0	0	v_{ba}	
0	1	1	0	1	0	0	0	1	1	0	v_{ca}	
1	0	1	0	0	0	0	0	1	1	0	v_{ca}	
1	0	1	1	0	0	0	0	0	1	1	v_{cb}	
1	0	0	1	0	1	0	0	0	0	1	v_{ab}	
1	0	0	1	1	1	1	0	0	0	0	v_{ac}	120–180°
1	0	0	0	1	0	1	1	0	0	0	v_{bc}	
1	0	1	0	1	0	0	1	1	0	0	v_{ba}	

the drift in α angle is eliminated. The choice of $N = 384$ gives 64 pulses per 60° interval and therefore matches to α angle resolution. The counter operation is illustrated in Fig. 10.2(a) for base interrupt at a 0° angle. The counter generates an interrupt when clear at the α angle, and then the interrupt service routine is executed. First, the registers used for computation are pushed into the stack and then the input signals in Table 10.1 are tested. The test of D_7 and D_6 status directs to the particular 60° segment and reading of the ϕ_A, ϕ_B, and ϕ_C logic signals identifies the thyristors to be fired. For example, when $\alpha = 40°$ for the segment shown in Fig. 10.2(a), the status of the signals are $D_7 = 0$, $D_6 = 0$, $\phi_A = 1$, $\phi_B = 0$, and $\phi_C = 0$, and therefore as the truth table indicates, thyristors Q_1 and Q_6 should be fired. The short firing pulses at the α angle can easily be converted to 120° long pulse train firing by external logic circuits and gating a high-frequency clock. Two thyristors (one from the positive group and another from the negative group) are fired at any instant to ensure current conduction. The line voltage source that contributes to output current in any segment is also indicated in the table. If $\alpha = 40°$ is maintained constant, only the upper one-third of the table is scanned in sequence. Upon completion of the interrupt service routine, execution of the main program begins. In this firing control scheme, the maximum delay to update the firing angle is 60°, because once the down-counter is loaded the firing angle has no control.

A gate pulse generation scheme where the α angle can be corrected instantaneously is shown in Fig. 10.4. The principle is explained for the phase a leg

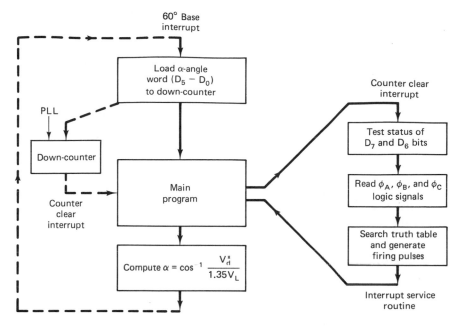

Figure 10.3 Flowchart for firing angle control.

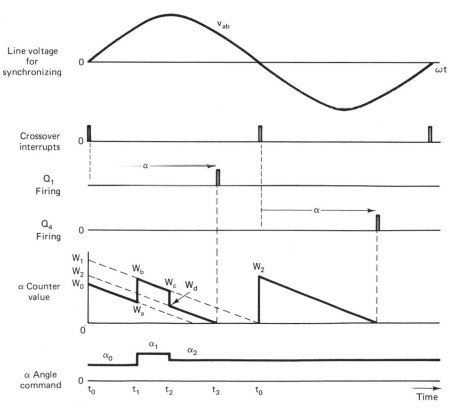

Figure 10.4 Gate pulse generation scheme with instantaneous angle correction capability.

only (consisting of thyristors Q_1 and Q_4) of a three-phase bridge converter. The interrupt pulses are generated when the line voltage v_{ab} crosses through zero. At this point, the microcomputer sends the firing angle data W_0 corresponding to the α_0 angle to a down-counter. If the command firing angle changes from α_0 to α_1 at time t_1, the microcomputer reads the present value W_a from the counter, calculates the new value $W_b = W_a + (W_1 - W_0)$, and loads the counter again. It is assumed that the task is completed within consecutive clock pulses of the counter so that its operation is not affected. If the firing angle command changes again to α_2 at t_2, the counter content is corrected instantaneously by the new value $W_d = W_c - (W_1 - W_2)$. When the counter clears at t_3 at the desired α angle, a firing pulse is generated for Q_1. In Fig. 10.4, the angle α_2 is assumed to remain constant in the negative half-cycle.

Feedback Control

A closed-loop power electronic system with digital control is a discrete-time system which may be either linear or nonlinear. The discrete-time nature arises because of periodic sampling in digital computation and switching of power semiconductor devices. If the sampling or switching interval is small compared to the system response time (typically by a factor of 10 or more), the discrete-time effect can be neglected. In such a case, a linear system can be represented by the Laplace transfer function, and the Bode or Nyquist method can be applied for the system stability study. Introducing a transport lag ($e^{-T_s s}$) in the transfer function for sampling time T_s will give better accuracy in the analysis. For a smaller ratio of response to sample time, the discrete-time effect should be considered in the analysis. A linear discrete-time system can be analyzed by the Z-transform method. The role of Z-transform to the discrete-time system is similar to that of Laplace transform to the continuous system. The dynamics of a discrete-time system can be described by difference equations in the time domain and then the Z-domain transfer functions can be derived for stability study. Instead of Z-transform analysis, the state variable method can be applied for the analysis of discrete-time system. The structure of state space equations in discrete time and continuous time are the same except that in the former case the system is represented by a set of first-order difference equations. The advantages of the state variable method are that the linear and nonlinear systems can be handled in a unified manner for both system analysis and design.

Figure 10.5 shows the block diagram of the digital speed control system of an inverter-fed ac machine, where the dashed area is under microcomputer control. The system has speed control in the outer loop and current control in the inner loop. It can be assumed that the torque is related to the current, and the flux is maintained constant. The system is shown with multirate sampling (i.e., the speed loop has sampling time T_{s1} and the current loop has sampling time T_{s2}). The sampling time of the inverter is neglected. Normally, the current loop has a larger bandwidth requirement than that of the speed loop, and therefore T_{s2} should be smaller than T_{s1}. In the speed loop, the microcomputer samples the command speed ω_r^*, feedback speed ω_r, and updates the current command I^* through the PI

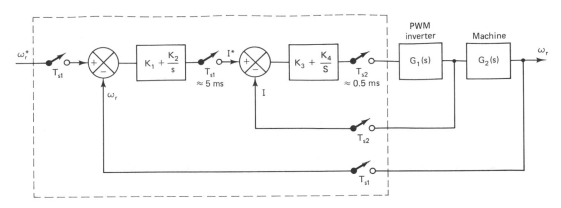

Figure 10.5   Block diagram of speed control system with multirate sampling.

compensator at a sampling rate of $1/T_{s1}$. The current loop performs similar computations and updates the inverter command at a higher rate $1/T_{s2}$.

In a multiloop feedback control system, if the sampling times are identical, Z-transform analysis can be easily applied. For example, a 60 Hz bridge-converter controlling a dc motor has a sampling interval of 60° (i.e., 2.77 ms). Here both the speed and current control loops can be synchronized to have the same sampling interval. The analysis and design with multirate sampling is usually quite complex and a computer simulation study is necessary. An approximate analysis where the inner loop is studied by Z-transform, but neglecting the discrete-time effect of the inner loop for study of the outer loop, may be adequate.

Digital Implementation of Compensators.   The control loops may contain PID, PI, or lag lead compensators, which are usually given in the form of Laplace transfer function. For microcomputer implementation, it is necessary to convert them into the time domain in the form of difference equations. Consider a PI compensator with transfer function

$$\frac{Y(S)}{X(S)} = K_1 + \frac{K_2}{S} \tag{10.2}$$

or

$$SY(S) = K_2X(S) + K_1SX(S) \tag{10.3}$$

If the sampling time T_s is small, a derivative can be represented in finite difference form, and therefore equation (10.3) can be written as

$$\frac{Y(N + 1) - Y(N)}{T_s} = K_2X(N) + K_1\left[\frac{X(N + 1) - X(N)}{T_s}\right] \tag{10.4}$$

where N and $N + 1$ are the consecutive sampling instants. Equation (10.4) can be written as

$$Y(N + 1) - Y(N) = K_2 T_s X(N) + K_1 X(N + 1) - K_1 X(N)$$

or   (10.5)

$$Y(N + 1) = Y(N) + K_1 X(N + 1) + (K_2 T_s - K_1) X(N)$$

Difference equation (10.5) can be expressed in state variable form as

$$Z(N + 1) = AZ(N) + BX(N) \tag{10.6}$$

$$Y(N) = CZ(N) + DX(N) \tag{10.7}$$

where $Z(N)$ is the state variable, $A = 1$, $B = K_2 T_s$, $C = 1$, and $D = K_1$. The digital implementations of equations (10.6) and (10.7) are given in the block diagram of Fig. 10.6. The computation flowchart can then be summarized as

$$X(N)K_2 T_s + Z(N) = Z(N + 1)$$

$$Z(N + 1) \xrightarrow{\hspace{2cm}} Z(N) \tag{10.8}$$

$$Z(N) + X(N)K_1 = Y(N)$$

In practical implementation, the gain constants K_1 and K_2 can be adjustable software parameters which may be fine tuned by experiment.

For a lag-lead compensator, the transfer function is

$$\frac{Y(S)}{X(S)} = K \frac{1 + \tau_1 S}{1 + \tau_2 S} \tag{10.9}$$

The transfer function can be manipulated as follows:

$$Y(S) + \tau_2 SY(S) = KX(S) + K\tau_1 SX(S)$$

or   (10.10)

$$SY(S) = -\frac{1}{\tau_2} Y(S) + \frac{K}{\tau_2} X(S) + K \frac{\tau_1}{\tau_2} SX(S)$$

Equation (10.10) can be converted to a difference equation as

$$Y(N + 1) = \left(1 + \frac{T_s}{\tau_2}\right) Y(N) + K\frac{\tau_1}{\tau_2} X(N + 1)$$

$$+ K \left(\frac{T_s}{\tau_2} - \frac{\tau_1}{\tau_2}\right) X(N) \tag{10.11}$$

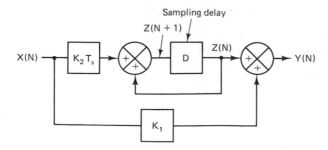

Figure 10.6 Digital implementation of PI compensator.

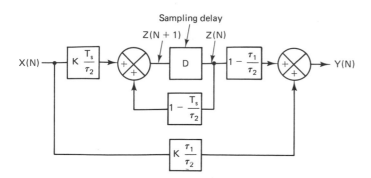

Figure 10.7 Digital implementation of lag-lead compensator.

Equation (10.11) can be written in state variable form as

$$Z(N + 1) = \left(1 - \frac{T_s}{\tau_2}\right)Z(N) + \frac{KT_s}{\tau_2}X(N) \qquad (10.12)$$

$$Y(N) = \left(1 - \frac{\tau_1}{\tau_2}\right)Z(N) + \frac{K\tau_1}{\tau_2}X(N) \qquad (10.13)$$

Equations (10.12) and (10.13) can be represented in block diagram as shown in Fig. 10.7.

Limiting Function

In a multiple-loop feedback control system, the control variables, such as position, speed, current, and flux, should have limited excursions. The limit values of each command signal can be precomputed and stored in memory. These limit values can easily be made adaptable to the operating conditions, if necessary.

Linearization

If the static transfer characteristics of a system are nonlinear and predictable, the system can be linearized by generating inversely nonlinear function with the help of a microcomputer. The method of nonlinear compensation by $\cos^{-1}$ look-up table was indicated for firing angle control. Another example of a linearization problem is the discontinuous conduction of a phase-controlled converter, as discussed below.

Converter Discontinuous Conduction

A phase-controlled converter can operate in a discontinuous conduction mode, especially with a counter emf type of load. At discontinuous conduction, the converter transfer characteristics become nonlinear, giving variable gain at different operating points. With closed-loop current control, if the loop gain is designed for continuous conduction, then as the conduction becomes discontinuous, the loop gain deteriorates, giving a sluggish response time. The nonlinearity at discontinuous conduction can be compensated by one of the following methods:

- Provide an inner voltage control loop with a faster response time.

- Provide an inversely nonlinear look-up table for gain compensation so that the system loop gain remains constant irrespective of continuous or discontinuous conditions. A modification of this method is to generate an auxiliary firing angle $\Delta\alpha$ (Ref. 2) to compensate the nonlinearity, which is discussed below.
- Provide a model referencing adaptive control (MRAC) (Ref. 10) system where the "plant" is forced to follow a fixed-parameter reference model.

Figure 10.8(a) shows the load characteristics of a three-phase bridge converter with counter emf load for both continuous and discontinuous conductions, where the load parameter X/R is assumed as constant. Consider, for example, the case of continuous conduction at fixed $\alpha = 70°$ where the dc current I_d is gradually decreased by increasing the counter emf until the conduction becomes discontinuous

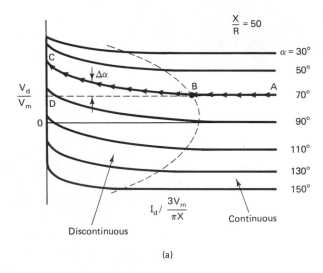

(a)

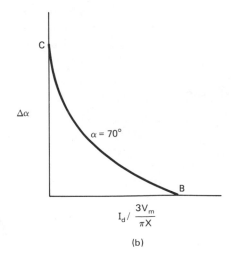

(b)

Figure 10.8 (a) Load characteristics of a three-phase bridge converter under continuous and discontinuous conductions, and (b) $\Delta\alpha - I_d$ relation to compensate nonlinear load characteristics at $\alpha = 70°$.

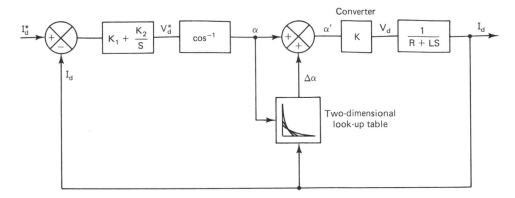

Figure 10.9 Block diagram of nonlinear compensation of converter at discontinuous conduction.

at point B. At continuous conduction, the dc voltage V_d is independent of load condition, but from the point B to point C, V_d increases with a decrease in I_d. The voltage V_d can be maintained constant to the level of continuous conduction by adding a compensating angle $\Delta\alpha$ so that the BC curve is forced to stay on line BD. The $\Delta\alpha$–I_d relation to compensate the nonlinear load characteristics is shown in Fig. 10.8(b), and Fig. 10.9 shows the corresponding control block diagram. A two-dimensional look-up table for the $\Delta\alpha$–I_d relation corresponding to each segment of the α angle is precomputed and stored in memory. The microcomputer senses the α angle to identify the table segment and then retrieves $\Delta\alpha$ as a function of I_d. The modified angle $\alpha' = \alpha + \Delta\alpha$ fires the converter, which is represented by the invariant gain constant K.

Feedback Signals

The synthesis of various feedback signals, such as flux, torque, and unit vector, were discussed in Chapters 7 and 8. The analog/digital dedicated hardware or microcomputer can be used for processing feedback signals from the raw sensor outputs. The analog feedback signals, such as voltage and current, which are directly measurable by sensors, can be converted to digital form by A/D converters. The feedback signals that are not directly measurable can then be computed by knowing the mathematical relations. The slowly varying analog signals can be converted to digital form by voltage-controlled oscillators and counters.

The principles of digital speed measurement from incremental-type pulse tachometers which are commonly used in microcomputer control systems are illustrated in Fig. 10.10. The digital pulse train is usually obtained from an optical-type tachometer mounted on a machine shaft. For speed measurement in the high-speed range, the pulse train is accumulated in a counter and the speed in rpm is computed from the following relation:

$$N = 6 \times 10^4 \, \frac{m}{MT_s} \qquad (10.14)$$

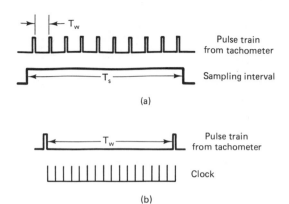

Figure 10.10 Digital speed measurement principles from pulse tachometer: (a) high speed; (b) low speed.

where T_s is the sampling interval in ms, m the number of tach pulses in the interval T_s, and M the tach pulses per revolution. For accuracy of speed measurement, it is desirable to have a large sampling interval, or for a given sampling time the pulses per revolution should be high. The sampling time is dictated by the bandwidth of the speed loop, and the pulses per revolution are limited in a practical speed sensor. Therefore, a compromise is made between the accuracy of the speed signal and the bandwidth of the speed loop.

Again, as the speed decreases the pulse train frequency decreases, introducing more inaccuracy in the pulse integration type of speed measurement. This can be avoided by pulse interval measurement, as shown in Fig. 10.10(b). In this method, the pulses from a fixed-frequency clock are integrated over the pulse interval T_w and the speed is given by the relation

$$N = 6 \times 10^4 \frac{f_c}{Mn} \qquad \text{rpm} \qquad (10.15)$$

where f_c is the clock frequency in KHz, M the pulses per revolution, and n the clock pulses for interval T_w. In this method, the speed loop sampling time corresponds to time T_w and the accuracy of the speed signal is improved by increasing the clock frequency.

A position signal can be obtained from an incremental encoder with the help of a counter(s) and a reference pulse(s) (see, for example, Fig. 8.4). This encoder, however, does not give absolute position information at a standstill condition. The absolute position can be obtained by an analog-type resolver with a resolver-to-digital converter or a digital absolute position encoder. The position signal can be processed to get the speed signal, if desired.

Digital Filtering

Digital filters of specified transfer functions can be implemented in the microcomputer by first describing the functions in discrete-time state variable form. The feedback-loop compensators described early in this section are examples of simple digital filters. The digital filters can be used in feedback loops to modify poles and zeros of the system or simply with the intention of attenuating noise and harmonics from feedback signals. For a slowly varying signal, a simple form of digital filter consists of sampling the signal consecutively several times and then computing the mean value.

Programmable Delay

It is easy to generate a programmable time delay in a control system using either a hardware or a software timer. A software timer consists of preloading a digital word to a memory location and then decrementing it periodically. When the location clears at the desired delay time, an event can start. Sampling times of various lengths can be generated by software timers in multitasking operation, which is discussed later.

Phase-Locked Loop

As discussed early in this section, the PLL frequency for gate firing control can be generated either by hardware or by software. Figure 10.11 illustrates sine reference

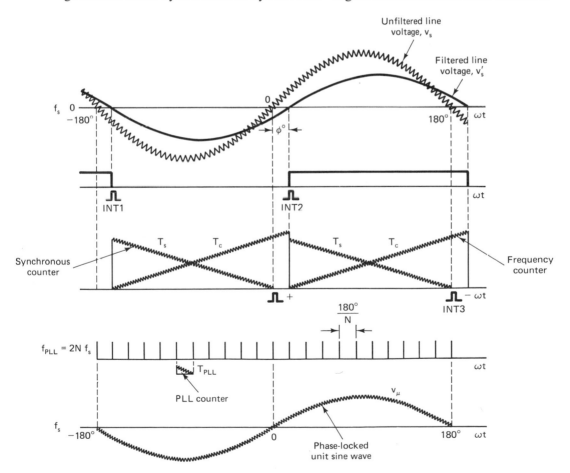

Figure 10.11   Phase-locked-loop synthesis of sine wave.

wave synthesis by the PLL principle where the reference wave is in phase with the unfiltered line voltage. The scheme can be considered as a low-pass filter without phase shift. The unfiltered line voltage v_s is filtered by a low-pass RC filter, and correspondingly, a square wave is generated from the filtered wave v_s' through a zero-crossing detector. The v_s' wave lags the v_s wave by ϕ^0, given by the relation

$$\phi = \tan^{-1} 2\pi f_s RC \tag{10.16}$$

where f_s is the line frequency and RC the filter parameter. The unit amplitude sine reference wave v_μ is to be locked in phase with the v_s wave irrespective of line frequency variation. The phase locking is achieved by a set of three counters: a frequency counter T_c, a synchronous counter T_s, and a PLL counter T_{PLL}. Counter T_c is an up-counter, whereas T_s and T_{PLL} are down-counters, and all are clocked by the same oscillator frequency f_{os}. The T_c counter tracks the line frequency and it is gated every half-cycle by the interrupt pulse, as shown. The count at the end of the half-cycle is given by

$$W_c = \frac{f_{os}}{2f_s} \tag{10.17}$$

The interrupt pulse also enables the synchronous counter, which is loaded by the word

$$W_s = \frac{\pi - \phi}{\pi} W_c \tag{10.18}$$

where ϕ is given by equation (10.16). The angle ϕ can be generated as a function of f_s by a look-up table. The PLL counter is of the auto-reload type and its buffer is updated every half-cycle of line frequency by the word

$$W_{\text{PLL}} = \frac{W_c}{N} \tag{10.19}$$

so that

$$f_{\text{PLL}} = 2Nf_s \tag{10.20}$$

where $2N$ is the desired ratio between the PLL frequency and the line frequency. The v_μ wave can be generated from a sine-wave look-up table which is retrieved by the interrupts of the PLL counter. The 0° pointer of the sine look-up table is set by the interrupt of the synchronous counter, which also clears the PLL counter and identifies a positive or negative half-cycle. The analog v_μ wave can be generated by a D/A converter and its amplitude can be modulated by a multiplying-type D/A converter. The frequency f_{PLL} can be used for gate firing control.

Pulse Width Modulation

For a chopper or inverter-fed drives, the PWM signals can be generated by a microcomputer if the carrier frequency does not exceed a few KHz. There are several techniques for inverter pulse width modulation, which have been reviewed in Section 4.4. All these methods can be implemented by a microcomputer resulting in considerable hardware simplification. The look-up-table oriented methods appear especially attractive for microcomputer implementation. The look-up table of digital words corresponding to pulse and notch widths can be converted to the time domain by down-counters. The clock frequency of the down-counter should relate to the fundamental frequency so that time intervals correspond to angular intervals. A computation-intensive method where the carrier and the signal waves are solved for the point of intersection to generate the pulse and notch widths is also possible. The computation-oriented method tends to be very time critical at higher frequency. Therefore, a PWM method that adopts a computation-intensive technique at low frequency but transitions to a look-up table oriented method at higher frequencies is desirable (Ref. 3).

Monitoring and Warning

Monitoring of variables by digital/analog display is very convenient with microcomputer control, since the variables already reside in memory. The monitoring of signals, constants, and different modes of control for system debugging can be conveniently done with the help of a diagnostic panel, which is discussed later.

If a number of signals, such as speed, stator current, and so on, exceed a safe limit, the microcomputer can easily generate audio and/or visual warning signals. If the operator fails to respond, the microcomputer can take appropriate action to ensure the safety of the equipment.

Protection and Fault Overriding Control

A sophisticated protection and fault overriding control can be designed with a microcomputer. The protective function (i.e., tripping a breaker or suppressing gate signals) can be decided on the basis of a Boolean function synthesized in software. Similarly, fault overriding control or a degraded mode of system operation can be designed by careful analysis of the system. For example, in the HVDC converter the thyristors can be fired selectively to overcome commutation or misfiring failure, or a shoot-through fault in an inverter can be cleared by switching all the thyristors simultaneously. Similarly, a three-phase machine can be operated in a single-phase mode, if permitted by the system. A fast fault where the microcomputer delay cannot be tolerated can be cleared by dedicated control hardware.

Programmable Set Point Commands

Programmable set point commands can be generated by the host microcomputer for dedicated hardware or microcomputer-controlled power electronic systems. This function is non-time critical and therefore is popular with dedicated hardware control systems. The programmed profile can be generated for test purposes, such as evaluation on a dynamometer of electric vehicle drive system performance for an urban driving cycle.

Sequencing Control

Sequencing control is an important function in which the microcomputer allows the drive system to acquire the various modes of operation and causes a smooth transition between the modes. The possible modes of an ac drive system can be identified as:

- Start neutral
- Forward constant-torque motoring
- Forward constant-torque regeneration
- Reverse constant-torque motoring
- Reverse constant-torque regeneration
- Forward field weakening motoring
- Forward field weakening regeneration
- Reverse field weakening motoring
- Reverse field weakening regeneration
- Shutdown
- Diagnostic tests

Figure 10.12 shows a typical sequencing diagram in which the circles are the modes and the arrows indicate the permissible transition between the modes. The transition from one mode to the other is dictated by a set of conditionals represented in the form of Boolean variables. When the particular Boolean function is satisfied, the transition is initialed by executing an action routine. If the transition is unsuccessful, it falls back repetitively in the same mode. For example, when the operator presses the START button, the microcomputer is powered up, hardware and software initialization occur, and after checking favorable system conditions, the microcomputer closes the power circuit breaker and acquires the NEUTRAL mode. Then it goes to the FORWARD or REVERSE modes as commanded by the operator. For static diagnostic tests, the system transitions directly to this mode from the NEUTRAL mode, as shown. In any mode, if a fault develops in the system, it goes to the SHUTDOWN mode. For a complex system there may be many more modes of operation, and careful system analysis and computer simulation may be required for a satisfactory design of sequencing control. Bumpy transition and chattering between the modes are some of the common problems in sequencing design.

Data Acquisition

The data in a digital control system can be acquired for the following purposes:

- Data recording and creating archives for future evaluation.
- Test or diagnostic data. The diagnostic data may be stored in continuous memory or passed on to the host computer in case of malfunction of the drive

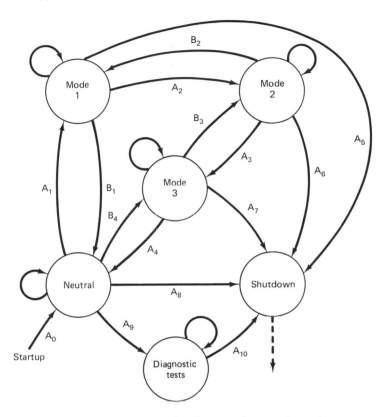

Figure 10.12 Typical sequencing diagram of power electronic system.

system components.   Ring buffer storage of various software parameters can be created in microcomputer memory for diagnostic purposes.

- Data acquisition for control and monitoring purposes.

Tests and Diagnostics

The microcomputer-based tests and diagnostics of a power electronic system have the following advantages:

- Tests can be performed by a semiskilled technician.
- The duration of tests can be reasonably short—saving costly equipment down-time.
- The test procedure is methodical; there is no chance of faulty diagnosis.
- There is a minimal possibility of damaging healthy components through human error.
- Sophisticated test equipment is not needed.
- Safety problems are minimal during the tests.
- The test procedure can be formulated in a general way such that it can easily be extrapolated to other systems.

The tests and diagnostics can be performed in the following modes:

- Highly automated without user intervention
- Semiautomated with partial user intervention
- Heavy user involvement assisted by the microcomputer

The procedures can be exercised in the following modes of system operation:

- *On-line:* when the system is operating and the microcomputer is executing the application program
- *Off-line:* when the system is not operating but the microcomputer is active

Nondiagnostic tests.   Various tests can be performed on power electronic systems with the help of a microcomputer.   The test data can be processed and performance curves can be plotted directly.   The examples are speed, efficiency, and temperature characteristics of the machine at different loadings, and open-circuit and locked-rotor tests of induction motors for estimation of parameters.

On-line tests, such as structure and parameter identification tests, can also be done on the system.   At different steady-state operating points of a nonlinear system, small sinusoidal or step signals can be injected and correspondingly, the response can be analyzed by the microcomputer to identify the system.   The identification of poles, zeros, and gain permits the controller to be adaptive, giving optimal system performance.

Diagnostic tests.   The microcomputer hardware, software, and components of the power electronic system can have various types of faults during the development stage and regular running conditions.   Some amount of diagnostics is usually incorporated in a microcomputer to identify the faults and possibly to take remedial measures.   A separate diagnostic computer can be used if more sophisticated diagnostic tests are desired.   Test routines can be exercised in the following modes:

- During startup of the system, when the microcomputer hardware, software, and system components are tested for healthiness.   The system will successfully start if everything is healthy or otherwise will stop giving an error message.
- The diagnostic routine can be exercised periodically as a low-priority task during system operation or can be user invoked as desired.   The error messages are given in a display panel, or for unsafe operation the system will gracefully shut down, causing minimal damage.   The possible causes of failure can be identified by error messages or by retrieving a ring buffer continuous memory containing software parameters.
- The diagnostic routine in part can be embedded with the application program of the computer.   For example, if a contactor does not successfully open as indicated by the "acknowledge" signal, an error message will be displayed and the operation will be continued if considered safe.

- At system shutdown under fault conditions, the microcomputer performs an automated or semiautomated test procedure to identify the fault condition.
- As above, but the microcomputer gives instructions in the test procedure, and actual tests are made with heavy user involvement.

The following hardware/software diagnostics can easily be implemented:

- *Watchdog timer:* A retriggerable monostable timer can be strobed periodically at a task timing interval.   If the microcomputer operation is healthy, a green light is indicated; otherwise, a red light glows.
- *RAM write/read test:* Shows successful RAM memory operation.
- *ROM checksum test:* Shows integrity of application software and successful operation of ROM memory.
- *I/O write/read test:* Indicates healthy operation of I/O.
- *A/D converter tests:* An A/D converter with a known reference voltage input is tested to match a digital word.
- *A/D–D/A converters wraparound test:* A D/A converter is loaded with a digital word and its output is impressed as input to an A/D converter to indicate the same digital word.   Calibration can be performed in the whole range.
- *Timer test:* A hardware timer can be tested in parallel with a software timer to verify the timing interval.
- *Warm reset:* Warm reset is in contrast to cold or initial reset of a microcomputer.   In case of suspected malfunction of a microcomputer, a warm reset is activated and the program is restarted.   The warm reset bypasses hardware and software initialization.

Diagnostic design methodology.   A systematic diagnostic design methodology is indicated as a flowchart in Fig. 10.13.   The flowchart assumes that the faults are concentrated primarily in the power electronic system, and diagnostic tests will be exercised on an off-line basis.   For a comprehensive diagnostic test, a large number of sensors for identification of the system condition may be required. Figure 10.14 shows the possible I/O signals of a general thyristor rectifier–transistor inverter ac drive system.   The microcomputer exercises the test routines systematically in sequence, as indicated in Fig. 10.15, and tries to identify the fault conditions as defined by the Boolean functions.   The test procedure can be performed by user intervention, as discussed earlier.

10.2 DESIGN METHODOLOGY

A microcomputer-based control design is generally more involved and time consuming than that with dedicated hardware.   The design consists of the following four stages:

1. System design with algorithm development

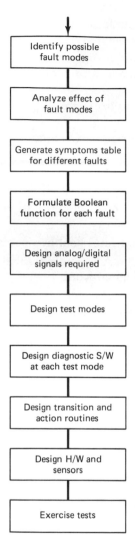

Figure 10.13 Flowchart for diagnostic design methodology.

2. Hardware design
3. Software design
4. System integration and test

Figure 10.16 shows a general flow diagram of a microcomputer control design. The heavy arrows indicate the main direction of flow, although many back-and-forth iterations and overlapping of stages are not uncommon.

System Design

The system design starts after receiving the performance specifications. The control system functions are first identified, followed by detailed analysis and design of each function. Computer simulation of some functions may be desirable. The

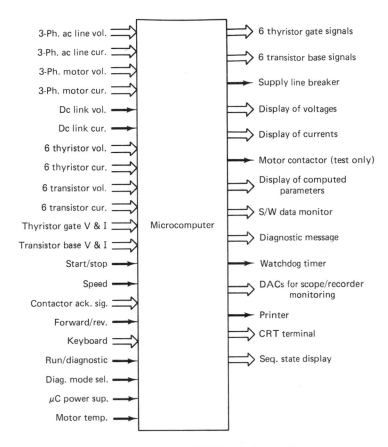

Figure 10.14 Microcomputer I/O signals for ac drive system.

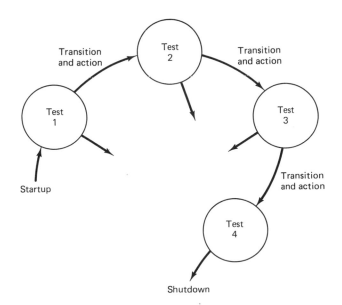

Figure 10.15 Diagnostic tests sequencing.

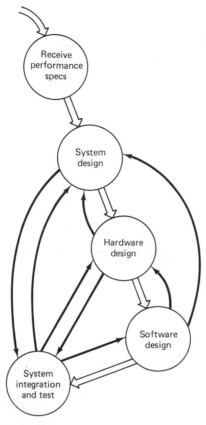

Figure 10.16 Flow diagram of microcomputer control project.

design of various control functions was discussed in the preceding section. The system design stage terminates with a description of control algorithms in the form of flowcharts which become input to the software design.

Hardware Design

The first step in hardware design is to analyze the hardware–software tradeoff and decide on the boundary between dedicated hardware and software. For very time-critical functions, such as a high-bandwidth current control loop, PWM signal generation above a carrier frequency of a few KHz, or a fast-protection requirement, a microcomputer may be too slow, and therefore dedicated hardware control may be necessary. Independent testability and serviceability may also dictate the boundary between hardware and software. The important design considerations of a microcomputer hardware are reliability, economy, noise immunity, and debugging capabilities.

   Selection of microcomputer.   Several considerations that determine the selection of a microcomputer are:

- 8-bit or 16-bit
- Clock rate and instruction cycle time
- Single-chip or multichip computer

- Instruction set—does it support multiplication and division?
- Register set and memory addressability
- I/O, memory, and interrupt support
- Price
- Single or multisourcing
- Availability of chips in industrial and military grade
- Peripheral hardware support
- Software support
- Development system
- In-circuit emulation (ICE) capability
- Software development support in time-sharing computer

Microcomputers are available from a number of competitive manufacturers, but it is desirable to select one manufacturer only. The in-house availability of a development system and prior knowledge of a certain type may also dictate the selection of a microcomputer.

Single-board computers (SBCs) versus custom design.   Single-board computers in the form of SBCs or SDKs (developmental kits) may be available from the vendor, which may considerably simplify the hardware design task. The boards have areas for optional memory or I/O expansion, or auxiliary custom-designed hardware can be added to develop fully functional hardware. The SBC/ SDK boards are satisfactory for initial feasibility studies or design of one or several sample units, but for large-volume applications custom design proves more economical. Development of universal hardware for a particular type of power electronic system may prove very economical where only the software is customer modified.

Memory design.   The size of ROM and RAM memory has to be estimated, corresponding to program and data size, respectively. In most cases, estimation of memory size is difficult, and therefore it should be designed conservatively with provision for expansion. The assembly language program for a typical application can hardly exceed 4K bytes, but the size may be much higher in high-level language. The program should be segmented in several sections, if possible, so that a separate EPROM chip can be used for each segment. The segment which requires alteration or contains adjustable parameters would be located in a separate chip. The EPROM memory permits flexible program development and is economical except for large volume application. The access time of memory and I/O chips should be compatible with the processor speed to avoid WAIT states which slow down the program execution.

I/O design.   As a first step in I/O design, the signals, which may be analog, digital, or logic, should be identified. The signals and microcomputer ground should be isolated from the system to prevent noise coupling. The analog signals may be unipolar or bipolar, and the resolution requirement determines the bit size of A/D and D/A conversion systems.   Optical isolation of digital signals is cheaper than that of analog signals.   Adequate hardware noise filtering of analog and digital signals is necessary.   For logic signals obtained from mechanical switches, debouncers should be used.   It is desirable to provide some spare I/O channels in the initial design.

Methods of speed enhancement.   For time-critical computations it is necessary that the microcomputer should operate at maximum possible speed.   The various hardware-software techniques for speed enhancement can be summarized as follows:

- Use fast microcomputer.
- Formulate algorithm for time efficient computation.
- Use assembly language instead of high-level language.
- Use dedicated hardware counters, I/O signal processors, coprocessors, etc.
- Use look-up table method of computation.
- Use DMA techniques for large data transfer.
- Use hybrid analog and digital computation.
- Use multiprocessing.

The higher-bit-size computer is desirable for faster computation.   Besides the higher clock frequency, the instruction set with direct multiplication/division (both signed and unsigned), register structure, addressibility, and so on, should be considered for fast computation.

The mutiprocessing technique may considerably enhance the speed of digital processing and therefore requires further discussion.   In a multiple-function system, the functions can be appropriately distributed between several computers for parallel operation and thus enhance the speed of processing.   The processors may operate in a master–slave or master–master relationship.   In parallel processing, the microcomputers are required to communicate data among them.   The communication protocol adds complexity in hardware and software design and may adversely affect the execution speed.   The protocol design is simpler with loose coupling than with tight coupling.   The different communication protocols can be summarized as follows:

Serial link communication: Very slow but simple to design.

Parallel I/O transfer with handshake: Generally satisfactory for small number of bytes.

DMA transfer: Efficient typically above 100 bytes of data.

FIFO channel transfer: Asynchronously a block of data can be transferred in FIFO mode and stored into the RAM buffer.

Dual-port mailbox memory: A section of RAM memory that may be accessed by two processors for read/write operation is shown in Fig. 10.17.   The memory control logic decides which processor should have access in case of conflict.

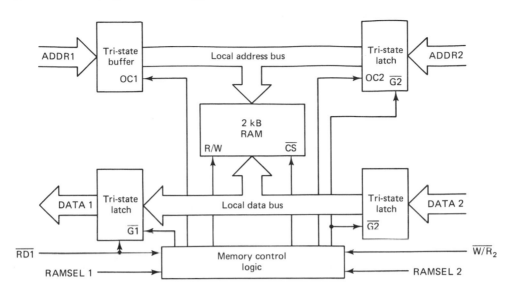

Figure 10.17   Dual-port mailbox memory.

A general multiprocessor architecture with multibus is shown in Fig. 10.18. Here, the global memory and I/O can be accessed by either microcomputer one or two through the multibus. A bus arbiter (not shown) determines on the basis of a priority technique, which microcomputer can have access on the multibus.

Direct memory control: This principle is explained for a central microcomputer linked to the peripheral microcomputers in a radial configuration, as shown in Fig. 10.19. Each microcomputer has its local ROM, RAM, and I/O. In this architecture, the central computer maintains a partitioned data base that is shared by all peripheral units. The peripheral units are designed to communicate only through the central computer but not directly among them. The peripheral units request communication from the central processor by halting its own operation and initiating an interrupt request. The central unit then services and restarts the unit. The central unit is also allowed to address the peripheral unit's memory as if it is a part of central's memory

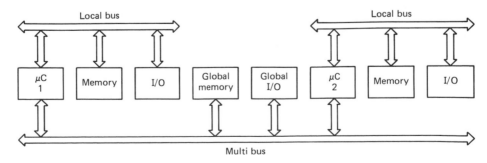

Figure 10.18   Multiprocessor architecture with multibus.

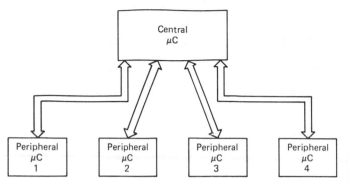

Figure 10.19   Radial configuration of multiprocessor system.

map, thus allowing data transfers to take place in parallel fashion in minimum time.

Diagnostic box.   The diagnostic or debug box is an important tool required during the project development stage.   It permits modification of software parameters and monitoring of software activity within the microcomputer when the application program is running.   The diagnostic box function can be summarized as follows:

- Simulates artifical analog/digital I/O signals of microcomputer for dynamic testing of software modules
- Modifies the software control parameters and constants during program operation
- Permits reading of software parameters and variables during program operation

The block diagram of a simple diagnostic box is shown in Fig. 10.20 and its operation is explained with the flowchart of Fig. 10.21.   The box permits writing data in any memory location and retrieving data from memory locations to DACs A and B and hex displays A and B.   The write/read operations can be done repetitively at a specified sampling interval determined by the diagnostic software.   The command is entered by the keys as shown in the figure.   The address and write data are set up by hex thumbwheel switches.   The command, address, and data are entered to the microcomputer by the address lines A_1, A_3, and A_2, respectively, through latch A.   Latch B updates the hex displays by address lines A_4 and A_5.   The microcomputer interrogates the command at every T_4 sampling period (see Fig. 10.24) by making A_1 line low and enabling latch A.   If, for example, it senses a write command, it will accept data from thumbwheel switches by lowering line A_2 and write to the address as specified on line A_3.   For display or DAC update command, the memory location specified by thumbwheel switches at address A_3 will be retrieved.   The DAC outputs, which may be feedback-loop variables, can be displayed repetitively on a scope or recorded on a chart recorder.

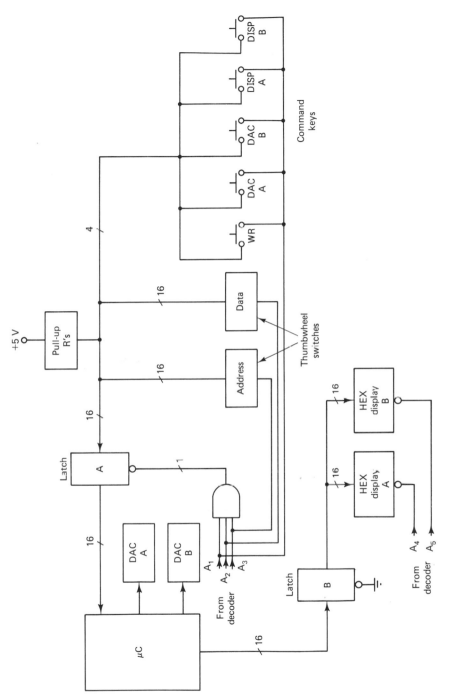

Figure 10.20 Block diagram of diagnostic box.

387

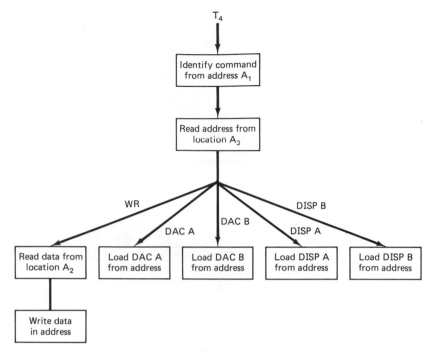

Figure 10.21   Flowchart explaining operation of diagnostic box.

Software design

The software of a microcomputer consists of a set of application program modules which are executed in real time in a time-sequenced manner. The salient features of software design can be reviewed as follows.

Selection of language. The control of power electronic systems, especially the feedback control loops and computations related to them, are very time critical, and therefore assembly language is normally used. The assembly language is nearest to machine code, and therefore a program written in assembly language has a fast execution time and occupies small memory size. However, a program development in assembly language is time consuming, tedious, and may require many iterations. The programs in high-level languages, such as PL/M, PASCAL, and FORTRAN, on the other hand, are slow in execution, require large memory size, but have the advantage of fast development time. The total application program can be judiciously partitioned according to the timing requirements and both high-level and low-level languages can be mixed appropriately. In such a case, the segments of source program can be separately compiled and assembled, then linked and located in ROM memory of prototype microcomputer. The steps in program development are discussed later.

Sampling time and bit resolution.   The sampling time and bit resolution are the most important considerations in the digital control system design. The performance of a digital system will approach that of an analog system with faster sampling intervals and higher bit resolution.

Since a microcomputer executes a control loop in a time-sequenced manner, a sampling interval, or the corresponding sampling rate of computation, should be specified. The sampling interval should provide spare time for multitasking operations, discussed later. A longer sampling interval is convenient for the microcomputer, which possibly permits a high-level language. On the other hand, the long sampling time introduces delay and distortion in the control loop, tending it to be unstable. The choice of a practical sampling time is determined by the transient response time (i.e., the closed-loop bandwidth) of a loop. A selection of sampling time that is one-tenth or less of the loop response time is typical. Generally, an inner control loop is required to have faster response than the outer loop, which means that the outer loop may have a slower sampling time. The sampling time of the loops may be dictated by the switching interval of the converter semiconducters, as discussed in Section 10.1.

The performance accuracy requirement of a system determines the bit resolution of the signals, which affects whether an 8-bit or 16-bit microcomputer should be selected. In a digital control system, the controlled variable oscillates with ± 1-bit jitter, known as quantization noise. The permissible jitter of the corresponding physical variable determines the bit size of the signal. For example, in a speed control loop with the range ± 3000 rpm, a bipolar 12-bit computation will give jitter of $3000/(2^{11} - 1) = 1.47$ rpm. The jitter in an outer loop can be amplified by the sensitivity of inner loop bit jitter. For a multiloop control system, the bit resolution of signals in different loops should be carefully determined by sensitivity analysis (Ref. 5). With the resolution of the outer loop normally specified, the resolution of inner loops should be equal or better so that the induced effect in the outer loop is negligible. The figure of merit of a loop is sometimes defined by the ratio of bit resolution and sampling time.

Scaling.   Another important consideration in digital signal manipulation is scaling. The scaling maintains resolution of the computed software parameters and prevents overflow or underflow of a parameter. Internally, a microcomputer handles addition and subtraction operations of bipolar signals in two's-complement arithmetic, and signed multiplication and division instructions may be used if available. If a memory location tends to overflow, it is clamped to the limit value. The concept of digital scaling is similar to analog scaling in analog computation. A simple example of digital scaling can be given as follows.

Consider a digital torque computation from the expression

$$T_e = \frac{VI}{N} \frac{60}{2\pi} \tag{10.21}$$

where T_e is the torque in Nm, V the voltage in volts, I the current in amps, and

N the speed in rpm. The variables can be related to software variables as follows:

$$V = \frac{V - V}{K_1} \tag{10.22}$$

$$I = \frac{V - I}{K_2} \tag{10.23}$$

$$N = \frac{V - N}{K_3} \tag{10.24}$$

$$T_e = \frac{V - T_e}{K_4} \tag{10.25}$$

where $V - V$, $V - I$, and so on, are the software variables and K_1, K_2, and so on, are the corresponding scale factors. Substituting in equation (10.21) yields

$$V - T_e = 9.55 \left(\frac{K_4 K_3}{K_1 K_2}\right) \frac{V - V \cdot V - I}{V - N} \tag{10.26}$$

Assume that V and I are bipolar signals which are sampled by 10-bit A/D converters and speed is available as a 10-bit unipolar signal. Assume also that $V_{max} = 200$ V, $I_{max} = 400$ A, and $N_{max} = 3000$ rpm. Then

$$K_1 = \frac{2^9 - 1}{200} = 2.56 \text{ bits/V}$$

$$K_2 = \frac{2^9 - 1}{400} = 1.28 \text{ bits/A}$$

$$K_3 = \frac{2^{10} - 1}{3000} = 0.34 \text{ bits/rpm}$$

Select $K_4 = 8$ bits/Nm (i.e., 1 bit $= 0.125$ Nm). Substituting all the scale factors in equation (10.26), we have

$$V - T_e = 7.93 \frac{V - V \cdot V - I}{V - N} \tag{10.27}$$

Equation (10.27) should be calculated to compute the software torque parameter $V - T_e$.

Software structure and timing. In the initial design stage of software, the software functional modules are to be identified and their interaction with the physical system through the I/O signals should be studied. Figure 10.22 shows such a block diagram, which includes diagnostic, monitoring, and warning functions. The modules are classified according to the priority levels of implementation and the corresponding sampling times are determined. Figure 10.23 shows a simplified structure chart of software where the tasks with the respective sampling time shown are under the control of a real-time scheduler (RTS). The RTS is an

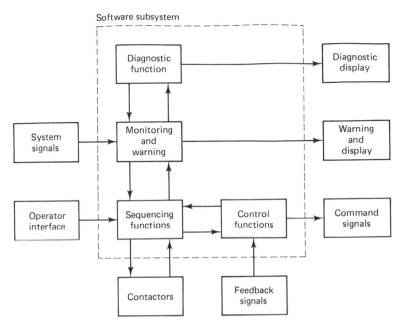

Figure 10.22 Software functional block diagram.

executive software which is responsible for orderly execution of tasks according to their level of priorities.   It is driven by a hardware-interrupt clock T_1, the interval of which is the sampling interval of the highest-priority task (task 1), as shown. The power-up reset, shutdown of the system by a pushbutton, and software idle-time counting features are also indicated.

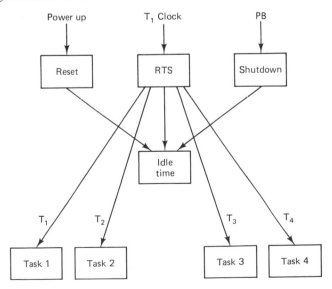

Figure 10.23 Simplified structure chart of software.

Figure 10.24 shows the tasks timing diagram where the typical sampling times are shown. Task 1 is always completed at a T_1 clock input within the interval T_1 before any low-priority task is executed. If a low-priority task was under execution, it is suspended until completion of task 1. In Fig. 10.24, task 2 is suspended once, as indicated by marker 1, whereas task 4 is suspended three times, as indicated by markers 2, 3, and 4.

The loading factor of a microcomputer can be defined as the fraction of time the computer remains busy during the longest sampling time T_4, that is,

$$\text{LF} = \frac{\sum \text{computing time}}{T_4} \tag{10.28}$$

Normally, the sampling intervals are chosen such that each is an integral multiple of the next-lower interval. If the execution time of the four tasks are t_1, t_2, t_3, and t_4, respectively, then

$$\text{LF} = \frac{1}{T_4}\left(\frac{T_4}{T_1}t_1 + \frac{T_4}{T_4}t_2 + \frac{T_4}{T_3}t_3 + t_4\right) \tag{10.29}$$

$$= \frac{t_1}{T_1} + \frac{t_2}{T_2} + \frac{t_3}{T_3} + \frac{t_4}{T_4}$$

The load factor is always less than unity and $(1 - \text{LF})$ indicates the per unit idle time. Each task has to be completed within the specified sampling time to prevent overflow; that is,

$$t_1 < T_1 \tag{10.30}$$

$$t_2 < \left(T_2 - \frac{T_2}{T_1}t_1\right) \tag{10.31}$$

$$t_3 < \left(T_3 - \frac{T_3}{T_1}t_1 - \frac{T_3}{T_2}t_2\right) \tag{10.32}$$

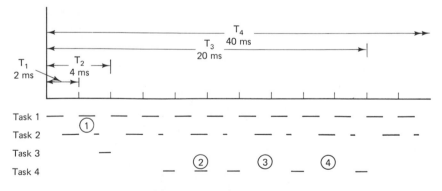

Task 1 (T_1 period): current loop control, diagnostics
Task 2 (T_2 period): flux and speed loop control
Task 3 (T_3 period): sequencing, diagnostics
Task 4 (T_4 period): background functions

Figure 10.24 Tasks timing diagram.

$$t_4 < \left(T_4 - \frac{T_4}{T_1} t_1 - \frac{T_4}{T_2} t_2 - \frac{T_4}{T_3} t_3 \right) \qquad (10.33)$$

A simplified flowchart of a real-time scheduler that controls the timing diagram is shown in Fig. 10.25. The timing intervals T_2, T_3, and T_4 are generated by the T_1 clock through software timers. Task 1 is always scheduled at the T_1 interrupt and the other tasks are executed in sequence as shown. Task 1 can reenter a lower-priority task by suspending it and saving its status in the stacks. Of course, any nonscheduled interrupt-driven task has higher priority and can be serviced during execution of the tasks 1, 2, 3, and 4.

Program development and emulation test. The development of source programs begins after the microcomputer function algorithms have been described in the form of detailed flowcharts. The programs can be developed either in high-level language, assembly language, or by a mixture of both, as discussed before.

Modular programming. Many programs are too long or complex to write as a single unit. Programming becomes much simpler when it is divided into small functional units. Modular programs are easier to code, debug, and alter than monolithic programs. The modular approach to programming is similar to the design of hardware that contains numerous functional circuits. The program is logically divided into "black boxes" with specific inputs and outputs. Once the interfaces between the units have been defined, detailed design of each module can proceed independently. Finally, the modules are integrated to develop the composite program.

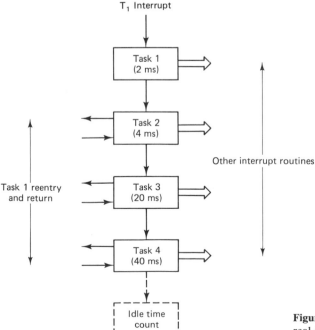

Figure 10.25 Simplified flowchart of real-time scheduler.

Microcomputer development system. A development system is an important tool for the user's software development. It permits development of the source programs, which can then be translated to an absolute object program, to be used in a prototype microcomputer. Alternatively, programs can be developed in a mini/personal computer, compiled/assembled, and the object-program can be downloaded to the memory of the microcomputer directly. A special advantage of the development system is that it permits system operation in the real time emulation mode. Figure 10.26 shows a block diagram of a microcomputer development system with the peripheral units. The system itself includes a CPU with RAM and ROM memory, a CRT, an ASCII keyboard with cursor controls, and a limited storage floppy disk drive. The development system is linked with a line printer, disk drive system, EPROM programmer, and in-circuit emulator (ICE), as shown in the figure. It can be linked to a mainframe or minicomputer for distributed program development. The disk operating system software permits compiling/assembling, debugging of programs, and efficient file-handling operations for the user.

Program development. The flowchart for the program development process is shown in Fig. 10.27. The rectangular boxes indicate the development tools and the ellipses are the user-coded software. Once the source program modules are written, these are entered into disk files using the text editor. The CRT-based text editor permits correction of source files by insertion, deletion, string search and so on. The assembler (or compiler) then translates the source code into object code. The assembly program has three constituent parts: machine instructions, assembler directives, and assembler controls. A machine instruction is a machine code that can be executed by the machine. Assembler directives are used to define program structure and symbols and to generate nonexecutable code (data, messages, etc.). Assembler controls set the assembly modes and direct the assembly flow.

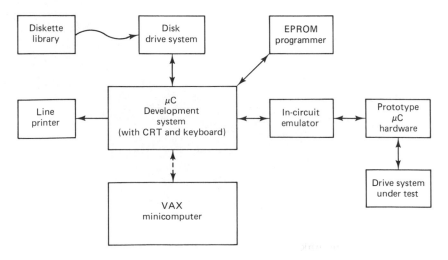

Figure 10.26 Microcomputer development system with peripheral units.

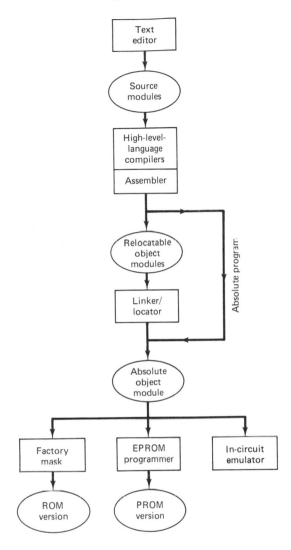

Figure 10.27   Flowchart for program development.

   The assembler produces an object file and a listing file showing the results of assembly. The object file may be relocatable or absolute. When the assembler invocation contains the DEBUG control, the listing file also receives the symbol table and other debug information for use in symbolic debugging of the program.

   The object modules contain machine language instructions and data that can be loaded into microcomputer memory for execution. The object modules which are relocatable are processed through the linker/locator. This assigns absolute memory locations to the relocatable segments and combining segments that have the same name and type. The linker/locator resolves all references between modules and outputs an absolute object module and a summary listing file showing the results of the link/locate process. The absolute object file can then be used in one of the following modes:

- Generate a factory interconnection mask for the ROM version of the program
- Burn an EPROM through an EPROM programmer
- Use the program for an in-circuit emulation test

If the development system is tied with the VAX minicomputer as shown in Fig. 10.26, the user can integrate the resources of both the systems. The VAX can be used for storage, maintenance, and management of program source and object files. The source program, developed by a number of programmers, can be down-loaded to a development system for compilation, assembly, linkage, and location. The linked modules can be transmitted and saved on VAX to be shared by all programmers. The VAX-based codes can then be down-loaded to the development system for EPROM programming or in-circuit emulation.

In-circuit emulation tests.   The ICE permits debugging of prototype hardware, software, and system operation under the real-time operating condition of software. The interface of ICE with the development system is shown in Fig. 10.25. With the help of ICE, the memory and I/O of the development system can be shared with a prototype microcomputer in the initial development stage. The prototype hardware and software can be developed and debugged interactively before integration with the system. The system can then be tested extensively in the emulation mode and the program resident in the development system can be iterated. Finally, the program can be loaded in EPROM or ROM.

REFERENCES

1. P. C. Tang, S. S. Lu, and Y. C. Wu, "Microprocessor-Based Design of a Firing Circuit for Three-Phase Full-Wave Thyristor Dual Converter," *IEEE Trans. Ind. Elec.*, Vol. IE-29, pp. 67–73, Feb. 1982.

2. T. Ohmae, T. Matsuda, T. Suzuki, N. Azusawa, K. Kamiyama, and T. Konishi, "A Microprocessor-Controlled Fast Response Speed Regulator with Dual Mode Current Loop for DCM Drives," *IEEE Trans. Ind. Appl.*, Vol. IA-16, pp. 388–394, May-June, 1980.

3. B. K. Bose and H. A. Sutherland, "A High Performance Pulse-Width Modulator for an Inverter-Fed Drive System Using a Microcomputer," *IEEE Trans. Ind. Appl.*, Vol. IA-19, pp. 235–243, Mar.–Apr. 1983.

4. B. K. Bose, P. Szczesny, and R. L. Steigerwald, "Microcomputer Control of a Residential Photovoltaic Power Conditioning System," *IEEE Trans. Ind. Appl.*, Vol. IA-21, pp. 1182–1191, Sept.-Oct. 1985.

5. T. Konishi, K. Kamiyama, and T. Ohmae, "A Performance Analysis of Microprocessor-Based Control Systems Applied to Adjustable Speed Motor Drives," *IEEE Trans. Ind. Appl.*, Vol. IA-16, pp. 378–387, May–June 1980.

6. B. K. Bose, *Microcomputer Control of Power Electronics and Drives*, IEEE Press, New York, 1986 (in press).

7. B. K. Bose, "A Microprocessor-Based Control System for a Near-Term Electric Vehicle," *IEEE Trans. Ind. Appl.*, Vol. IA-17, pp. 626–631, Nov.–Dec. 1981.

8. P. Katz, *Digital Control Using Microprocessors*, Prentice-Hall, Englewood Cliffs, N.J., 1981.

9. *Intel Development Systems Handbook*, May 1983.

10. H. Naitoh, "Model Reference Adaptive Control for Adjustable Speed Motor Drives," *Conf. Rec. Int. Power Elec. Conf.*, Tokyo, pp. 1705–1716, 1983.

11. R. G. Hoft, T. Khuwatsamrit, and R. McLaren, "Microprocessor Applications for Power Electronics in North America," *Conf. Rec. Microelectron. Power Electron. Electr. Drives*, Darmstadt, pp. 29–42, Oct. 1982.

12. B. K. Bose, "Motion Control Technology—Present and Future," *IEEE Trans. Ind. Appl.*, Vol. IA-21, pp. 1337–1342, Nov.-Dec. 1985.

INDEX